International Encyclopaedia of

NANO TECHNOLOGY, SCIENCE AND CHEMISTRY

International Encyclopaedia of

NANO TECHNOLOGY, SCIENCE AND CHEMISTRY

Volume 1

R.N. Parmar

ANMOL PUBLICATIONS PVT. LTD.

NEW DELHI - 110 002 (INDIA)

ANMOL PUBLICATIONS PVT. LTD.
H.O.: 4374/4B, Ansari Road, Daryaganj,
New Delhi-110 002 (India)
Ph.: 23278000, 23261597
B.O.: No. 1015, Ist Main Road, BSK IIIrd Stage
IIIrd Phase, IIIrd Block,
Bangalore - 560 085 (India)
Visit us at: www.anmolpublications.com

International Encyclopaedia of Nano Technology, Science & Chemistry

First Edition, 2008
ISBN 978-81-261-3681-0 (Set)

PRINTED IN INDIA

Printed at Mehra Offset Press, Delhi.

Contents

Preface vii

VOL. 1

Glossary 1

1. Introductory Overview of Nano Chemistry, Nanostructured Materials and Nanoparticles: Major Properties and Issues 56

2. An Overview of Relationships Among Nano Chemistry, Nanotechnology and Nano-velcro: An In-depth Analysis 133

3. Nanoscale Chemistry, Nano Synthesis and Assembly: Towards Knowing the Various Implications 161

4. Nanoscale Interfacial Materials and Molecular Manufacturing: Understanding the Dimensions and Processes 211

VOL. 2

5. Multiple Applications of Nanocomposites and Nano Chemistry with Special Focus on Industries 289

6. Towards Understanding the Global Contemporary Issues Regarding Nanoparticles, Nanostructured Materials' Chemistry and Polymer Nanotechnology 326

7. Nano Chemistry, Nanoworld Particles and Molecules: Functional Relationship Analysis 363

8. Soft Nanotechnology, Nano-Vasculoid and Biomimetic Self-Assembly: Industrial Applications and Processes 414

9. Fundamentals and Applications of Carbon Nanotubes, Nanotube Theory and Nanotube Rings 500

Bibliography 553

Index 573

Preface

Nano Chemistry is the branch of science used for analyzing ultra small systems and for this one needs ultra small tools. Whether dealing with single molecules or single cells, the precision and sensitivity of the manipulation and detection schemes used are absolutely critical. Nano Chemistry scientists have been interested in confining single molecules in solution to perform chemistry at the lowest level possible, as well as in delivering small volumes of chemicals to single sub-cellular organelles. By mimicking what nature has developed over billions of years to conduct chemistry in ultra small packets (cells and organelles), Nano Chemistry scientists have arrived at what is to date the smallest test tubes known - synthetic liposomes which are manipulated and electrofused to mix their contents. These liposomes have volumes of a few femtoliters or less, matching the probe volumes of standard single molecule detection techniques, and are made in minutes by a protocol developed in this lab. In sum, Nano Chemistry is the science of tools, technologies, and methodologies for chemical synthesis, analysis, and biochemical diagnostics, performed in nanolitre to femtolitre domains. Nano Chemistry is the use of synthetic chemistry to make nanoscale building blocks of desired shape, size, composition and surface structure, charge and functionality with an optional target to control self-assembly of these building blocks at various scale-lengths.

This book entitled "International Encyclopaedia of Nano Technology, Science & Chemistry" provides readers with an elaborate introduction to molecular nanotechnology, Nanostructured materials, Nanoparticles and Nanodevices. An overview of relationship between nanotechnology and Nano-velcro is presented herein. The modern global issues regarding Nanoparticles, Nanostructured materials and polymer nanotechnology are discussed in detail. The subject areas of Nanocomposites and Nanostructured materials are covered detailing about their respective properties and applications. The inter-relationship among today's Nanoworld, Nano Chemistry and the all important process of self-assembly is highlighted. An overview of nanotech products is presented. There are so many other applications of Nano Chemistry ranging from mundane to incredibly profound. Other applications include energy storage materials, photovoltaic cells, fuel cells, polymer Nanocomposites, super-tough Nanocoatings, landmine detectors, and thousands more. Although it is a field that has already seen its share of hype, there is a general consensus that over the long term, Nano Chemistry is likely to reconfigure everything. Nanostructured materials do not represent a new phenomenon. However, the ability to probe, manipulate, understand and engineer matter at atomic scales has only recently come within our grasp. Developments such as the invention of the Scanning Tunneling Microscope have since made nanoscale science a reality. Today, the term nanotechnology is used to describe specific construction and manipulation performed on an atomic and molecular scale. It could be one's strong opinion that Nano Chemistry will revolutionize many areas of chemistry, and is of strategic importance. In the first instance, the goals are focused around the future needs in pharmaceutical chemistry and clinical diagnostics, where paradigm shifts are very close. Novel approaches towards combinatorial Nano Chemistry are studied and developed. Tools like chip-based Nanovials, flow reaction channels, dispensing devices, mechanical and chemical microactuators, optical detectors etc. are being developed by joint expertise in micro structure technology, chemistry as well as specialists in surface and coating technology. Applications which are being studied are e.g. optimization of catalysis, asymmetric synthesis, combinatorial chemistry, novel photochemistry etc. In this context, new strategies for high throughput handling of very small amounts of materials of major concern, which will have important applications in biochemistry, molecular biology, clinical diagnostics, drug development and material science.

Generally, systems with a maximum size of less than a hundred nanometers are classified as nanotechnology. Nanotechnology is a completely new production process in a molecular world, opening up possibilities far beyond what we have known before. Nature provides the inspiration for this new technology – from time immemorial, nanomachines have been at work in all organic cells, building our world of plants, animals and people from a store of chemical elements. Experts world over believe nanotechnology will have a more far-reaching impact than the silicon integrated chip with many more applications than electronics. Nanotechnology is expected to revolutionize certain areas, such as semiconductors, pharmaceuticals and materials. The total societal impact of nanotechnology is expected to be much greater than that of the silicon integrated circuit because it is applicable in many more fields than electronics. Nanotechnology experts believe that because of Nano Chemistry, we would see more changes in the next 30 years than we saw in the entire last century. The main focus of research work in Nano Chemistry is on the synthesis and characterization of metal and semiconductor Nanoparticles. These particles have interesting size-dependent properties and can be used as building blocks for new materials. We use dynamic light scattering to study the formation of superstructures of Nanoparticles.

This book tends to cover the aforementioned subject areas in their broadest possible terms. All useful supplementary research and reference tools have been included for readers' further investigations in the subject area of Nano Chemistry. This book mainly focuses on the chemical aspects of Nanotechnology; Nanoparticles,

Nanomaterials and Nanostructures. This book "International Encyclopaedia of Nano Technology, Science & Chemistry" seeks to explain this complex field of studying its totality. It also seeks to provide an understanding of what is happening now in the field of Nano Chemistry, as well as what to expect in future by generations to come. Keeping in mind the applications and utilities of Nano Chemistry for meeting the contemporary global medical requirements, it seems that this subject will become the core area of future nanoscience and nanotechnology research. Developments such as the invention of the Scanning Tunneling Microscope have since made nanoscale science a reality. Today, the term nanotechnology is used to describe specific construction and manipulation performed on an atomic and molecular scale. Generally, systems with a maximum size of less than a hundred nanometers are classified as nanotechnology. This book tends to cover the aforementioned subject areas in their broadest possible terms. All useful supplementary research and reference tools have been included for readers' further investigations in the subject area of Nano Chemistry. This "International Encyclopaedia of Nano Technology, Science & Chemistry" is user friendly and future research oriented.

—EDITOR

Glossary

Acetylcholine: Neurotransmitter that functions at cholinergic chemical synapses, found both in the brain and in the peripheral nervous system. It is the neurotransmitter at vertebrate neuromuscular junctions.

Acrosomal Process: Long, thin, actin-containing spike produced from the head of certain sperm when they make contact with the egg. Seen in sea urchins and other marine invertebrates whose eggs are surrounded by a thick gelatinous coat.

Active Component: A device that adds intelligence in some manner to the signal or data that passes through it. For example, in networking, an active hub regenerates fading input pulses into new, strong output pulses. In contrast, a passive hub is just a junction box that does not affect the passing data. Non-mechanical circuit component that has gained or switched current flow, such as a diode, transistor, etc.

Active Device: Device requiring a source of energy for its operation and has an output that is a function of present and past input signals. Examples include controlled power supplies, transistors, LEDs, amplifiers, and transmitters.

Adsorption: Formation of molecules of a gas similar to a thin film that binds on the surface of a solid. Unlike absorption, the binding to the surface in adsorption is usually weak and reversible. Adsorption is a method for separating mixtures and purifying fluids.

Affinity: A thermodynamic measurement of the strength of binding between molecules, say between an antibody and antigen. Each antibody/antigen pair has an association constant, Ka, expressed in L/mol.

Akebono: A Japanese scientific studies satellite that became operational in 1989. It contained only one non-Japanese instrument: Canada's Suprathermal Ion Mass Spectrometer, sent to investigate the Earth's outer atmosphere.

Algorithm: In mathematics, computing, linguistics, and related disciplines, an algorithm is a definite list of well-defined instructions for completing a task; that given an initial state, will proceed through a well-defined series of successive states, eventually terminating in an end-state. The transition from one state to the next is not necessarily deterministic; some algorithms, known as probabilistic algorithms, incorporate randomness.

Amide: A molecule containing an amine bonded to a carboxyl group; the resulting bond has substantial double-bond character. Also termed a peptide; amide bonds link amino acids in proteins.

Amino Acid: A molecule containing both an amine and a carboxylic acid group; in the 20 genetically encoded amino acids in biology, both groups are bound to the same C. Amino acids joined by amide bonds form peptides and proteins; these do not contain amino acids as such, and are instead said to contain *amino acid residues*. Organic molecules that are the building blocks of proteins. There are some two hundred known amino acids, of which twenty are used extensively in living organisms.

Amperometric Sensor: Amperometric sensors involve a heterogeneous electron transfer as a result of an oxidation/reduction of an electro-active species at a sensing electrode surface. A current is measured at a certain imposed voltage of the sensing electrode with respect to the reference electrode. Analytical information is obtained from the current-concentration relationship at that given applied potential.

Analyte: An analyte is a substance or chemical constituent that is determined in an analytical procedure, such as a titration. For instance, in an immunoassay, the analyte may be the ligand or the binder, while in blood glucose testing, the analyte is glucose. In medicine, analyte often refers to the type of test being run on a patient, as the test is usually determining a chemical substance in the human body.

Anechoic Chamber: Anechoic chambers duplicate the silence of space, where there is no medium (like air or water) through which the sound waves can travel. The walls of the anechoic chambers are made of carbon compound pyramids which absorb any microwave and radio signals that hit them.

Angle of Inclination: The angle at which a satellite's orbit is tilted in relation to the Earth's equator. A 90 degree angle of inclination is a polar orbit. A zero degree angle of inclination is an *equatorial orbit*.

Anisotropic: Exhibiting different values of a property in different crystallographic directions.

Antenna: A piece of equipment that allows transmission and reception of radio signals. Satellites need antennas to communicate with Earth. A satellite may need to receive instructions and transmit the information it collects, or it may relay the information sent to it to another site on Earth. Since the information is transmitted using radio waves, which move at the speed of light, this method allows for very fast communications (only a very small time lag).

Aperture: In optics, an aperture is a hole or an opening through which light is admitted. More specifically, the aperture of an optical system is the opening that determines the cone angle of a bundle of rays that come to a focus in the image plane. An optical system typically has many openings, or structures that limit the ray bundles (ray bundles are also known as *pencils* of light).

Aromaticity: Aromaticity is a chemical property in which a conjugated ring of unsaturated bonds, lone pairs, or empty orbitals exhibit a stabilisation stronger than would be expected by the stabilisation of conjugation alone. It can also be considered a manifestation of cyclic delocalisation and of resonance. This is usually considered to be because electrons are free to cycle around circular arrangements of atoms, which are alternately single- and double-bonded to one another.

Artificial Intelligence (AI): A field of research that aims to understand and build intelligent machines; this term may also refer to an intelligent machine itself.

At the Core: A Non-Heavy Metal Quantum Dot. Research at Evident, which is focused on advancing product development, led to the first commercially available nanocrystals that comprise a III-V semiconductor material: Indium, Gallium and Phosphide core quantum dots, with a unique molecular metallic platting compound to form the shell or InGaP/ZnS EviDots-MP™.

Atmosphere: The many layers of gases that surround a planet. The Earth's atmosphere is composed of several layers of gases that separate our planet from space. The major gases in the Earth's atmosphere are nitrogen and oxygen. The air we breathe is part of the atmosphere.

Atmospheric Studies Satellite: A type of scientific satellite that studies the Earth's atmosphere. They were some of the very first satellites launched into space, including Canada's first satellite Alouette.

Atom: An atom is the smallest particle characterizing a chemical element. An atom consists of a dense nucleus of positively charged protons and electrically neutral neutrons, surrounded by a much larger electron cloud consisting of negatively charged electrons. When the number of protons matches the number of electrons, an atom is electrically neutral; otherwise it is an ion and has a net charge. An atom is classified according to its protons and neutrons: the number of protons determines the chemical element and the number of neutrons determines the isotope of that element. The concept of the atom as an indivisible component of matter was first proposed by early Indian and Greek philosophers.

Atomistic Simultations: Atomic motion computer simulations of macromolecular systems are increasingly becoming an essential part of materials science and nanotechnology. Recent advances in supercomputer simulation techniques provide the necessary tools for performing computations on nanoscale objects containing as many as 300,000 atoms and on materials simulated with 1,000,000 atoms. This new capability will allow computer simulation of mechanical devices or molecular machines using nanometer size components.

Avogadro's Number: The number of atoms in exactly 12 grams of pure 12C, equal to 6.022 × 1023.

Ballistic Magnetoresistance (BMR): Ballistic Magnetoresistance is yet another way in which spin orientation, encoding information on a storage medium such as a hard drive, can modify electrical resistance in a nearby circuit, thereby accomplishing the sensing of that orientation.

Band Gap Energy (Eg): For semiconductors and insulators, the energies that lie between the valence and conduction bands; for intrinsic materials, electrons are forbidden to have energies within this range.

Banoelectronics: Nanoelectronics in Japan and Korea tends to focus on next generation semiconductor devices and *single electron devices (SED).*

Banopositioning: The means of controlling motion on the nanometre scale—is now a key enabling technology in high- tech fields such as semiconductor test and measurement, photonics alignment, scanning microscopy and microlithography.

Banoscale: At the nanoscale, physics, chemistry, biology, materials science, and engineering converge toward the same principles and tools. The nanoscale is not just another step toward *miniaturisation*, but a qualitatively new scale. The new behaviour is dominated by quantum mechanics, material confinement in small structures, large interfacial volume fraction, and other unique properties, phenomena and processes. Many current theories of matter at the microscale have critical lengths of nanometer dimensions.

Banoscience: Nanoscience is primarily the extension of existing sciences into the realms of the extremely small (*nanomaterials, nanochemistry, nanobio, nanophysics*, etc.) while *nanoengineering* represents the extension of the engineering fields into the nano- scale realm (*nanofabrication, nanodevices*, etc.). The exponential growth of nanoscience is largely due to the development of new instruments and related techniques that are used to "routinely" probe and manipulate material at the atomic and molecular level. *Scanning probe microscopies,* analytical electron- beam techniques, epitaxial growth facilities, and *synchrotron* radiation sources are all opening huge opportunities.

Banoshells: Procedures that target cancer cells while leaving normal cells untouched, patients controlling the release of medicine in their bodies with an infrared light, and medical test results produced in seconds rather than days – these are three new medical technologies currently being tested by nanotechnology researchers at Rice University.... The research focuses on nanoshells, a new type of nanoparticle invented by Naomi Halas, Rice professor of electrical and computer engineering and chemistry. Nanoshells are layered nanoparticles whose ability to manipulate light and color can be designed into the nanoparticle by varying the thickness of the nanoparticle's layers.

Baronatation: In medical nanorobotics, locomotion through a frozen fluid by applying mechanical pressure along the path travelled to induce melting ahead, followed by regelation (refreezing) behind.

Bearing: A mechanical device that permits the motion of a component (ideally, with minimal resistance) in one or more degrees of freedom while resisting motion (ideally, with a stiff restoring force) in all other degrees of freedom.

Bilayer Lipid Membrane (BLM): The structure found in most biological membranes, in which two layers of lipid molecules are so arranged that their hydrophobic parts interpenetrate, whereas their hydrophilic parts form the two surfaces of the bilayer. The reduction in the free energy of a system that occurs when a ligand binds to a receptor. Generally used to describe the total energy required to remove something, or to take a system apart into its constituent particles—for example, to separate two atoms from one another, or to separate an atom into electrons and nuclei.

Binding: The process by which a molecule (or ligand) becomes bound, that is, confined in position (and often orientation) with respect to a receptor. Confinement occurs because structural features of the receptor create a potential well for the ligand; van der Waals and electrostatic interactions commonly contribute.

Biocatalysis: The application, both actual and potential, of biological catalysts (including whole cells or isolated components thereof, natural and modified enzymes and catalytic antibodies) for the synthesis, interconversion or degradation of chemical species. In addition to papers describing the synthetic applications of biotransformations the journal particularly welcomes papers focusing on the mechanistic principles, kinetics and thermodynamics of biocatalytic processes, the chemical or genetic modification of biocatalysts and the activity and stability of biocatalysts in non- aqueous and multi- phasic environments, including the design of large scale biocatalytic processes. The scope of the journal also encompasses biomimetic systems and environmental applications of biocatalysis where the mechanistic principles are understood or novel biocatalytic activities are involved.

Biochauvinism: The prejudice that biological systems have an intrinsic superiority that will always give them a monopoly on self-reproduction and intelligence.

Biocompatible Coated Materials: Biocompatible materials usually used in dental and bone implants that enhance biologic fixation, thereby increasing the bond strength between the coated material and bone, and minimize possible biological effects that may result from the implant itself. MeSH, 1999

Biodefense: The need for improved national defenses against biological attacks are urgently needed. A real-time reporting system needs to be developed, whereby officials can be informed about an emerging threat before an agent has a chance to affect thousands of people. The development of integrated systems for detecting and monitoring biological agents is instrumental to this goal. The development of vaccines and novel detection platforms will enable preparedness, while also benefiting human health.

Biomaterials: The development of *biomaterials* is not a new area of science, having existed for around half a century. The study of biomaterials is called biomaterial science. It is an exciting field of science, having experienced steady and strong growth over its history with companies such as Smith and Nephew investing large amounts of money in new products. Biomaterials science encompasses elements of medicine, biology, chemistry and materials science. While a definition for biomaterials has been difficult to formulate, a widely accepted definition for biomaterials is that: "A biomaterial is any material, natural or man-made, that comprises whole or part of a living structure or biomedical device which performs, auguments, or replaces a natural function".

Biomechanics: Biomechanics is mechanics applied to biology (Fung). This includes research and analysis of the mechanics of living organisms and the application of engineering principles to and from biological systems. This research and analysis can be carried forth on multiple levels, from the molecular, wherein biomaterials such as collagen and elastin are considered, all the way up to the tissue and organ level. Some simple applications of Newtonian mechanics can supply correct approximations on each level, but precise details demand the use of continuum mechanics.

Biomedical Nanotechnology: The science and technology of diagnosing, treating, and preventing disease and traumatic injury, of relieving pain, and of preserving and improving human health, using molecular tools and molecular knowledge of the human body.

BioMEMS Biological MicroElectro Mechanical Systems: BioMEMs & Biomedical Nanotechnology WORLD 2000 encompasses all interfaces and intersections of the life sciences and clinical disciplines with microsystems and nanotechnology. Areas of interest include, but are not limited to, micro- and nanotechnology for drug delivery, tissue engineering, harvesting, manipulation, amplification, and sequencing of nucleic acids, proteomics, microfluidics and miniaturized total analysis systems (microTAS), biosensors, molecular assembly, nanoscale imaging, and integrated systems. Contributions addressing all stages of research and development are welcome, from basic science fundamentals and technology concepts to product development, clinical investigations, and business and ethical considerations. BioMEMs and Nanotechnology World 2001 Sept. 22-25 2001, Columbus OH. Broader term MEMs.

Biomimetics: Imitating, copying, or learning from nature. Nanotechnology already exists in nature; thus, nanoscientists have a wide variety of components and tricks already available. In other words, the design of systems, materials, and their functionality to mimic nature. Current examples include layering of materials to achieve the hardness of an abalone shell or understanding why spider silk is stronger than steel.

Biomolecular Materials: An emerging discipline, materials whose properties are abstracted from biology. They share many of the characteristics of biological materials but are not necessarily of biological origin. For example, they may be inorganic materials that are organised or processed in a biomimetic fashion. A key feature of biological and biomolecular materials is their ability to undergo *self- assembly*.

Bionanotechnology: Bionanotechnology is the intersection of biology and nanotechnology. Bionanotechnology is a broad and somewhat vague term which is sometimes used interchangeably with nanobiotechnology, which usually refers more specifically to the use of nanotechnological devices for applications in biotechnology. Bionanotechnology may also refer to the use of biomolecules for applications in nanotechnology. A major example of this is DNA nanotechnology, which uses self-assembling nucleic acid structures to control matter at the nanoscale. In a wider sense, bionanotechnolgy refers to synthetic technology based on the principles and chemical pathways of living organisms. In other words, includes molecular motors, biomaterials, single molecule manipulation technologies, biochip technologies, etc.

Biorobotics: Our research focuses on the role of sensing and mechanical design in motor control, in both robots and humans. This work draws upon diverse disciplines, including biomechanics, systems analysis, and neurophysiology. The main approach is experimental, although analysis and simulation play important parts. In conjunction with industrial partners, we are developing applications of this research in biomedical instrumentation, teleoperated robots, and intelligent sensors.

Biotin: A vitamin of the B complex. It is a co-enzyme for various enzymes that catalyse the incorporation of carbon dioxide into various compounds. It is essential for the metabolism of fats. Biotin is attached to pyruvate carboxylase by a long, flexible chain like that of lipoamide in the pyruvate dehydrogenase complex. Adequate amounts are normally produced by the intestinal bacteria in animals.

Biotinylated-DNA: A DNA molecule labelled with biotin by incorporation of biotinylated -dUTP into a DNA molecule. It is used as a non- radioactive probe in hybridisation experiments, such as Southern transfer. The detection of the labelled DNA is achieved by complexing it with streptavidin (an antibiotic with a high affinity for biotin) to which is attached a colour- generating agent such as horseradish peroxidase that gives a fluorescent green colour upon reaction with various organic reagents. Biotinylation: To label a probe with biotin.

Bond: Two atoms are said to be bonded when the energy required to separate them is substantially larger than the van der Waals attraction energy. Ionic bonds result from the electrostatic attraction between ions; covalent and metallic bonds result from the sharing of electrons among atoms; hydrogen bonds are weaker and result from dipole interactions and limited electron sharing. When used without modification, "bond" usually refers to a covalent bond.

Bone Substitutes: Synthetic or natural materials for the replacement of bones or bone tissue. They include hard tissue replacement polymers, natural coral, hydroxyapatite, beta- tricalcium phosphate, and various other biomaterials. The bone substitutes as inert materials can be incorporated into surrounding tissue or gradually replaced by original tissue.

Bottom-up Nanotechnology: Mostly chemists attempting to create structure by connecting molecules.

Broadband Mode: In this mode of operation the error signal E representing bending of the cantilever not corrected by the feedback is scaled and added to the Z signal to provide a mathematically correct image (I=E+Z).

Brownian Assembly: Brownian motion in a fluid brings molecules together in various position and orientations. If molecules have suitable complementary surfaces, they can bind, assembling to form a specific structure. Brownian assembly is a less paradoxical name for self-assembly (how can a structure assemble itself, or do anything, when it does not yet exist?).

Brownian Motion: Motion of a particle in a fluid owing to thermal agitation. Motion of a particle in a fluid owing to thermal agitation, observed in 1827 by Robert Brown. Stochastic motion of a particle suspended in a surrounding gas or liquid of other particles, molecules or atoms, which is in thermodynamical equilibrium. The motion of the particle is due to the collisions with the particles from its surrounding and as such is 'random', i.e. the change of particle's speed in any of the encounters is random. Although the Brownian motion is stochastical, it has some statistical properties which are fixed.

Buckminsterfullerene: A fullerene with the chemical formula (C60). A broad term covering the variety of buckyballs and carbon nanotubes that exist. Named after the architect Buckminster Fuller, who is famous for the geodesic dome, which buckyballs resemble.

Cantilever Vibration Mode: An atomic force microscope (AFM) combined with ultrasonic frequency vibration of a cantilever excited at its support. This method enables both topography and elasticity imaging of stiff samples such as metals and ceramics, without a need for bonding a transducer to the sample.

Capacitance: The charge-storing ability of a capacitor, defined as the magnitude of charge stored on either plate divided by the applied voltage. A 1-F capacitor charged to 1 V contains C of charge and 1 C is an amount of charge equal to that of about 6.24 × 1018 electrons.

Capillaries: Capillaries are the smallest of a body's blood vessels, measuring 5-10 ìm in diameter, which connect arterioles and venules, and enable the interchange of water, oxygen, carbon dioxide, and many other nutrient and waste chemical substances between blood and surrounding tissues.

Carbon Black: Carbon black is a powdered form of elemental carbon. The primary use of carbon black is in rubber products, mainly tyres and other automotive products, but also in many other rubber products such as hoses, gaskets and coated fabrics. Much smaller amounts of carbon black are used in inks and paints, plastics and in the manufacture of dry-cell batteries.

Carbon Nanofoam: Carbon nanofoam is the fifth known allotrope of carbon discovered in 1997 by Andrei V. Rode and co-workers at the Australian National University in Canberra. It consists of a low-density cluster-assembly of carbon atoms strung together in a loose three-dimensional web. The large-scale structure of carbon nanofoam is similar to that of an aerogel, but with 1% of the density of previously produced carbon aerogels — only a few times the density of air at sea level. Each cluster is about 6 nanometers wide and consists of about 4000 carbon atoms linked in graphite-like sheets that are given negative curvature by the inclusion of heptagons among the regular hexagonal pattern. This is the opposite of what happens in the case of buckminsterfullerenes, in which carbon sheets are given positive curvature by the inclusion of pentagons. Unlike carbon aerogels, carbon nanofoam is a poor electrical conductor. The nanofoam contains numerous unpaired electrons, which Rode and colleagues propose is due to carbon atoms with only three bonds that are found at topological and bonding defects.

Carbon Nanotubes: (AKA. Bucky™ tubes), a member of the fullerene family. Individual single wall nanotubes have the electrical conductivity of copper or silicon, the thermal conductivity of diamond, and also happen to be the stiffest and strongest fibers known. Their molecular perfection enables them to be manipulated using the versatile chemistry of carbon. Carbon nanotube tips have several advantages [as atomic force microscopy tips] , including high aspect ratio for imaging deep and narrow crevices, low tip- sample adhesion for gentle imaging, the ability to elastically buckle rather than break when large forces are applied, and the potential to achieve resolutions in the range of 1.0 nm or less. In addition, carbon nanotubes have well defined molecular structures so that it is possible to control their synthesis to make every tip with an identical structure and resolution. Carbon nanotubes can be selectively modified at their ends with organic or biological molecules to allow functional sensitive imaging. These tiny straw-like cylinders of pure carbon have useful electrical properties. They have already been used to make tiny transistors and one-dimensional copper wire. They were developed by using nanotechnology, a relatively new field that involves building electronic circuits and devices from single atoms and molecules. Nano means one thousand millionth of a unit. A nanometer is therefore one thousand millionth of a meter. The first nanofabrication experiments occurred in 1990 when individual xenon atoms were placed on a nickel substrate and used to spell out a company logo. One primary goal of nanotechnology is to build computer chips and other devices that are thousands of times smaller than they are now Carbon nanotubes have enormous theoretical possibilities but have not lived up to the hype surrounding their development. Researchers have continued to look for ways to use them, however, as successful applications have the potential to be highly lucrative. Scientists have recently succeeded in altering carbon nanotubes so that they supply electrons when exposed to light. This was done by having two flat rings of carbon molecules sandwich a ferrocene (iron) molecule. Ferrocene is known for its tendency to relinquish electrons. When exposed to visible light, the carbon atoms accepted the ferrocene molecule.

Carbonyl: A chemical moiety consisting of O with a double bond to C. If the C is bonded to N, the resulting structure is termed an amide; if it is bonded to O, it is termed a carboxylic acid or an ester linkage.

Carboxylic Acid: Carboxylic acids are organic acids characterized by the presence of a carboxyl group, which has the formula -C(=O)OH, usually written -COOH or -CO2H. Carboxylic acids are Bronsted/Lowry acids — they are proton donors. Salts and anions of carboxylic acids are called carboxylates. The simplest series of carboxylic acids are the alkanoic acids, R-COOH, where R is a hydrogen or an alkyl group. Compounds may also have two or more carboxylic acid groups per molecule. Carboxylic acids are polar, and form hydrogen bonds with each other.

Cathode: The electrode in an electrochemical cell or galvanic couple at which a reduction reaction occurs; in other words the electrode receiving electrons from an external circuit.

Cation: An ion consists of one or more atoms and carries a unit charge of electricity. Those that are positively electrified (hydrogen and the metals) are called cations (cf. anion).

Cell Repair Machine: Nanocomputers including system and molecular scale sensors and tools, programmed to repair damage to cells and tissues. Molecular and nanoscale machines with sensors, nanocomputers and tools, programmed to detect and repair damage to cells and tissues, which could even report back to and receive instructions from a human doctor if needed.

Charge Coupled Device: A highly sensitive camera, this counts individual photons of light to put together an image. It can take very detailed photographs of small areas.

Charged Particle: In physics, a charged particle is a particle with an electric charge. It may be either a subatomic particle or an ion. A collection of charged particles, or even a gas containing a proportion of charged particles, is called a plasma, which is called the *fourth state of matter* because its properties are quite different from solids, liquids and gases (plasma is the most common state of Matter in the universe). Particles either have a positive, negative or no charge (being neutral). Their effects go beyond the laboratory though: see aurora borealis

Chemotactic Nanosensor: In medical nanorobotics, a nanosensor used to determine the chemical characteristics of surfaces, possibly configured as a pad coated with an array of reversible, perhaps reconfigurable, artificial molecular receptors.

Chromatography: The physical method of separation in which the components to be separated are distributed between two phases, one of which is stationary while the other moves in a definite direction. Chromatography is a widely used for the separation, identification, and determination of the chemical components in complex mixtures.

Chronocyte: In medical nanorobotics, a theorized mobile, mass-storage (nanorobotic) device, similar to a communicyte, that may be used as a mobile source of precisely synchronized universal time inside the human body.

Clays: A family of finely crystalline materials, composed primarily of silicates, with layered crystal structures. Clays may be refined from geological materials or may be synthetic.

Cobots: Collaborative robots designed to work alongside human operators. Prototype cobots are being used on automobile assembly lines to help guide heavy components like seats and dashboards into cars so they don't damage auto body parts as workers install them.

Cold Plasma Analyzer: An instrument on board the Swedish scientific satellite Freja. By sweeping its sensor through plasma, it obtains measurements of the charged particles in a layer of the atmosphere, the magnetosphere and upper ionosphere, and thus helps scientists understand Earth's atmosphere and how to protect it from pollutants.

Combinatorial Biology: Involves genetic manipulation of bacteria and fungi that produce complex natural products. This technology includes construction of large libraries of recombinant microbes capable of generating novel organic molecules and engineering secondary metabolite biosynthetic pathways to modify valuable biologically active microbial metabolites.

Communications Satellite: A type of satellite used for communications on Earth by allowing radio, television, and telephone transmissions to be sent live anywhere in the world. Before satellites, transmissions were difficult or impossible at long distances. The signals, which travel in straight lines, could not bend around the round Earth to reach a destination far away. Because communications satellites are in orbit, the signals can be sent instantaneously into space and then redirected to another satellite or directly to their destination.

Communicyte: In medical nanorobotics, a theorized mobile, mass-storage (nanorobotic) device that can be used for information transport throughout the human body.

Composites: Combinations of metals, ceramics, polymers, and biological materials that allow multi-functional behaviour. One common practice is reinforcing polymers or ceramics with ceramic fibres to increase strength while retaining light weight and avoiding the brittleness of the monolithic ceramic. Materials used in the body often combine biological and structural functions (e.g., the encapsulation of drugs).

Computational Nanotechnology: Permits the modeling and simulation of complex nanometer-scale structures. The predictive and analytical power of computation is critical to success in nanotechnology: nature required several hundred million years to evolve a functional "wet" nanotechnology; the insight provided by computation should allow us to reduce the development time of a working "dry" nanotechnology to a few decades, and it will have a major impact on the "wet" side as well.

Computronium: A highly (or optimally) efficient matrix for computation, such as dense lattices of nanocomputers or quantum dot cellular automata.

Confocal Microscopy: A light microscopic technique in which only a small spot is illuminated and observed at a time. An image is constructed through point-by- point scanning of the field in this manner. Light sources may be conventional or laser, and fluorescence or transmitted observations are possible.

Conjugation: In medical nanorobotics, the docking of two or more nanorobots for the purpose of exchanging information, energy or materials, or to establish a larger multirobotic structure; in biology, the union of two unicellular organisms accompanied by an interchange of nuclear material, as in Paramecium.

Coupling: A connection between more than one object or energy pathway so that together they function as a single unit.

Covalent Bond: A bond formed by sharing a pair of electrons between two atoms. A primary interatomic bond that is formed by the sharing of electrons between neighboring atoms. An attachment between two atoms resulting from the sharing of a pair of electrons

Covalent Radius: Given a set of N elements that can form covalent single bonds in molecules, with $N(N - 1)$ possible elemental pairings, it has proved possible to define a covalent radius for each element such that the actual bond length between any two elements that form a covalent single bond is roughly equal to the sums of their covalent radii.

Creep: The time-dependent permanent deformation that occurs under stress; for most materials it is important only at elevated temperatures.

Crosslinked Polymer: A polymer in which adjacent linear molecular chains are joined at various positions by covalent bonds.

Cryogenic: In physics or engineering, cryogenics is the study of the production of very low temperatures (below –150 °C, –238 °F or 123 K) and the behaviour of materials at those temperatures. (Rather than the familiar temperature scales of Fahrenheit and Celsius, cryogenicists use the Rankine and Kelvin scales.) The word *cryogenics* literally means "the production of icy cold"; however the term is used today as a synonym for the low-temperature state. It is not well-defined at what point on the temperature scale refrigeration ends and cryogenics begins. The workers at the National Institute of Standards and Technology at Boulder, Colorado have chosen to consider the field of cryogenics as that involving temperatures below –180 °C (93.15 K). This is a logical dividing line, since the normal boiling points of the so-called permanent gases (such as helium, hydrogen, neon, nitrogen, oxygen, and normal air) lie below -180 °C while the Freon refrigerants, hydrogen sulfide, and other common refrigerants have boiling points above -180 °C.

Crystal Structure: For crystalline materials, the manner in which atoms or ions are arrayed in space. It is defined in terms of the unit cell geometry and the atom positions within the unit cell.

Cyclotron: A type of particle accelerator in which an ion introduced at the center is accelerated in an expanding spiral path by use of alternating electrical fields in the presence of a magnetic field.

Dendrimer: A polymer having a regular branched structure; If suitably functionalized may be used as a soluble support, in which case the desired, dendrimer-supported, material may be isolated by size- exclusion chromatography. Dendrimers may also be attached to a polymer and used as a solid support, with significantly increased loading over the initial resin. Dendrimers consist of interconnected monomeric subunits that hybridize to form a tree-like structure. Each monomer is a double- stranded DNA molecule where the two strands share a region of sequence complementarity in the middle of molecule. A dendrimer (from Greek *dendra* for tree) is an artificially manufactured or synthesized molecule built up from branched units called monomers. Such processes involve working on the scale of nanometers (a nanometer is 10^{-9} meter or a millionth of a millimeter). Technically, a dendrimer is a polymer, which is a large molecule comprised of many smaller ones linked together. Dendrimers have some proven applications, and numerous potential applications. They have been used in the production of industrial adhesives. They are expected to serve as components in a variety of nanomachines. Dendrimers are of interest to researchers in medical technology, where they might help carry and deliver drugs in the body, or serve as replacements for plasma components. Dendrimers might also prove useful in the manufacture of nanoscale batteries and lubricants, catalysts, and herbicides. Dip Pen Nanolithography DPN: A new AFM [Atomic Force Microscopy] -based soft- lithography technique which was recently discovered in our [Chad Mirkin, Northwestern Univ.] labs. ... We have found that an important requirement for creating stable nanostructures is that the transported molecules anchor themselves to the substrate via chemisorption. When T-substituted alkanethiols are patterned on a gold substrate, a monolayer is formed in which the thiol headgroups form relatively strong bonds to the gold and the alkane chains extend roughly perpendicular to surface. The thiol lattice formed is identical to that of a monolayer obtained via solution deposition of alkanethiols on gold. Creating nanostructures using DPN is a single step process which does not require the use of resists.

Using a conventional Atomic Force Microscope it is possible to achieve ultra-high resolution features - as small as 15 nm linewidths and ~ 5 nm spatial resolution, Figure 3a. For nanotechnology applications, it is not only important to pattern molecules in high resolution, but also to functionalize surfaces with patterns of two or more components. One of the most important attributes of DPN is that because the same device is used to image and write a pattern, patterns of multiple molecular inks can be formed on the same substrate in very high alignment, Figure 3b. With the aid of software created in- house, we have devised a DPN nanoplotter with which to write any type of complicated pattern. Just for fun we have reproduced Richard Feynmann's famous speech in which he predicts a bright future for nanotechnology.

Directed-Assembler: A specific type of assembler that makes use of directed-assembly, such that the assembly process requires external energy or information input.

Disassembler: In molecular nanotechnology, a nanomachine or system of nanomachines able to take an object apart while at each step recording the structure and composition of that object at the molecular level.

Disasterbation: Idly fantasizing about possible catastrophes (ecological collapse, full-blown totalitarianism) without considering their likelihood or considering their possible solutions and preventions.

Dissolution: Deterioration in an organism such that its original structure cannot be determined from its current state.

Distributed Intelligence: An intelligent entity which is distributed over a large volume (or inside another system, like a computer network) with no distinct center. This is the opposite to the strategy of Concentrated intelligences. Distributed intelligences have much longer communications lags, but are more flexible in their structure and can survive damage to their parts.

DNA Diagnostics - Miniaturisation of: In the areas of sample preparation and assay, it is clear that miniaturisation is key. To reduce the size of samples by a factor of 10 or greater, barriers in microfluidics, micromachining, robotics, microchemistry, nucleic acid chemistry, and surface chemistry must be overcome. To implement miniaturized protocols accurately and efficiently, substantial automation of the process will be required. In the development of miniaturized systems, it is essential that the system can be adapted for high levels of parallelisation. Miniaturisation poses significant technological risks. Currently, there exists no universally accepted precedent for the handling, replication, amplification, or cloning of DNA in nanoliter volumes. Due to the size and charge of the DNA molecule, and the relative instability of many of the enzymes involved in the sample preparation processes, nanoliter and less volumes may pose substantial challenges. In addition, interactions of the biological molecules with the surfaces of the reaction chambers must be minimized. For some methodologies, it is not clear what the optimal sample will be, so substantial improvement in DNA fragmentation technologies or DNA cloning vectors may be required for the ultimate efficient application to diagnostics. Improvements in any of these areas are likely to be of value to other non- DNA based diagnostic applications such as antibody screening protocols and enzyme based diagnostics, because miniaturized robotic or micro- electro mechanical systems developed for DNA could be modified to be used for these purposes.

DNA Probes: A DNA or nucleic acid probe is a short strand of DNA that locates and binds to its complementary sequence in samples containing single strands of DNA or RNA enabling identification of specific sequences. Nucleic acid probe assays exploit the fundamental hybridisation reaction that occurs spontaneously between two complementary DNA.

DNA Sequencing: There are two main classical methods for sequencing DNA: The first method, developed by Allan Maxam and Walter Gilbert, involves chemicals used to cleave the DNA at certain positions, generating a set of fragments that differ by one nucleotide. The second method, developed by Fred Sanger and Alan Coulson, involves enzymatic synthesis of DNA strands that terminate in a modified nucleotide. Analysis of fragments is similar for both methods and involves gel electrophoresis and autoradiography or fluorescence. The enzymatic method has largely replaced the chemical method as the technique of choice, although there are some situations where chemical sequencing can provide data more casily than the enzymatic method.

DNA: A molecule encoding genetic information, found in the cell's nucleus. Deoxyribose Nucleic Acid. DNA is a molecule used in biological systems to store genetic information. It has a long, double stranded structure with the two strands wrapped around each other to form a right handed helix. There are about 10 nucleotide pairs per helical turn. Each strand is composed of a sugar phosphate backbone and attached bases. It is connected to a complementary strand by hydrogen bonding (non-covalent) between paired bases, adenine (A) with thymine (T) and guanine (G) with cytosine (C). DNA is an abbreviation of DeoxyriboNucleic Acid. It is the genetic 'blueprint' of all life on this planet. DNA typing is often applied to Human Identification which has also been called DNA Fingerprinting. DNA Fingerprinting has been attributed to Alex Jeffries, who with his colleagues in 1985, discovered a sequence of nucleotides, subsequently termed short tandem repeats (STR) when comparing the DNA sequence of the human á-globin chromosomes isolated from different individuals.

Dnhancer: Regulatory DNA sequence to which gene regulatory proteins bind, influencing the rate of transcription of a structural gene that can be many thousands of base pairs away.

Domain: A region of a ferromagnetic or ferrimagnetic material in which all atomic or ionic magnetic moments are aligned in the same direction.

Dopeyballs: Superconducting Buckyballs (they) have the highest critical temperature of any known organic compound.

Doppler Effect: An apparent shift in the *frequency* of a wave. For example, when someone is listening to the sound of an ambulance siren, and that person is staying still but the ambulance is driving by, the person will hear a change in pitch of the ambulance siren. That change in pitch is caused by the doppler effect. The frequency of a sound wave determines the pitch, and the distance of the source of the sound from the sound's observer determines the amount that the frequency seems to have shifted, known as the doppler shift.

Drug Design: Includes not only ligand design, but also pharmacokinetics (Pharmacogenomics) toxicity, which are mostly beyond the possibilities of structure- and/ or computer- aided design. Nevertheless, appropriate chemometric (Chemoinformatics) tools, including experimental design and multivariate statistics, can be of value in the planning and evaluation of pharmacokinetic and toxicological experiments and results. Drug design is most often used instead of the correct term "ligand design".

Drug Discovery and Molecular Imaging: Molecular imaging's core role for the pharmaceutical industry is in drug discovery and development. Pure research is already profiting from cell- based molecular imaging, which will continue to be based on fluorescence, bioluminescence, and confocal microscopy. Applications in small animal imaging, lead characterisation, and lead optimisation are also discussed. The insights into basic cell biology that this research is yielding today will form the basis of drug development during the second half of the decade, as the results are absorbed by the pharmaceutical industry.

Drug Ontology: A three year project to develop a highly structured drug knowledge base. Unlike existing reference sources, it is intended solely for use by software applications.

Druggable: Able to be modulated by a small molecule to produce a desired phenotypic change in cell targets.

Dry Nanotechnology: Derives from surface science and physical chemistry, focuses on fabrication of structures in carbon silicon, and other inorganic materials. Unlike the 'wet' technology, 'dry' techniques admit use of metals and semiconductors. The active conduction electrons of these materials make them too reactive to operate in a 'wet' environment, but these same electrons provide the physical properties that make 'dry' nanostructures promising as electronic, magnetic, and optical devices. Another objective is to develop 'dry' structures that possess some of the same attributes of the self-assembly that the wet ones exhibit.

Emergence: A complex whole created by simple parts, as in the brain where billions of neurons work individually, but collectively make up our consciousness and give us the ability to think, rationalize, and create.

Encryption Device: An instrument often used in spy satellites. It encodes digital images or data before it is sent back to Earth.

Endocrine Cell: Specialized animal cell that secretes a hormone into the blood; usually part of a gland, such as the thyroid or pituitary gland.

Endocytosis: Uptake of material into a cell by an invagination of the plasma membrane and its internalisation in a membrane-bounded vesicle.

Engulf Formation: In medical nanorobotics, a configuration that may be adopted by a metamorphic nanorobot, in which the nanorobot reshapes itself to create an interior cavity capable of trapping a living cell, virion, or other biological particle.

Enhancer: Regulatory DNA sequence to which gene regulatory proteins bind, influencing the rate of transcription of a structural gene that can be many thousands of base pairs away.

Entanglement: From quantum mechanics, entanglement is a relationship between two systems in which they exist in more than one state simultaneously (by superposition), and the state of one system determines the state of the other.

Enternal Computer: The satellite's method of storing and analyzing data collected by the satellite, and a way of controlling its various systems. The technical term for this satellite subsystem is *TT&C* (Telemetry, Tracking, and Control).

Enzyme immunoassay/EIA: In an enzyme immunoassay (EIA), an enzyme-labeled antibody or antigen is used for the detection and quantification of the antigen-antibody reaction. In an electrochemical EIA, the enzyme-catalyzed reaction is monitored electrochemically (amperometric, potentiometric, voltametric or conductometric). In EIA, the antibody-antigen reaction furnishes the needed specificity. The enzyme label provides the sensitivity via chemical amplification. A large molecule, usually a protein, that catalyzes biological reactions. A protein that acts as a catalyst in a biochemical reaction. Protein that catalyzes a specific chemical reaction.

Equator: The equator is an imaginary circle drawn around a planet (or other astronomical object) at a distance halfway between the poles. The equator divides the planet into a Northern Hemisphere and a Southern Hemisphere. The latitude of the equator is, by definition, 0°. The length of Earth's equator is about 40,075.0 km, or 24,901.5 miles. The equator is one of the five main circles of latitude based on the relationship of the Earth's rotation and plane of orbit around the sun. Additionally, the equator is the only line of latitude which is also a great circle. The Sun, in its seasonal movement through the sky, passes directly over the equator twice each year on the March and September Equinoxes. At the equator, the rays of the sun are perpendicular to the surface of the earth on these dates. Places near the equator experience the quickest rates of sunrise and sunset in the world, taking minutes. Such places also have a relatively constant amount of day/night time on every day throughout the year compared with more northerly or southerly places.

Fatty Acid: Compound such as palmitic acid that has a carboxylic acid attached to a long hydrocarbon chain. Used as a major source of energy during metabolism and as a starting point for the synthesis of phospholipids.

Fault-tolerant: Describes a system that can suffer failure in a component or subsystem, yet continue to function correctly.

Femtosecond: Femtosecond is one quadrillionth of a second, and is to a second what a second is to 32,700,000 years. At 186,000 miles per second, in one femtosecond light travels only far enough to traverse about 1,000 silicon atoms. When used to time a laser pulse, it allows for ultra-precise micromachining, with virtually no damage to surrounding material.

Femtotechnology: Femtotechnology the art of manipulating materials on the scale of elementary particles (leptons, hadrons, and quarks). The next step smaller after picotechnology, which is the next step smaller after nanotechnology.

Fermi Energy: The energy level in a solid at which the probability of finding an electron is 1/2. For a metal, the energy corresponding to the highest filled electron state in the valence band at 0 K.

Ferroelectric Material: A dielectric material such as Rochelle salt and barium titanate with a domain structure containing dipoles (asymmetric distributions of electrical charge) which spontaneously align. Their domain structure makes them analogous to ferromagnetic materials. They exhibit hysteresis and usually the piezoelectric effect.

Ferromagnetism: Permanent and large magnetisations found in some metals (e.g. Fe, Ni and Co), resulting from the parallel alignment of neighboring magnetic moments.

Fertilisation: Fusion of a male and a female gamete (both haploid) to form a diploid zygote, which develops into a new individual.

Fiber-optic: Relates to transmission of information as modulated light in tiny transparent fibers instead of copper wires.

Fibrin Sealant: A biological tissue glue derived from human plasma. Sometimes referred to as fibrin glue, it contains the components necessary for the last step of coagulation: i) fibrinogen and other coagulation factors and ii) thrombin. It is used in surgical procedures to arrest bleeding and as an adjunct to wound healing.

Field-Effect Transistor (FET): Field-Effect Transistor — semiconductor device whose insulated gate electrode controls current flow.

Fluidic System: Device for synthesis or screening in which fluids such as reagents or assay buffers may be directed to specified locations by the opening and closing of valves in a stationary network of tubes and wells.

Fluorescence Microscopy: Microscopy of specimens stained with fluorescent dye (usually fluorescein isothiocyanate) or of naturally fluorescent materials, which emit light when exposed to ultraviolet or blue light. Immunofluorescence microscopy utilizes antibodies that are labeled with fluorescent dye.

Fluorescence Polarisation: Based on the observation that emission signals from small fluorescent molecules are relatively depolarized, while binding to a larger molecule reduces the tumbling rate of the fluorescer, resulting in a relatively polarized emission signal. First described in 1926 (Perrin) and has been a powerful tool in the study of molecular interactions. When fluorescent molecules are excited with plane polarized light, they emit light in the same polarized plane, provided that the molecule remains stationary throughout the excited state (4 nanoseconds in the case of fluorescein). However, if the excited molecule rotates or tumbles out of the plane of polarized light during the excited state, then light is emitted in a different plane from that of the initial excitation.

Fluorescent dye: Molecule that absorbs light at one wavelength and responds by emitting light at another wavelength; the emitted light is of longer wavelength (and hence of lower energy) than the light absorbed.

Fluorescent Dye: Molecule that absorbs light at one wavelength and responds by emitting light at another wavelength; the emitted light is of longer wavelength (and hence of lower energy) than the light absorbed.

Focal Length: The distance needed between a lens and an object to make the object visible and in focus. The human eye has different focal lengths depending on how much the muscles in the eye around the lens are contracted. That is how the human eye focuses.

Footprint: The area with which a satellite in *geostationary orbit* can communicate. A footprint can be as large as an entire country; for example, many Canadian satellites have footprints almost the entire size of Canada, from coast to coast. Click here for a picture of a satellite's footprint.

Free Energy: Energy that can be extracted from a system to drive reactions. Takes into account changes in both energy and entropy.

Free-energy Change: Change in the free energy during a reaction: the free energy of the product molecules minus the free energy of the starting molecules. A large negative value of DG indicates that the reaction has a strong tendency to occur.

Fullerene: A new allotrope of carbon characterized by a closed cage structure consisting of an even number of three coordinate carbon atoms devoid of hydrogen atoms. This class was originally limited to closed-cage structures with twelve isolated five- membered rings, the rest being six- membered rings.

Geostationary Orbit: An orbit in which a satellite appears to remain in the same spot in the sky all the time. When a satellite is in geostationary orbit, it travels at exactly the same speed as the Earth is rotating below it. A satellite in geostationary orbit is very high up, at 35 850 km above the Earth.

Geostationary satellites are always located directly above the equator. The area with which a satellite in geostationary orbit can communicate is called its *footprint*.

Green Ceramic Body: A ceramic piece, formed as a particulate aggregate, that has been dried but not fired.

Henry: Unit of inductance. One henry (H) is the inductance of a closed circuit in which an electromotive force of 1 volt is produced when the electric current in the circuit varies uniformly at the rate of 1 ampere per second.

Heterochromatin: Region of a chromosome that remains unusually condensed and transcriptionally inactive during interphase.

Highest Occupied Molecular Orbital (HOMO): The highest energy molecular orbital of an atom or molecule that contains an electron.

Histonatation: In medical nanorobotics, locomotion (swimming) through tissues by a nanorobot.

Histone: A protein present in the eukaryotic nucleus that is bound to the DNA at regularly spaced intervals.

Immunoglobulin: An antibody molecule. Higher vertebrates have five classes of immunoglobulin—IgA, IgD, IgE, IgG, and IgM—each with a different role in the immune response.: : immunoglobulin like (Ig-like) domain: : Characteristic protein domain of about 100 amino acids that is found in antibody molecules and in many other proteins that form the Ig superfamily.

Immunohistochemistry: Histochemical localisation of immunoreactive substances using labeled antibodies as reagents. Immunohistochemistry involves using antibodies (typically visualized via an enzyme- linked antibody assay) that specifically bind to proteins of interest. This method allows one not only to assess levels of a protein but also to localize the protein within cells in the tissue sample.

Inductance: The property of an electric circuit which tends to oppose change in current in the circuit. One henry (H) is the inductance of a closed circuit in which an electromotive force of 1 volt is produced when the electric current in the circuit varies uniformly at the rate of 1 ampere per second.

Induction (embryonic): Change in the developmental fate of one tissue caused by an interaction with another tissue.

Inductor: Energy storage circuit component consisting of a coil of wire and possibly a magnetic material.

Infrared Spectromicroscopy: Synchrotron Radiation is a powerful tool in research, broadly applied in the VUV and x-ray spectral regions. The three main advantages of synchrotron radiation are: broadband characteristics, high collimation and the possibility to calculate its properties with Schwinger's equation. Synchrotron radiation can also be superior in the infrared spectral region.

Infrared: Invisible electromagnetic radiation having a longer wavelength, and lower frequency, than visible red light.

Inhibitor: A chemical substance that, when added in relatively low concentrations, slows down a chemical reaction.

Inline Universities: (As opposed to online universities), nanocomputer implants serving to increase intelligence and education of their owners, essentially turning them into walking universities.

Inmessaging: In medical nanorobotics, conveyance of information from a source external to the human body, or external to working nanodevices, to a receiver located inside the human body.

Input Device: A digitizer, scanner, line art, clip art, keyboard, video camera or any other device that is used to generate and send a design or other instruction into a computer for eventual production on an output device.

Integrated Microdevices IM: DARPA/MTO wishes to capitalize on the investments in the manufacture, design and test of **semiconductor** integrated circuits by applying them to the emerging mixed technology systems area. The tight integration of microdevices in mixed technology systems, however, requires more than just an electronic domain analysis to understand and optimize the functionality of the design. Issues such as coupled energy domain simulation; three dimensional shape analysis before, and possibly after, integration; and mixed technology interconnect design and analysis need to be addressed as part of the design process. Key to enabling mixed technology systems is the development of a design environment that supports design and manufacture based upon many available mixed technology and electronic building blocks. In addition, design trade- offs, optimisations and synthesis need to be explored from an overall systems perspective in a mixed domain design and layout environment. The ultimate goal of the Composite CAD program is to provide a validated design environment that enables the reliable and predictable design of mixed technology systems. A new term, integrated microdevices (or IM's), is being introduced to identify this new type of monolithic system. DARPA seeks innovative proposals in the following areas: I. Process and shape based analysis for the process and device engineer, II. Circuit and physical layout synthesis and verification environments for the systems designer, and III. Applications which utilize the resulting design technology on integrated microdevices (IM's). The primary focus is on linking under- development shape based analysis tools into a functional circuit and physical layout environment. The circuit and layout environment could be enhanced from standard mixed signal design tool set but needs to allow the design and incorporation of mixed technology devices such as microelectromechanical (MEMS) resonators, optoelectronic transmitters and receivers, and microfluidic valves, channels and chambers.

Interface: In physical terms, an interface is the boundry between two phases, for instance between a solid and liquid or between a liquid and gas.

Intermolecular: Between more than one molecule. Describes an interaction (e.g., a chemical reaction) between different molecules.

Intrinsic: Characterizes pure undoped semiconductor; electrical conductivity depends only on temperature and the band gap energy.

Ion Exchange: A chemical reaction involving the exchange of hydrated ions in a solid for similarly charged but different element ions in solution.

Ion Microscopy: Use of the Secondary Ion Mass Spectrometry SIMS technique to obtain micrographs of the elemental (or isotopic) distribution at the surface of a sample with a spatial resolution of 2 mm or better.

Ion: An atom or a molecule that has lost or gained an electron. An ion is also known as a charged particle. Ions are usually more reactive than neutral, uncharged particles. They cause all sorts of things from static electricity and chemical reactions like smog, to body processes like cellular respiration and digestion.

Ionic bond: A chemical bond resulting chiefly from the electrostatic attraction between positive and negative ions. A coulombic interatomic bond existing between two adjacent and oppositely charged ions.

Ionophore: A macro-organic molecule capable of specifically solubilizing an inorganic ion of suitable size in organic mediums.

Ketone: Organic molecule containing a carbonyl group linked to two alkyl groups.

Kevlar (TM): A synthetic fiber made by E. I. du Pont de Nemours & Co., Inc. Stronger than most steels, Kevlar is among the strongest commercially available materials and is used in aerospace construction, bulletproof vests, and other applications requiring a high strength-to-weight ratio.

Kilobyte (kB): 210 (= 1024, or about one thousand) bytes of information.

Laser Tweezers: Laser tweezers and related approaches are important for manipulation and isolation of subcellular **organelles** and structural components.

Leucine Zipper: Structural motif seen in many DNA-binding proteins in which two a helices from separate proteins are joined together in a coiled-coil, forming a protein dimer.

LIF Laser Induced Fluorescence: The optical emission from molecules that have been excited to higher energy levels by absorption of electromagnetic radiation. The main advantage of fluorescence detection compared to absorption measurements is the greater sensitivity achievable because the fluorescence signal has a very low background. For molecules that can be resonant excitated, LIF provides selective excitation of the analyte to avoid interferences. LIF is useful to study the electronic structure of molecules and to make quantitative measurements of analyte concentrations. Analytical applications include monitoring gas-phase concentrations in the atmosphere, flames, and plasmas; and remote sensing using light detection and ranging (LIDAR).

Ligand: In protein chemistry, a small molecule that is (or can be) bound by a larger molecule is termed a ligand. In organometallic chemistry, a moiety bonded to a central metal atom is also termed a ligand; the latter definition is more common in general chemistry.

Ligand-directed Drug Discovery: The first goal is to select a small number of likely targets (which at least in some cases, such as targets in pathogens, are validated targets) rapidly, without the need for costly and time- consuming biologically based target validation and functional determination. The second goal is to obtain lead compounds for these targets, instead of starting with a biologically validated target and then attempting to screen libraries for lead compounds that modulate it.

Light Emitting Diodes: Light Emitting Diodes (LEDs) work on a completely different concept. Traditionally LEDs are created from two semiconductors. By running current in one direction across the semiconductor the LED emits light of a particular frequency (hence a particular color) depending on the physical characteristics of the semiconductor used. The semiconductor is covered with a piece of plastic that focuses the light and increases the brightness. These semiconductors are very durable, there is no filament, they donít require much power, theyíre brighter and they last a long time. By densely packing red, blue and green LEDs next to each other on a substrate one can create a display.

Limit of Detection: The smallest measurable input. This differs from resolution, which defines the smallest measurable change in input. For a temperature measurement, this would provide an indication of the lowest temperature a sensor could generate an output in response to.

Limited Assembler: An assembler with built-in limits that constrain its use (for example, to make hazardous uses difficult or impossible, or to build just one thing). Assembler capable of making only certain products; faster, more efficient, and less liable to abuse than a general-purpose assembler.

Linde Scenario: A scenario for indefinite survival of intelligent life. It assumes it is possible to either create basement universes connected to the original universe with a wormhole *or* the existence of other cosmological domains. Intelligent life continually migrates to the new domains as the old grow too entropic to sustain life.

Linear: Aside from its geometric meaning, linear describes systems in which an output is directly proportional to an input. In particular, a linear elastic system is one in which the internal displacements are (at equilibrium) directly proportional to applied forces.

Liposome: Type of nanoparticle made of lipids, or fat molecules, surrounding a water core. Liposomes, several of which are widely used to treat infectious diseases and cancer, were the first type of nanoparticle to be used to create therapeutic agents with novel characteristics.

Macroscale: Larger than nanoscale; often implies a design that humans can directly interact with; too large to be built by a single assembler (one cubic micron of diamond contains 176 billion atoms).

Macrosensing: In medical nanorobotics, the detection of global somatic states and extrasomatic states (sensory data originating outside of the human body) by in vivo nanorobots.

Magnetic Force Microscopy (MFM): A method for observing local magnetic fields near a surface by scanning the surface with a magnetic probe.

Magnetic Resonance Force Microscopy MRFM: A new microscopic imaging technique that combines aspects of atomic force microscopy AFM with magnetic resonance imaging MRI. Interest in this technique is driven by the possibility of reaching the "Holy Grail" of microscopy: true three- dimensional, sub- surface imaging with atomic resolution and chemical specificity.

Magnetic Susceptibility (÷m): The proportionality constant between the magnetisation M (see under"magnetisation") and the magnetic field strength H. The magnetic susceptibility is unitless.

Magnetisation (M): The total magnetic moment per unit volume of material. Also, a measure of the contribution to the magnetic flux by some material within an H field. The magnitude of M is proportional to the applied field as: M = ÷m × H, with ÷m the magnetic susceptibility.

Magnetostrictive Material: A material that changes dimension in the presence of a magnetic field or generates a magnetic field when mechanically deformed.

Manufacturing: The ability to make products, in this case ranging from clothing, to electronics, to medical devices, to books, to building materials, and much more.

Martensite: A metastable iron phase supersaturated in carbon that is the product of a diffusionless (athermal) transformation from austenite.

Mask: Pattern on glass, like a photographic negative, for producing integrated-circuit elements on semiconductor wafer.

Mass Spectrometry: This technique can be used to both measure and analyze molecules under study. It involves introducing enough energy into a target molecule to cause its ionisation and disintegration. The resulting fragments are then analyzed, based on the mass/ charge ratio to produce a "molecular fingerprint.

Mass to Charge Ratio: A number defining how a particle will respond to an electric or magnetic field that can be calculated by dividing the mass of a particle by its charge. For instance, mass spectrometry provides information on the mass to charge ratio of a molecular species. In this case, the ratio is expressed by m/z where m is the molecular weight and z is the elementary charge.

Massometer: In medical nanorobotics, a nanosensor device for measuring the mass of individual molecules or small physical objects to single-proton resolution.

Materials Science: Science of ceramics, glass, metals, plastics, semiconductors.

Matter as Software: "Autonomous, motile microdevices clearly are on the horizon. They may be regarded as the first step in the evolution of a technology for "programming" the structure and properties of material objects at the microscopic and the submicroscopic levels. As this evolution progresses, *the physical and economic properties of such programmable matter are likely to become much like those of present day software.*"

MCM: MultiChip Module; the interconnection of two or more semiconductor chips in a semiconductor-type package.

Measurand: A physical quantity, condition, or property that is to be measured.

Mechatronics: The study of the melding of AI and electromechanical machines to make machines that are greater than the sum of their parts. The synergistic combination of precision mechanical engineering with electronic control.

Megabyte (MB): 220 (= 1,048,576, or about one million) bytes of information.

MEMS MicroElectro Mechanical Systems: A technology that combines computers with tiny mechanical devices such as **sensors**, valves, gears, mirrors, and **actuators** embedded in semiconductor chips. Paul Saffo of the Institute for the Future in Palo Alto, California, believes MEMS or what he calls analog computing will be "the foundational technology of the next decade." MEMS is also sometimes called **smart matter**.

Meridian Lines: The imaginary lines that were drawn on the globe for navigational purposes. Meridian lines mark longitude and run through both the north and the south poles. They separate the Earth into time zones and they are used to indicate a coordinate on the Earth. The Greenwich meridian is the central meridian line from which all time zones are set.

Messenger Molecule: A chemically recognizable molecule which can convey information after it is received and decoded by an appropriate chemical sensor.

Metal Nanoshells: A new type of nanoparticle composed of a semiconductor or dielectric core coated with an ultrathin conductive layer.. By adjusting the relative core and shell thicknesses, metal nanoshells can be fabricated that will absorb or scatter light at any wavelength across the entire visible and infrared range of the electromagnetic spectrum.

Metastable: A classical system is metastable if it is above its minimum-energy state, but requires an energy input before it can reach a lower-energy state; accordingly, a metastable system can act like a stable system, provided that energy inputs (e.g., thermal fluctuations) remain below some threshold. Systems with strong metastability are commonly described as stable. Quantum mechanical effects can permit metastable states to reach lower energies by tunneling, without an energy input; an associated, broader definition of *metastable* embraces all systems that have a long lifetime (by some standard) in a state above the minimum-energy state.

Meteosat: A geostationary weather satellite launched by the European Space Agency (ESA) and now operated by the company Eumetsat. The most recent version of Meteosat was launched in June 1988. It provides weather imaging of the Earth at both visible light and infrared wavelengths.

Micellar Nanocontainers: Block copolymer micelles are water- soluble biocompatible nanocontainers with great potential for delivering hydrophobic drugs. An understanding of their cellular distribution is essential to achieving selective delivery of drugs at the subcellular level.

Micelles: Micelles are small, spherical structures composed of molecules that attract one another to reduce surface tension. The head of the molecule is hydrophilic, meaning it likes water, while the interior portion is hydrophobic, meaning it avoids water.

Microbubbles: Very small encapsulated gas bubbles (diameters of micrometers) that can be used in diagnostic and therapeutic applications. Upon exposure to sufficiently intense ultrasound, microbubbles will cavitate, rupture, disappear, release gas content, etc. Such characteristics of the microbubbles can be used to enhance diagnostic tests, dissolve blood clots, and deliver drugs or genes for therapy. MeSH 2004

Microchemistry: The study of chemical reactions, using small quantities of materials, frequently less than 1 milligram or 1 milliliter, and often requiring special small apparatus and microscopical observation. In other words, the application of chemical tests to minute objects or portions of matter, magnified by the use of the microscopy; distinguished from macrochemistry. [OMD]

Microchip: A microchip (sometimes just called a "chip") is a unit of packaged computer circuitry (usually called an integrated circuit) that is manufactured from a material such as silicon at a very small scale. Microchips are made for program logic (logic or microprocessor chips) and for computer memory (memory or RAM chips). Microchips are also made that include both logic and memory and for special purposes such as analog-to-digital conversion, bit slicing, and gateways.

Microdevices: Narrower terms integrated microdevices, nanodevices, SED Single Electron Devices. Related terms microelectronics, microfluidics, micro- TAS.

Microelectronics: Narrower terms MEMS, nanoelectronics, optoelectronics, SED Single Electron Devices. Related term molecular electronics, semiconductors

Microfabrication: This technology includes techniques used to manufacture integrated circuits (ICs), discrete microelectronic devices, MEMS devices such as sensors and actuators, and various electro-optic devices. [University of Louisville Microfabrication Lab] is currently serving as a center for research activity in the areas of micromachined sensors and actuators, electro- optic devices, special- purpose microelectronic devices, planar waveguides, chemical transducers, microstrip and microgap radiation detectors, micromachined nozzles, and micromachined ink- jet printheads. Microfabrication holds the key to developing monolithically integrated devices capable of extremely high throughput. ... Microfabrication of surface characteristics provides a means of controlling the cell- environment interaction. Microfabricated structures also present new opportunities for cell, protein and nucleic acid separations and analysis- including high throughput single protein molecule sequencing. New approaches to analysis utilizing conductance techniques or tunneling probe microscopy merely emphasize the breadth of the potential of **nanofabrication** for future technology development relevant to biological issues.

Microfluidics: Within the microelectromechanical systems (MEMS) and biological and chemical detection communities, microfluidics refers to the research and development of microscale devices that handle very small volumes of fluids (nano- and picoliter volumes). These devices may perform tasks such as DNA analysis or the separation of human blood cells.

Microinjection: The insertion of a substance into a cell through a microelectrode. Typical applications include the injection of drugs, histochemical markers (such as horseradish peroxidase or lucifer yellow) and RNA or DNA in molecular biological studies. To extrude the substances through the very fine electrode tips, either hydrostatic pressure (pressure injection) or electric currents (ionophoresis) is employed. A technique for introducing a solution of DNA, protein, or other soluble material into a cell using a fine microcapillary pipet. [Life Sciences Dictionary]

Micromachining: Techniques used to fabricate MEMS. Broader term microengineering.

Microparticles: Applications include calibration of flow cytometers, particle and hematology analyzers, confocal laser scanning microscopes and zetapotential measuring instruments; flow measurements in gases and liquids like Laser Doppler Anemometry (LDA); Particle Dynamics Analysis (PDA), and Particle Image Velocimetry (PIV); medical diagnostics; separation phases for chromatography; support for immobilized enzymes; spacer in liquid crystal displays (LCD's); peptide synthesis; cell separation; tracers in environmental science; model systems in medicine, biochemistry, colloid chemistry, and aerosol research.

Microphone: A microphone, sometimes referred to as a mike or mic (pronounced "mike"), is an acoustic to electric transducer that converts sound into an electrical signal. Microphones are used in many applications such as telephones, tape recorders, hearing aids, motion picture production, live and recorded audio engineering, in radio and television broadcasting and in computers for recording voice, VoIP and numerous other computer applications.

Microplate Reader: Created from the tube spectrophotometer designs of the 1970s to save precious antibody samples. At first clumsy and inaccurate, absorbance microplate readers have evolved to pack unbelievable power and precision, replacing cuvette spectrophotometers for most multisample applications. Continuous improvement is enhancing the classic designs to embrace the world of high- throughput (HT) screening and to allow complete analytical automation. To handle the HT range (more than 10 microplates a day or 1,000 assays), many instruments now allow robotic handling of plates "stacked" in accessory plate handlers.

Microprocessor: A microprocessor incorporates the functions of a central processing unit (CPU) on a single integrated circuit (IC). The first microprocessors used a word size of only 4 bits, so that the transistors of its logic circuits would fit onto a single part. One or more microprocessors typically serve as the processing elements of a computer system, embedded system, or handheld device. Microprocessors made possible the advent of the microcomputer in the mid-1970s. Before this period, CPUs were typically made from small-scale integrated circuits containing the equivalent of only a few transistors. By integrating the processor onto one or a very few large-scale integrated circuit packages (containing the equivalent of thousands or millions of discrete transistors), the cost of processing capacity was greatly reduced.

Microspectrophotometry: Analytical technique for studying substances present at enzyme concentrations in single cells, in situ, by measuring light absorption. Light from a tungsten strip lamp or xenon arc dispersed by a grating monochromator illuminates the optical system of a microscope. The absorbance of light is measured (in nanometers) by comparing the difference between the image of the sample and a reference image. MeSH 1990

Microstructures: The last decade has seen rapid developments in the fabrication, characterisation and conceptual understanding of synthetic microstructures in many different material systems including silicon, III-V and II-VI semiconductors, metals, ceramics and organics. The objective of this journal [Superlattices and Microstructures] is to provide a common interdisciplinary platform for the publication of the latest research results on all such "nanostructures" with dimensions in the range of 1 - 100 nm; the unifying theme here being the dimensions of these artificial structures rather than the material system in which they are fabricated.

Microtubule: A cylindrical biological structure made of tubulin (which is a specific type of protein) dimers, and found in cells. Their role is to mechanically support cellular structure (cell rigidity) and the cell movement. Additionally, biological molecular motors, such as kinezin, use microtubules to move along them while transporting proteins in cells. Tubulin dimers in a microtubule are organised in protofilaments along which they bond relatively strongly.

Microwave: A form of electromagnetic radiation that is beyond the range of the *visible light* spectrum. Microwaves have very high frequencies and wavelengths of 1 mm to 50 cm.

Miller Indices: A set of 3 integers (4 for hexagonal) that designate crystallographic planes, as determined from reciprocals of fractional axial intercepts.

Miniaturisation: Desirable for many technologies for overall cost reduction (including reduction in the amount of reagents and analytes). Important to remember that building space is often the least available and most expensive component of an overall laboratory budget.

Molecular Assembler: A molecular assembler has been defined as a single molecule machine that would assemble individual atoms or molecules according to specific instructions to construct a desired product. Some biological molecules the such as a ribosomes fit this definition since working within cell's environment, it receives instructions from mRNA and then assembles specific sequences of amino acids to construct protein molecules. However, the term "molecular assembler" usually refers to theoretical man-made or synthetic devices.

Molecular Beam Epitaxy (MBE): Process used to make compound (multi-layer) semiconductors. Consists of depositing alternating layers of materials, layer by layer, one type after another (such as the semiconductors gallium arsenide and aluminum gallium arsenide).

Molecular Combing: A method ... which can straighten and align molecules of genomic DNA on a solid surface. The technology also includes a battery of novel statistical methods developed for analyzing the large amounts of data obtained from FISH analyses made on individual DNA molecules. Molecular combing relies on the action of a receding air/ water interface, or meniscus, to uniformly straighten and align DNA molecules on a solid surface.

Molecular Design: The application of all techniques leading to the discovery of new chemical entities with specific properties required for the intended application.

Molecular Electronics: Seeks to use individual molecules to perform functions in electronic circuitry now performed by semiconductor devices. Individual molecules are hundreds of times smaller than the smallest features conceivably attainable by semiconductor technology. Because it is the area taken up by each electronic element that matters, electronic devices constructed from molecules will be hundreds of times smaller than their semiconductor-based counterparts. Moreover, individual molecules are easily made exactly the same by the billions and trillions. The dramatic reduction in size, and the sheer enormity of numbers in manufacture, are the principle benefits promised by the field of molecular electronics. Research is currently focused on the development of methodologies for self- assembly of metal and semiconductor nanocrystals for use as nanoscale electronic devices. Electrical transport properties of layers of gold nanoparticles self- assembled between metal electrodes on silicon oxide surfaces is being studied in detail. The assembled nanoparticles are separated by specific molecular linkers and the layers are typically a few monolayers thick...l Another active area of molecular electronics research is the development of strategies for fabrication of molecular devices using biological assembly principles.

Molecular Machine: A mechanical device that performs a useful function using components of nanometer scale and defined molecular structure; includes both artificial nanomachines and naturally occurring devices found in biological systems. Any machine with atomically precise parts of nanometer dimensions; can be used to describe molecular devices found in nature.

Molecular Manipulator: A device combining a proximal probe mechanism for atomically precise positioning with a molecule binding site on the tip; can serve as the basis for building complex structures by positional synthesis. A programmable device able to position molecular tools with high precision, for example, to direct a sequence of mechanosynthetic steps; a molecular assembler.

Molecular Manufacturing: Manufacturing using molecular machinery, giving molecule-by-molecule control of products and by-products via positional chemical synthesis. Exponential general-purpose molecular manufacturing — that's a mouthful, but what does it mean? Let's take the phrase apart to see why it is so important. The building of complex structures by mechanochemical processes. The production of complex structures via nonbiological mechanosynthesis.

Molecular Medicine: A variety of pharmaceutical techniques and gene therapies that address specific molecular diseases or molecular defects in biological systems. A variety of pharmaceutical techniques and therapies in use today. Studying molecules as they relate to health and disease, and manipulating those molecules to improve the diagnosis, prevention, and treatment of disease.

Molecular Mimicry: The process in which structural properties of an introduced molecule imitate or simulate molecules of the host. Direct mimicry of a molecule enables a viral protein to bind directly to a normal substrate as a substitute for the homologous normal ligand. Immunologic molecular mimicry generally refers to what can be described as antigenic mimicry and is defined by the properties of antibodies raised against various facets of epitopes on the viral protein.

Molecular Motors: Protein based machines that are involved in or cause movement such as the rotary devices (flagellar motor and the F1 ATPase) or the devices whose movement is directed along cytoskeletal filaments (myosin, kinesin and dynein motor families). [MeSH]

Molecular Nanoscience: An emerging interdisciplinary field that combines the study of molecular/ biomolecular systems with the science and technology of **nanoscale** structures. The potential applications are very broad and include such possibilities as: (i) the use of **biomolecules** and cellular systems to self-assemble nanoelectronic circuitry and other nanoscale structures and (ii) the use of lamellar host frameworks containing **nanopores** that can be tailored to include guest molecules for separation of chemicals for pharmaceutical and other applications.

Molecular Nanotechnology (MNT): Molecular nanotechnology (MNT) is the engineering of functional systems at the molecular scale. An equivalent definition would be "machines at the molecular scale designed and built atom-by-atom". This is distinct from nanoscale materials. Based on Richard Feynman's vision of miniature factories using nanomachines to build complex products (including additional nanomachines), this advanced form of nanotechnology (or *molecular manufacturing*) will make use of positionally-controlled mechanosynthesis guided by molecular machine systems.

Molecular Recognition: The ability of biological macromolecules such as proteins and DNA to recognize selectively and to bind to other species to form larger supramolecular complexes is a key element in the extraordinarily diverse and controlled chemistry exhibited by nature. Chemists today are increasingly interested in mimicking these processes which involve 'molecular recognition' of one molecule by another via formation of specific nonequivalent bonds between them and spontaneous binding together of two to many thousands of molecules into well defined supramolecular systems with new chemical properties. Industrially important applications already include drug design, synthesis of stereo regular polymers, affinity chromatography, crystallisation, epitaxial growth and liquid crystal displays.

Molecular Sorting Rotor: A class of nanomechanical device capable of selectively binding (or releasing) molecules from (or to) solution, and of transporting these bound molecules against significant concentration gradients.

Molecular Surgery: In medical nanorobotics, the analysis and physical correction of molecular structures in the body using medical nanomachines. Analysis and physical correction of molecular structures in the body using medical nanomachines.

Molecular Systems Engineering: Design, analysis, and construction of systems of molecular parts working together to carry out a useful purpose.

Molecular Wire: A molecular wire - the simplest electronic component - is a quasi-one-dimensional molecule that can transport charge carriers (electrons or holes) between its ends.

Molecularly Imprinted Polymers MIPs: A new class of materials that have artificially created receptor structures. Since their discovery in 1972, MIPs have attracted considerable interest from scientists and 3engineers involved with the development of chromatographic absorbents, membranes, sensors and enzyme and receptor mimics.

Moledular Technology: Moledular technology promises to create a new class of imaging agents that offer distinct advantages and can be used for the early detection and diagnosis of disease. Diagnostic molecules have the potential to act as biomarkers in drug development and diagnostics, and can be used in the imaging of cancer in living subjects. This meeting will address the challenges in implementing nanotechnology for drug delivery systems and imaging agents, and promote dialogue between diagnostic and therapeutic development. The production and application of structures, devices and systems by controlling shape and size at nanometre scale.

Monkeywrenching: In medical nanorobotics, the mechanical or chemical jamming of cellular equilibrium processes, with a cytocidal objective.

Nanoarray: An ultra-sensitve, ultra-miniaturized array for biomolecular analysis. BioForce Nanosciences' Nanoarrays utilize approximately 1/10,000th of the surface area occupied by a conventional microarray, and over 1500 nanoarray spots can be placed in the area occupied by a single microarray domain.

Nanoassembler: The Holy Grail of nanotechnology; once a perfected nanoassembler is availble, building anything becomes possible, with physics and the imagination the only limitation (of course each item would have to be designed first, which is another small hurdle).

Nanobiology: Many fundamental biological functions are carried out by molecular machineries that have the sizes of 1-100 nm. You find many examples in molecular biology and cell biology: single enzymes, transcription complex, ribosome, transport complex, nuclear pore, and so on. To understand the functions of these machineries, one has to describe their movements, changes in their shapes, and their localisation. This means the mechanistic study is equivalent to dynamic morphology at this level of size, making a new field that merges mechanistic biology and morphology. The emergence of nanobiology depended on the invention of nano- technology: scanning probe microscopy, modern optical techniques, and micro- manipulating techniques. This concept of nanobiology was first proposed in a group study named "Biological Nano- Mechanisms", which was supported by Japanese Agency of Science and Technology (1992-1997).

Nanobubbles: Nanobubbles are tiny gas bubbles that have been found on some surfaces in liquids. Typically composed of atmospheric gases, they occur on hydrophobic (water repellent) surfaces in water. The existence of nanobubbles has also been inferred in other solvents, (in this case the surfaces are called lyophobic). The nanobubbles are from 10-100nm in diameter, and, although they appear to be forbidden by thermodynamics, their lifetime is at least of the order of hours. They likely arise by nucleation of gas at the surface from a supersaturated solution. Nanobubbles appear to be responsible for the long-range attraction that occurs between hydrophobic colloid particles and surfaces. Nanobubbles bridge apposite surfaces and draw the particles into contact, leading to a capillary adhesion that makes aggregation of such particles practically irreversible.

Nanocantilever: The simplest micro-electro-mechanical system (MEMS) that can be easily machined and mass-produced via the same techniques used to make computer chips. The ability to detect extremely small displacements make nanocantilever beams an ideal device for detecting extremely small forces, stresses and masses. Nanocantilevers coated with antibodies, for example, will bend from the mass added when substrate binds to its antibody, providing a detector capable of sensing the presence of single molecules of clinical importance.

Nanocentrifuge: In medical nanorobotics, a proposed nanodevice that can spin materials at very high speed, imparting rotational accelerations of up to one trillion gravities (g's), thus permitting rapid sortation.

Nanochemistry: To analyze ultrasmall systems you need ultrasmall tools. Whether dealing with single molecules or single cells, the precision and sensitivity of the manipulation and detection schemes used are absolutely critical. We have been interested in confining single molecules in solution to perform chemistry at the lowest level possible, as well as in delivering small volumes of chemicals to single subcellular organelles. By mimicking what nature has developed over billions of years to conduct chemistry in ultrasmall packets (cells and organelles), we have arrived at what is to date the smallest test tubes known - synthetic liposomes which are manipulated and electrofused to mix their contents. Our liposomes have volumes of a few femtoliters or less, matching the probe volumes of standard single molecule detection techniques, and are made in minutes by a protocol developed in this lab. In other words, the goal of the Nanochemistry group is to develop chemistry beyond the realm of individual molecules by exploring how systems of several molecules or other electronically confined entities self-organise into distinct structures on the nanometer scale. The new insight is used to study and manipulate biological nanosystems and to create novel nanoarchitectures with unique structural, electronic, optical and magnetic properties. Specific research areas include bionanotechnology and self- assembling molecular electronics.

Nanochip: In Friday's issue [Aug. 1, 2001] of the journal *Science,* physicists from IBM's Thomas J. Watson Research Center [Philip G. Collins, Michael S. Arnold and Phaedon Avouris] announce their fabrication of the world's first array of transistors made from carbon nanotube. In other words, nanochip is developing the next generation of removable, portable memory chips for consumer, computer, and server applications. Using MEMs technology along with arrays of atomic force probe tips, Nanochip is developing storage chips that will store tens of Gigabytes per chip. Our technology is not bound by lithography constraints so we will be able to offer storage chip products through our customers that will exceed 100 GB per chip set in the beginning and grow to Terabytes per chip set in the future. Nanochip is an integrated circuit (IC) that is so small, in physical terms, that individual particles of matter play major roles. Miniaturisation of electronic and computer components has always been a primary goal.

Nanoclusters: A nanocluster or *nanocrystal* is a fragment of solid comprising somewhere between a few atoms to a few tens of thousands of atoms. Nanoclusters are therefore a novel state of matter with properties that are neither those of a bulk crystal nor those of individual atoms and molecules. Over the past 10 years huge advances have been made both in the synthesis of size- tunable, monodisperse (i.e. size- selected) nanoclusters of various chemical compositions and in the development of techniques for their assembly into nanostructured solids (facilitating the synthesis of what have been termed "designer materials").

Nanocones: Nonplanar graphitic structures. Carbon-based structures with five-fold symmetry that form due to disclination defects in two-dimensional graphene sheets. They have been observed as nanotube caps and as freestanding structures.

Nanocrystals: A nanocrystal typically has a diameter of between 1 and 10 nm and may contain as few as a hundred or as many as tens of thousands of atoms. Many fundamental properties of nanocrystals depend strongly on their size in smooth and predictable ways. Examples include the external field required to switch a magnetized particle of great importance in magnetotactic bacteria and in hard disk drives and the color of light emission from a semiconductor used for the fluorescent labeling of cells and in lasers. This facile tuning of properties by size variation is one reason why nanocrystals are widely viewed as promising components for new artificial optical and electrical materials..

Nanodevices: Scientists believe nanoscale devices may lead to computer chips with billions of transistors, instead of millions - which is the typical range in today's semiconductor technology. The more transistors crammed on a chip, the more powerful it is. "This technology has the potential to replace existing manufacturing methods for integrated circuits, which may reach their practical limits within the next decade when Moore's Law eventually hits a brick wall," said physicist Bernard Yurke of Bell Labs. DNA, which provides the molecular blueprints for all living cells, is an ideal tool for making nanoscale devices. "We took advantage of how pieces of DNA - with its billions of possible variations - lock together in only one particular way, like pieces of a jigsaw puzzle," Yurke said.

Nanodots: Nanodots, also known as quantum dots, consist of 100s-1000s of atoms of inorganic semiconductor nanoparticles and are approximately one billionth of a meter in size. Developed in the mid-1980s for optoelectronic applications, they have interesting structural, electronic, and optical properties—they strongly absorb light in the near UV range and re-emit visible light that has its color determined by both the nanodot size and surface chemistry. And as the size of nanodots can be controlled during synthesis with nanoscale precision, so the optical properties can be manipulated. In addition, nanodots have a longer life than organic fluorophores, and have a broad excitation spectrum. These factors combined make the use of quantum dots as light-emitting phosphors a strong candidate for a major application of nanotechnology in the future.

Nanoelectronics: Nanoelectronics in Japan and Korea tends to focus on next generation semiconductor devices and single electron devices (SED).

Nanoengineering: Nanoengineering and nanotechnology are concerned with developing structures and systems that use and enhance the significantly improved properties of their nanoscale components. By learning how to control feature size and to assemble appropriate "building blocks," it should be possible to enhance the properties of materials and to create functional devices with greatly improved or entirely new functions. This goal, however, requires both discovering the underlying principles and developing the tools needed to apply them systematically. We use chemistry to construct nanostructures and their composites, then focus our attention on the electronic, optical, and transport properties of these nanostructures and the macroscopic films and materials that can be constructed from them. This research lies at the common frontier of chemistry, condensed matter physics, optics, and bioengineering.

Nanofabrication: Nanofabrication methods can be divided into two categories: top-down methods, which carve out or add aggregates of molecules to a surface, and bottom- up methods, which assemble atoms or molecules into nanostructures. [George M. Whitesides and J. Christopher Love "The art of building small" Scientific American 285 (3): 39-47, Sept. 2001]

Nanofluidics: Researchers at Cornell University are using nanotechnology to build microscopic silicon devices with features comparable in size to DNA, proteins and other biological molecules — to count molecules, analyze them, separate them, perhaps even work with them one at a time.

Nano-genomics: The main objective of the group is to develop experimental tools that will accelerate the finding of genomic factors (e.g. genes, polymorphisms, methylation patterns) that contribute to complex traits. In the long term this may lead to predictive medicine based on personalized genomic information. Currently available molecular tools are too slow, cumbersome and expensive to allow comprehensive genomic information to be easily obtained from a large number of individuals in a population (e.g. for case/control studies). A radical re-invention of the way genes and DNA are analyzed is needed. We believe that the analysis of single DNA molecules can overcome many of the present limitations and in addition will reveal information which is inaccessible by conventional methods.

Nanoguitar: Smallest guitar, about the size of a human blood cell, carved out of crystalline silicon, illustrates new technology for nanosized electromechanical devices. The world's smallest guitar is 10 micrometers long — about the size of a single cell — with six strings each about 50 nanometers, or 100 atoms, wide. Just one of several structures that Cornell researchers believe are the world's smallest silicon mechanical devices. Researchers made these devices at the Cornell Nanofabrication Facility, bringing microelectro-mechanical devices, or MEMS, to a new, even smaller scale — the nano-sized world."

Nanoharvesting: Nanoharvesting agents are being designed to act as molecular mops for these biomarkers and can be directly queried by mass spectrometry. Dr. Emanuel F. Petricoin, Co-Director, NCI-FDA Clinical Proteomics Programme, Senior Principal Investigator, Center for Biologics Evaluation and Research, U. S. Food and Drug Administration.

Nanohorns: One of the single walled carbon nanotube types, with an irregular horn-like shape, which may be a critical component of a new generation of fuel cells. "The main characteristic of the carbon nanohorns is that when many of the nanohorns group together an aggregate (a secondary particle) of about 100 nanometers is created. The advantage being, that when used as an electrode for a fuel cell, not only is the surface area extremely large, but also, it is easy for the gas and liquid to permeate to the inside. In addition, compared with normal nanotubes, because the nanohorns are easily prepared with high purity it is expected to become a low-cost raw material."

Nanoimprinting: Sometimes called soft lithography. A technique that is very simple in concept, and totally analogous to traditional mould- or form-based printing technology, but that uses moulds (masters) with nanoscale features. As with the printing press, the potential for mass production is clear. There are two forms of nanoimprinting, one that uses pressure to make indentations in the form of the mould on a surface, the other, more akin to the printing press, that relies on the application of "ink" applied to the mould to stamp a pattern on a surface.

Nanomachines: Nanomachines somewhat similar to those envisaged by Molecular Nanotechnology already exist: they are biomacromolecules, which are nanoengines acting both as thermal engines and as informational engines like the so- called "assemblers" (cf. Molecular Nanotechnology), but the latter are not "self- programmed" like nanobiomachines are. The autopoietic character of nanobiomachines involves a physics of very different kind, largely unknown at present.

Nanomachining: Like traditional machining, where portions of the structure are removed or modified, nanomachining involves changing the structure of nano-scale materials or molecules.

Nanomanufacturing: The science that is carried out within the CNSI [California NanoSystems Institute] may be classified according to: The development of the chemical, biological, and materials approaches toward the fabrication and assembly of the fundamental nanoscale building blocks. This science includes the synthesis of traditional materials (e.g. semiconductors) that are designed to exhibit nontraditional behavior (e.g. ferromagnetism). It also includes the development of manufacturing routes for traditional devices (e.g. quantum well lasers) from nontraditional routes (e.g., biomimetic synthesis). It further includes the development of manufacturing routes for nontraditional devices (e.g. photonic band gap materials) from nontraditional routes (e.g., 3D assembly of modulated dielectrics). At the heart of these types of approaches is the concept of bottom- up manufacturing. Such an approach to manufacturing, which is inherent to all biological systems, is in contrast to all modern manufacturing approaches, such as semiconductor processing, which are classified as top- down manufacturing.

Nanomedicine: Nanomedicine is the medical application of nanotechnology and related research. It covers areas such as nanoparticle drug delivery and possible future applications of molecular nanotechnology (MNT) and nanovaccinology. Current problems for nanomedicine involve understanding the issues related to toxicity and environmental impact of nanoscale materials. Direct funding for nanomedicine projects has begun, and the US National Institute of Health received funding in 2005 to set up four nanomedicine centres. In April 2006, the journal Nature Materials estimated that 130 nanotech-based drugs and delivery systems were being developed worldwide.

Nanoparticles: Both synthetic (bottom- up) and transformative (top- down) fabrication rely on the availability of building block materials and artifacts such as quantum dots, nanotubes and nanofibers, ultrathin films and nanocrystals. These also include their assemblies in coatings, dispersions, colloids, aerogels and nanoporous structures, as well as organic dendrimers, block copolymers and nanocomposites, and also biomolecules such as proteins, nucleic acids etc. Despite the multitude of nanoparticles that have been generated in the laboratory, there is a need for fabrication and proliferation of a much larger variety of such nanomaterials. ... Serial direct structuring methods include robotic nanoassembly by the atomic force microscope (AFM) and strain-directed assembly, while parallel soft lithography techniques comprise various nanoimprinting, molding, embossing processes, as well as DNA-directed assembly etc. Alternatives to such technologies are offered by miniaturisation of hard microlithography, including X-ray (LIGA), electron or ion lithography, and *molecular beam epitaxy (MBE)*. Other processes address 3-D templating and growth of aerogels, sol- gel methods, laser or plasma vaporisation and condensation, laser- guided particle transport and electrophoretic techniques. ... Instrumentation and Equipment for Characterisation and Processing - As already mentioned, analogous technologies are presently used for both analysis (sensing) and synthesis (actuation) of nanostructures. These instruments include the various scanning probe (SPM, such as AFM) and near- field microscopes (NFM); electron beam (SEM, TEM), ion beam, and molecular beam epitaxy (MBE) equipment; optical (laser) tweezers, X-ray spectroscopy and nuclear magnetic resonance (NMR) instruments; large- scale synchrotron, neutron and photon sources etc. To improve throughput and efficiency, parallel probe arrays, distributed/

scanned beams and lab- on- a- chip MEMS, including chemical and biological array sensors and actuators, must be developed as robust, miniaturized alternatives to serial, large-scaled instrumentation. Integration of equipment across multiple scales, exemplified by a chain of a manipulator with a MEMS micro- manipulator as the end effector, to process nano- electro- mechanical structures (NEMS) is necessary to process multi- scale nanostructured parts. Instrumentation research must also strive to improve resolution and bandwidth, commensurate to the scale and dynamics of nanostructure interaction phenomena that must be monitored and regulated. Last, teleoperation of large and costly equipment for telecharacterisation and telefabrication via the internet is desirable for collaborative and shared use. Nanoparticles, including nano- clusters, [nano]- layers, [nano]- tubes, and two- and three- dimensional structures in the size range between the dimensions of molecules and 50 nm (or in a broader sense, submicron sizes as a function of materials and targeted phenomena), are seen as tailored precursors for building up functional nanostructures.

Nanopens & Nanopencils: (AKA: Atomic Pencil) *"Analogous to using a quill pen but on a billionth the scale"*, and may transform dip-pen nanolithography. Allows for drawing electronic circuits a thousand times smaller than current ones. The "pen" is an atomic force microscope (AFM).

Nanopipettes: "Cantilevered/Straight Nanopipettes can be used as *nanopens* for controlled chemical delivery or removal from regions as small as 100 nanometers. They can also be used as vessels for containing molecules whose optical properties change in response to their chemical environment." Other uses include "controlled chemical etching with the precision of atomic force microscopy; chemical imaging of surfaces; delivering femtosecond laser pulses; and performing NSOM/SNOM imaging using a UV excimer laser."

Nanoplotter: A multi-tip nanopen. "A device that can draw patterns of tiny lines just 30 molecules thick and a single molecule high. ... produces eight identical patterns at once and extends ... dip-pen nanolithography towards mass producing nanoscale devices and circuits by converting what was a serial process to a parallel one. May be use to "... miniaturize electronic circuits, pattern precise arrays of organic and biomolecules such as DNA and put thousands of different medical sensors on an area much tinier than the head of a pin."

Nanopore Technology: Nanopore technology is an elegant concept. A membrane with very small channels (a few nanometers in diameter), called nanopores, separates two solutions. When a voltage is applied across the membrane, charged biomolecules migrate through the pores in a controlled manner.... As each nucleotide passes through the nanopore, an electronic signature is produced that can be used to characterize it. The nanopore's size is such that it permits only a single nucleic acid strand to pass through it at any one time. Therefore the technology can be used to sequentially measure a biopolymer's properties along its length. Research in the field of *computational biology* will play a significant role in the development of nanopore technology. It will be used to help optimize experimental strategies and designs, and to create the mathematical methods needed to interpret the data generated. The potential capabilities of nanopore technology are very broad. Nanopore technology can distinguish between and count a variety of different molecules in a complex mixture. For example, nanopores could discriminate between hybridized or unhybridized unknown RNA and DNA molecules that differ by a single nucleotide only. Compared to existing techniques, nanopore technology is expected to provide direct characterisation of individual nucleic acid and protein molecules directly derived from biological samples, thereby making it applicable to a wide set of analyses.

Nanopore: Nanopore technology is an elegant concept. A membrane with very small channels (a few nanometers in diameter), called nanopores, separates two solutions. When a voltage is applied across the membrane, charged biomolecules migrate through the pores in a controlled manner. ... As each nucleotide passes through the nanopore, an electronic signature is produced that can be used to characterize it. The nanopore's size is such that it permits only a single nucleic acid strand to pass through it at any one time. Therefore the technology can be used to sequentially measure a biopolymer's properties along its length. Research in the field of computational biology will play a significant role in the development of nanopore technology. It will be used to help optimize experimental strategies and designs, and to create the mathematical methods needed to interpret the data generated. The potential capabilities of nanopore technology are very broad. Nanopore technology can distinguish between and count a variety of different molecules in a complex mixture. For example, nanopores could discriminate between hybridized or unhybridized unknown RNA and DNA molecules that differ by a single nucleotide only. Compared to existing techniques, nanopore technology is expected to provide direct characterisation of individual nucleic acid and protein molecules directly derived from biological samples, thereby making it applicable to a wide set of analyses. Because nanopore technology is in the very early stages of its development, its advantages are not fully characterized.

Nanorobotics: Nanorobotics is the technology of creating machines or robots at or close to the scale of a nanometres (10-9 metres). More specifically, nanorobotics refers to the still largely hypothetical nanotechnology engineering discipline of designing and building nanorobots. Nanorobots (nanobots, nanoids or nanites) would be typically devices ranging in size from 0.1-10 micrometers and constructed of nanoscale or molecular components. As no artificial non-biological nanorobots have so far been created, they remain a hypothetical concept at this time.

Nanoscale: 1 to 100 billionths of a meter. At the nanoscale, physics, chemistry, biology, materials science, and engineering converge toward the same principles and tools. The nanoscale is not just another step toward miniaturisation, but a qualitatively new scale. The new behavior is dominated by quantum mechanics, material confinement in small structures, large interfacial volume fraction, and other unique properties, phenomena and processes. Many current theories of matter at the microscale have critical lengths of nanometer dimensions. These theories will be inadequate to describe the new phenomena at the nanoscale. ... Innovative nanoscale properties and functions will be achieved through the control of matter at its building blocks: atom- by- atom, molecule- by- molecule, and nanostructure- by- nanostructure. Nanotechnology will include the integration of these nanoscale structures into larger material components, systems, and architectures. However, within these larger scale systems the control and construction will remain at the nanoscale.

Nanoscience: Nanoscience is primarily the extension of existing sciences into the realms of the extremely small (nanomaterials, nanochemistry, nanobio, nanophysics, etc.) while nanoengineering represents the extension of the engineering fields into the nano- scale realm (nanofabrication, nanodevices, etc.). In other words, the prefix "nano" originates from the Greek word for dwarf, and thus refers to something small. As a prefix for a unit of time or length, it means one billionth of that unit. Thus, a nanometer (nm) is 10-9 meter. The dot over this letter "i" is approximately one million nanometers in diameter. The fundamental building blocks of nature, atoms and molecules, have dimensions in the nanometer domain (i.e., the nanoscale). Many water molecules can easily occupy a sphere 1 nm in diameter. The DNA double helix is approximately 2 nm wide. The way molecules, for example, assemble into larger, supramolecular entities on the nanoscale determines important material properties (e.g., electrical, optical, and mechanical properties). Of course, it has long been recognized that nature performs this assembly very well in the creation of the sophisticated molecular machinery that supports our life on earth. In short, by controlling structure on the scale of ~ 1 – 100 nm, one can, in principle, ultimately design new materials with specific properties. Scientists have long imagined the possibility of manipulating individual atoms and molecules. Nanoscience is primarily the extension of existing sciences into the realms of the extremely small (nanomaterials, nanochemistry, nanobio, nanophysics, etc.) while nanoengineering represents the extension of the engineering fields into the nanoscale realm (nanofabrication, nanodevices, etc.).

Nanoshells: At Rice University in Texas, however, tiny constructions called nanoshells have shown promise for fighting cancer and administering drugs. The devices are simple enough: beads about three millionths of an inch wide, with an outer metal wall and an inner silicon core. But by varying the size ratio between wall and core, electrical and computer engineer Naomi Halas and her Rice colleagues can tune the shells precisely to absorb or scatter specific wavelengths of light.

Nanospheres: Self-assembling nanospheres that fit inside each other like Russian dolls are one form of a broad range of submicroscopic spheres ... The durable silica spheres, which range in size from 2 to 50 nanometers, form in a few seconds, are small enough to be introduced into the body, and have uniform pores that could enable controlled release of drugs. The spheres can absorb organic and inorganic substances including small particles of iron, which means they can be controlled by magnets and the contents released as needed.

Nanosprings: A nanowire wrapped into a helix. Speculation is that they "may someday make highly sensitive magnetic field detectors, perhaps finding application in hard drive read heads. Alternatively, nanosprings could serve as positioners, or even as tiny conventional springs, for nanomachines of the future."

Nanostructures: The NanoStructures Laboratory (NSL) at MIT (formerly the Submicron Structures Laboratory) develops techniques for fabricating surface structures with feature sizes in the range from nanometers to micrometers, and uses these structures in a variety of research projects. The NSL includes facilities for lithography (photo, holographic electron beam, ion beam, and x-ray), etching (chemical, plasma and reactive- ion), lift- off, electroplating, sputter deposition and e- beam evaporation. Much of the equipment and nearly all of the methods utilized in the NSL are developed in house. Generally, commercial IC processing equipment cannot achieve the resolution needed for nanofabrication, and it lacks the required flexibility. The research projects fall into four major categories: development of submicron and nanometer fabrication technology; short- channel semiconductor devices, quantum- effect electronics, and optoelectronics; periodic structures for x-ray optics, spectroscopy and atomic interferometry, crystalline films on non-lattice- matching substrates.

Nanosystems: There is currently an intense exploration to determine patterns of gene expression that encode for normal biological processes such as cell replication and signal transduction and to understand disease that results from alterations in normal gene expression. Such alternations can result from environmental factors, hereditary defects, developmental errors and aging. In the future, symptom- based descriptions of disease will give way to a molecular mechanistic nomenclature with treatments aimed at molecular modification of alterations that produce cell phenotypes of disease. Out of this research will arise molecular diagnostics and molecular therapeutic treatments. ... Molecular probes are used to study basic biology and the resultant knowledge will lead to the development of molecular tools for molecular diagnostics and therapeutics.

Nanotechnology: The technology arising from efforts to exploit the novel and improved properties that phenomena and processes exhibit at the scale between single atoms/ molecules and bulk behavior (approximately 10^{-9} to 10^{-7} meters). Emerging as a new field enabling the creation and application of materials, devices, and systems at atomic and molecular levels and the exploitation of novel properties that emerge at the nanometer scale. Many areas of biomedicine are expected to benefit from nanotechnology including sensors for use in the laboratory, the clinic, and within the human body; new formulations and routes for drug delivery; and biocompatible, high-performance materials for use in implants. Examples of potential uses of nanotechnology in biomedicine include the early detection and treatment of disease and the development of "smart", rejection- resistant implants that will respond appropriately as the body's needs change. Although research in this field dates back to Richard P. Feynman's classic talk in 1959, the term nanotechnology was first coined by K. Eric Drexler in 1986 in the book *Engines of Creation*. In the popular press, the term nanotechnology is sometimes used to refer to any sub- micron process, including lithography. Because of this, many scientists are beginning to use the term molecular nanotechnology when talking about true nanotechnology at the molecular level.

Nanotubes: A one dimensional fullerene (a convex cage of atoms with only hexagonal and/ or pentagonal faces) with a cylindrical shape. Carbon nanotubes discovered in 1991 by Sumio Iijima resemble rolled up graphite, although they can not really be made that way. Depending on the direction that the tubes appear to have been rolled (quantified by the 'chiral vector'), they are known to act as conductors or semiconductors. Nanotubes are a proving to be useful as molecular components for nanotechnology. [about.com] Narrower term carbon nanotubes (or is this equivalent?) Broader term fullerenes

Nanowires: Molecular wires millions of times smaller in diameter than a human hair. Described in a paper appearing in the February 23, 2001 issue of the journal Science, these "nanowires," so called because they have dimensions on the order of a nanometer (a billionth of a meter), have high rates of electron transfer with very low resistance. "That means less impedance to the flow of current, with little or no loss of energy," says chemist John Smalley, the lead Brookhaven researcher on the study.

National Nanotechnology Initiative: US federal government agencies participating include the National Science Foundation, the Department of Defense, the National Institute of Health, NASA, and NIST.

Navigation Satellite: A type of satellite that gives ships and aircraft their coordinate positions on the Earth. Navigation satellites were developed in the 1950's, and they rely on the *doppler effect* to calculate the position of vessels emitting a radio signal. Navigation satellites are also widely used by the military.

NEMS Nano ElectroMechanical Systems: The time is ripe for a concerted exploration of nanoelectromechanical systems (NEMS) - i.e. machines, sensors, computers and electronics that are on the nanoscale. Such efforts are under way in my group at Caltech, and in several others around the world. The potential payoffs are likely to be enormous and could benefit a diverse range of fields, from medicine and biotechnology to the foundations of quantum mechanics. [Michael Roukes "Nanoelectromechanical systems face the future" Physics World 14 (2) Feb. 2001] http://www.bioon.com/book/physicsweb.org/article/world/14/2/8

Operator: Short region of DNA in a bacterial chromosome that controls the transcription of an adjacent gene.

Optical Trapping: Optical trapping of small particles by forces exerted by laser radiation pressure has recently been introduced into the study of biological systems. The single- beam gradient trap (optical tweezers) employs a single strongly focused laser beam. In this case, particle size is much less than and the laser light exerts a force pulling the particle toward the high focus part of the beam. Such manipulation of micron- sized (particle size larger than) particles is also feasible. Manipulations of cells and intracellular organelles have extended to laser cutting (scissors) and to use of two- photon systems. A method for non- mechanical manipulation of small particles. Small transparent objects (such as glass- spheres or yeast cells) can be trapped using focused laser light. The method has many possible applications in various fields of physics and biology. The optical trap is based on the concept of photon momentum transfer. Photon momentum was proposed by Planck and used by Einstein to explain the photo-electric effect.

Optical Tweezers: Developed at Bell Labs in the 1980s ... The method relies on the fact that light waves exert minute forces as well as transmitting energy. If an object is small enough - in this case about 5 micrometers across — it can be "trapped" or held still by a focused beam of laser light. When the laser's position changes, the trapped object moves too ... The new technique allows researchers to make measurements not possible with conventional methods and should make it easier to judge the effectiveness of inhibitors and other medicines at a variety of concentrations. The single- beam gradient trap, or the optical tweezers, is a unique tool for micro- manipulation, enabling its user to access tiny objects without any mechanical contact. It uses a focussed laser beam to trap and manipulate microscopic particles in the size range 100 nm - 100 mm. The trap consists of a single laser beam which attracts transparent dielectric particles towards its focal region. The laser beam is introduced into a conventional optical microscope, so that the same objective is used to view and trap particles immersed in a liquid.

Organic Moledule: A molecule containing carbon; the complex molecules in living systems are all organic molecules in this sense.

Organotypic: Despite their wide use, the physiological relevance of organotypic slices remains controversial. Such cultures are prepared at 5 days postnatal. Although some local circuitry remains intact, they develop subsequently in isolation from the animal and hence without plasticity due to experience.

Development of synaptic connectivity and morphology might be expected to proceed differently under these conditions than in a behaving animal.

Pathogen (adjective pathogenic): An organism or other agent that causes diseases.

PCR (polymerase chain reaction): Technique for amplifying specific regions of DNA by multiple cycles of DNA polymerisation, each followed by a brief heat treatment to separate complementary strands.

Peptide Bond: Chemical bond between the carbonyl group of one amino acid and the amino group of a second amino acid - a special form of amide linkage.

Peptide Map: Characteristic two-dimensional pattern (on paper or gel) formed by the separation of the mixture of peptides produced by the partial digestion of a protein.

Peptide Nanotubes: NANOTUBES formed from cyclic peptides (PEPTIDES, CYCLIC). Alternating D and L linkages create planar rings that self assemble by stacking into nanotubes. They can form pores through CELL MEMBRANE causing damage. MeSH 2004

Peroxisome: Small membrane-bounded organelle that uses molecular oxygen to oxidize organic molecules. Contains some enzymes that produce and others that degrade hydrogen peroxide (H2O2)

Phagocyte: General term for a professional phagocytic cell - that is, a cell such as a macrophage or neutrophil that is specialized to take up particles and microorganisms by phagocytosis.

Phagocytosis: Process by which particulate material is endocytosed ("eaten") by a cell. Prominent in carnivorous cells, such as *Amoeba proteus*, and in vertebrate macrophages and neutrophils. (From Greek *phagein*, to eat.)

Pharmacodelivery: Site-directed pharmacodelivery is a desirable but elusive goal. Endothelium and epithelium create formidable barriers to endogenous molecules as well as targeted therapies *in vivo*.

Pharmacophore Generation: A procedure to extract the most important common structural features relevant for a given biological activity from a series of molecules with a similar mechanism of action.

Pharmacyte: In medical nanorobotics, a theorized (nanorobotic) device capable of delivering precise doses of biologically active chemicals to individually-addressed human body tissue cells (e.g. cell-by-cell drug delivery).

Phase Shift: A time difference between the input and output signals.

Phase Space: A classical system of N particles can be described by its $3N$ position and $3N$ momentum coordinates. The phase space associated with the system is the $6N$ dimensional space defined by these coordinates.

Phase Transformation: A change in the number and/or character of the phases that make up the microstructure of an alloy.

Phonon: A particle of sound. The energy E of a phonon is given by the Einstein relation,

$E = hf$. Here f is the frequency of the sound and h is Planck's constant. The momentum p of a photon is given by the de Broglie relation, $p = h/ë$. Here ë is the wavelength of the sound. A quantum of acoustic energy, analogous to the quantum of electromagnetic radiation, the photon. Thermal excitations in a crystal or in an elastic continuum can be described as a population of phonons (analogous to blackbody electromagnetic radiation). In highly inhomogeneous solids, a description in terms of phonons breaks down and localized vibrational modes become important.

Photonic Band Gap (PBG): An energy range (and corresponding wavelength range) for which a material neither absorbs light nor allows light propagation.

Photonic Crystal: Composite material made as periodical arrangement of pieces of material (e.g. spheres) embedded in a background matrix of another material that has very different (much lower) dielectric constant and/or magnetic permittivity (electromagnetic properties of material). Such structures do not allow for propagation of light of specific frequencies, due to a destructive interference of photons of such frequencies (this is a wave phenomenon also known as photonic band gap). Photonic crystals offer a possibility of manipulating the emission, propagation and splitting of electromagnetic waves which is expected to be useful for technologies of all-photon telecommunications. The production of such materials relies on nanotechnological methods (e.g. lithography, self assembly approaches...).

Pico Technology: (trillionth of a meter) — The next step smaller, after Nano-technology. The art of manipulating materials on a quantum scale.

Piezoelectric Material: A ferroelectric material in which an electrical potential difference is created due to mechanical deformation, or conversely, in which the application of a voltage causes dimensional changes in the material.

Piezoelectric: A piezoelectric material expands or contracts when a voltage is applied across it. Piezoelectric tubes are often used to move the tip in an AFM or STM. If a piezoelectric crystal is compressed, a voltage appears across it. This effect is sometimes used to measure displacements. A commonly used piezoelectric is PZT (Pb,Zr)TiO3.

Pinhole: The term pinhole embraces a wide variety of oxide defects and is used in a broad sense today. Listed in this category are cracks caused by thermal contraction after oxidation or by handling, and regions of oxide with low dielectric strength caused by dust particles, inadequate masking, contamination, or poor resist adhesion.

Pin-out: Diagram showing for electronic components the relations between connecting pins and internal components.

Pitting: A form of very localized corrosion wherein small pits or holes form, usually in a vertical direction.

Pixel: Picture Element — smallest element of an image, such as a dot on a computer monitor screen.

Plastic Deformation: Permanent or nonrecoverable deformation, accompanied by permanent atomic displacement.

Plasticizer: A low molecular weight polymer additive that enhances flexibility and workability and reduces stiffness and brittleness.

Point Defect: A crystalline defect associated with one or several atomic sites.

Poisson's Ratio: A bar of an isotropic, elastic material ordinarily shrinks laterally when it is stretched longitudinally. The lateral contracting strain divided by the applied tensile strain is Poisson's ratio, which varies from material to material.

Polar Orbit: Usually has an angle of inclination of 90 degrees to the equator. On every pass around the Earth, it passes over both the north and south poles. Therefore, as the Earth rotates to the east underneath the satellite which is travelling north and south, it can cover the entire Earth's surface. A polar orbiting satellite covers the entire globe every 14 days.

Polar: Satellite launched on February 24, 1996 by NASA in the Global Geospace Science project. Polar is an atmospheric studies satellite in polar orbit. One purpose of Polar is to gather information that will help scientists protect future satellites from radiation and other atmospheric dangers.

Polarisation (P): The total electric dipole moment per unit volume of dielectric material.

Polyclonal Antibodies: Antibodies produced by an animal's white blood cells (lymphocytes, specifically) in response to an antigen. This response occurs naturally or can purposely be created by injecting an animal, such as a rabbit or goat, with a specific antigen. More than one kind of anti-body is produced since more than one lymphocyte is producing antibodies. This is referred to as"polyclonal". The polyclonal antibodies are isolated from the animal and can be used for detection purposes. Because the antibodies are actually a mixture with different affinities (binding capability) for the antigen of interest, some variability in performance can occur from one test to another or one batch of antibodies to another.

Polycyclic: A cyclic structure contains rings of bonds; a structure having many such rings is termed polycyclic. In the polycyclic structures of interest in this volume, a large fraction of the atoms are members of multiple small rings, resulting in considerable rigidity.

Polymer: A molecule consisting of many smaller sub-units covalently linked together.

Polymerase Chain Reaction (PCR): A chemical reaction that uses the polymerase enzyme to carry out *in vitro* replication of DNA.

Polymeric Materials: Materials composed of large molecules, generally based on carbon, that have been formed from the chemical bonding of smaller units (monomers).

Polymerisation: A chemical reaction resulting in the bonding together of small molecular units (monomers) to form a much larger molecule (polymer). There are many different routes, or reaction mechanisms, for polymerisation.

Polymers Biomedical: More and more therapeutic problems are relevant to the use of polymer- based therapeutic aids for a limited period of time, namely the healing time related to the outstanding capacity of living systems to self-repair, e.g. bone fracture fixation with screws and plates, of wound closure by sutures and also of drug delivery from implants or similar systems based on polymeric matrices, or on aqueous dispersions or solutions of polymers.

Posthuman: Persons of unprecedented physical, intellectual, and psychological capacity, self-programming, self-constituting, potentially immortal, unlimited individuals.

Potentiometric Device: Monitors the voltage between a sensing electrode and a reference electrode. A high input impedance voltmeter is used to minimize current flow. The voltage typically is proportional to the logarithm of the analyte concentration.

Precision: The degree of reproducibility among several independent measurements of the same true value under specified conditions.

Presentation Semaphore: In medical nanorobotics, a mechanical device used to display specific antigens, chemical ligands, or other molecular objects to the external environment, with the purpose of selectively modifying the chemical or other surface characteristics of a nanorobot exterior.

Probability Density Function: Consider an uncertain physical property and a corresponding space describing the range of values that the property can have (e.g., the configuration of a thermally excited *N* particle system and the corresponding 3*N* dimensional configuration space). The probability density function associated with a property is defined over the corresponding space; its value at a particular point is the probability per unit volume that the property has a value in an infinitesimal region around that point.

Protein Design, Protein Engineering: The design and construction of new proteins; an enabling technology for nanotechnology.

Protein Folding: "The process by which proteins acquire their functional, preordained, three-dimensional structure after they emerge, as linear polymers of amino acids, from the ribosome." [The Scientist]

Protein: Living cells contain many molecules that consist of amino acid polymers folded to form more-or-less definite three-dimensional structures; these are termed proteins. Short polymers lacking definite three-dimensional structures are termed peptides. Many proteins incorporate structures other than amino acids, either as covalently attached side chains or as bound ligands. Molecular objects made of protein form much of the molecular machinery of living cells.

Proteomics: The study of the full expression of proteins in an organism The term proteome refers to all the proteins expressed by a genome, and thus proteomics involves the identification of proteins in the body and the determination of their role in physiological and pathophysiological functions. ... Ultimately it is believed that through proteomics new disease markers and drug targets can be identified that will help design products to prevent, diagnose and treat disease. [e-proteomics.net]

Proximal Probes: A family of devices capable of fine positional control and sensing, including scanning tunneling and atomic force microscopes; an enabling technology for nanotechnology.

Pyroelectricity: The property of certain crystals, such as tourmaline, of acquiring opposite electrical charges on opposite faces when heated.

Quantisation: The concept that energy can occur only in discrete units called quanta.

Quantum Computer: A computer that exploits quantum mechanical phenomena such as superposition and entanglement.

Quantum Confined Atoms (QCA): atoms caged inside nanocrystals. May find uses in clear-glass sunglasses, bio-sensors, and optical computing.

Quantum Corral: Structures where electrons are corralled (confined) by atoms placed on the surface of a metal.

Quantum Cryptography: A system based on quantum- mechanical principles. Eavesdroppers alter the quantum state of the system and so are detected. Developed by Brassard and Bennett, only small laboratory demonstrations have been made.

Quantum Physics: Describes fundamental electronic and optical properties of matter at microscopic level and wave and interference phenomena in particular . Quantum electronics, quantum optics and optoelectronics are important areas of application - atomic clocks, lasers, light emitting diodes, optical fibers, tunnel diodes and superconducting systems are important and well-known examples. The trend towards faster and more complicated microprocessors and microelectronics has resulted in electronic components now approaching the domains of quantum physics at research level. In about 20 years time, miniaturisation will also be halted in commercial applications. It will then probably utilize electronic wave phenomena and single- electron effects in semiconductor components and systems and also to create more complicated transistor components connected in more complicated ways where quantum phenomena, cooperative phenomena and even superconductivity may be important for function. "QNANO" — quantum nanophysics - will probably provide a target for research and development in semiconductor physics, molecular electronics and bioelectronics in the foreseeable future.

Quantum Teleportation: 'Teleportation' (or copying) of the quantum state from one particle to another. Not really spectacular as in Star Trek serial where macroscopic objects (humans, weapons and similar) are 'teleported'. Present technology can 'teleport' quantum objects such as photons, electrons ans similar. It is important to note that the object itself is not teleported during this process. What actually happens is that the quantum state (e.g. spin number) is exactly copied from one object to another, both spatially and temporally present *prior* to 'teleportation'. The technique used is based on the Einstein-Podolsky-Rosen effect (EPR) or paradox, which relies on the so-called nonlocality of the wavefunction. An EPR source produces pair of entangled particles, one of which serves as a 'carbon copy'. The third particle (original, source particle to be copied, introduced besides the EPR pair) is 'entangled' with one of the particles from the EPR pair which enables the copying of its state to another EPR particle that has never interacted with the original, yet it becomes its copy. The experimental setup requires one measuring apparatus that entangles the original with one of the EPR particles and another apparatus that acts on another EPR particle which becomes the copy. The experimentalists operating the two apparatuses must be in contact i.e. the one that acts on the copy particle must know the results of the entanglement measurement on the source (original) particle and another EPR particle which is 'waisted' in the process. It should also be noted here that none of the experimentalists does not know the quantum state of the original particle, even when the whole process finishes, neither is it measured during the process.

Quaternary Structure: Three-dimensional relationship of the different polypeptide chains in a protein complex.

Qubit: The quantum computing analog to a bit. Qubits exhibit superposition. Thus, unlike normal bits, qubits can be both 1 and 0 at the same time.

Reactant: A substance that is consumed in the course of a chemical reaction. It is sometimes known, especially in the older literature, as a reagent, but this term is better used in a more specialized sense as a test substance that is added to a system in order to bring about a reaction or to see whether a reaction occurs (e.g. an analytical reagent).

Reagent Device: A large reagent structure (or a large structure that binds a smaller reagent) serving as a component of a mechanochemical system. A reagent device exists chiefly to hold, position, and manipulate the environment of a reagent moiety.

Recessive: Refers to the member of a pair of alleles that fails to be expressed in the phenotype of the organism when the dominant member is present. Also refers to the phenotype of an individual that has only the recessive allele.

Recombinant DNA: Any DNA molecule formed by joining DNA segments from different sources. Recombinant DNAs are widely used in the cloning of genes, in the genetic modification of organisms, and in molecular biology generally.

Recombination: Process in which chromosomes or DNA molecules are broken and the fragments are rejoined in new combinations. Can occur in the living cell—for example, through crossing-over during meiosis—or in the test tube using purified DNA and enzymes that break and ligate DNA strands.

Reconnaisance Satellite: Also called a spy satellite because it is used to spy on other countries. It can provide intelligence information on military activities, detect missile launches or nuclear explosions, and pick up and record radio and radar transmissions while passing over a country. It can also be used as an orbital weapon by placing warheads on a low orbit satellite to be launched at a ground target, but this is not a recommended or frequent use of satellites.

Remote Sensing Satellite: A type of satellite that performs remote sensing from space. Remote sensing satellites generally monitor important resources for humans. For example, they might be used to track animal migration, locate mineral deposits, watch agricultural crops for weather damage, or see how fast the forests are being cut down. Because they are in space, remote sensing satellites are ideal for monitoring areas with harsh climates or difficult terrains.

Reproducibility: A property of an experiment or process where one tends to receive consistent results from following a specific procedure.

Resistor: Energy dissipative element consisting of a poor conductor in series with connecting wires.

Resolution: The minimum distance between two objects that can be distinguished in microscopy or the minimum spacing between two features that can be fabricated with lithography. The smallest measurable change in input that will produce a small but noticeable change in the output. In the context of chemical separations, defines the completeness of separation.

Resonant Frequency: The frequency at which a moving member or a circuit has a maximum output for a given input.

Respirocrit: In medical nanorobotics, the volume-fraction or bloodstream concentration of respirocyte nanorobots, expressed as a percentage.

Reverse bias: The insulating bias for a p-n junction rectifier; electrons flow into the p side of the junction.

ROM: Read only memory; memory used for permanent, storage of unalterable data; nonvolatile memory.

Sacrificial Layer: A thin film that is later removed to release a microstructure from its substrate.

Salmonella: Rod-shaped, motile, aerobic genus of bacteria. Includes species that cause food poisoning.

Salt Bridge: An ionic bond between charged groups that are part of larger covalent structures; salt bridges occur in many proteins.

SAMFET: (self assembled monolayer field effect transistor). Where a few molecules act as FETs, exhibiting both very strong gain, and extraordinarily rapid response. [Mark Ratner & MT 5(2) p. 20]

Sanguinatation: In medical nanorobotics, locomotion (especially swimming by a nanorobot) through the bloodstream.

Sarcomere: Repeating unit of a myofibril in a muscle cell, composed of an array of overlapping thick (myosin) and thin (actin) filaments between two adjacent Z discs.

Sarcoplasmic Reticulum: Network of internal membranes in the cytoplasm of a muscle cell that contains high concentrations of sequestered Ca2+ : that is released into the cytosol during muscle excitation.

Satellite DNA: Regions of highly repetitive DNA from a eucaryotic chromosome, usually identifiable by its unusual nucleotide composition. Satellite DNA is not transcribed and has no known function.

Saturated Molecule: Molecule containing carbon-carbon bonds that has only single covalent bonds.

Scanning Tunneling Microscope (STM): An instrument able to image conducting surfaces to atomic accuracy; has been used to pin: molecules to a surface. A device in which a sharp conductive tip is moved across a conductive surface close enough to permit a substantial tunneling current (typically a nanometer or less). In a common mode of operation, the voltage is kept constant and the current is monitored and kept constant by controlling the height of the tip above the surface; the result, under favorable conditions, is an atomic-resolution map of the surface reflecting a combination of topography and electronic properties. The STM has been used to manipulate atoms and molecules on surfaces. A type of scanning probe microscopy that maps the local electrical properties of an interface.

Science Court: A name (originally applied by the media) for a government-conducted fact forum.

Science: The process of developing a systematized knowledge of the world through the variation and testing of hypotheses. The pursuit of knowledge and understanding, from the Latin term *scientia*, which means 'knowledge'.

Sealed Assembler Laboratory: A work space, containing assemblers, encapsulated in a way that allows information to flow in and out but does not allow the escape of assemblers or their products.

Secretory Vesicle: Membrane-bounded organelle in which molecules destined for secretion are stored prior to release. Sometimes called secretory granule because darkly staining contents make the organelle visible as a small solid object.

Self-Assembler: A specific type of assembler that makes use of self-assembly such that the assembly process would theoretically not require external energy or information input.

Semiconductor: Material whose conductivity, due to charges of both signs, is normally in the range between that of metals and insulators and in which the electric charge carrier density can be changed by external means.

Sensors: An equipment which detects, and may indicate, and/or record objects and activities by means of energy or particles emitted, reflected, or modified by objects.

Signal to Noise: Ratio which can interfere with detection. Can also refer to data analysis. Biological data is often very "noisy". This is particularly seen when trying to look at low abundance biomolecules.

Single Crystal: A crystalline solid for which the periodic and repeated atomic pattern extends throughout its entirety without interruption.

Single Electron Devices (SED): Nanoscale devices that control the movement of individual electrons, may one day make it possible for integrated circuits to have as many as 10 billion electronic devices in a square centimeter, a density 1000 times greater than that believed feasible for conventional integrated circuits. In development since the mid- 1980s, these devices consist of two electrodes (typically 30 nm wide) separated by a 1 nm- deep insulating layer through which single electrons can tunnel. These devices have many potential applications, from building more sensitive measurement devices to understanding fundamental problems in physics. In the last several years, researchers have built two- junction devices that share a middle electrode. These devices are called "single- electron transistors," because, like conventional transistors, their current can be controlled by modifying the surface charge on the middle electrode, making it an ideal element for an integrated circuit. A circuit made of single- electron devices, however, would have to be operated at a temperature of 4 K or below to reduce thermal effects which disturb the movements of single electrons in the solid.

Sky Hook: A long, very strong, cable in orbit around a planet which rotates around its center of mass in such a way that when one end is closest to the ground, its relative velocity is almost zero. It would function as a kind of space elevator; shuttle craft would anchor to the end and then be lifted into orbit where they would be released. It is closely related to the idea of a beanstalk.

Slip Casting: A forming technique used to shape ceramic materials. A slip or suspension of solid particles in water is poured into a porous mold. A solid layer forms on the inside wall as water is absorbed by the mold, leaving a shell (or a solid piece) in the shape of the mold.

Small Molecule Therapeutics: Low molecular-weight drug. Compared to larger molecular weight pharmaceuticals such as proteins, peptides, and carbohydrates, small molecules can more easily penetrate cell membranes and the blood brain barrier. Can be delivered orally or intravenously. These molecules tend to incur lower process development and manufacturing costs.

Smart Materials and Products: Here, materials and products capable of relatively complex behaviour due to the incorporation of nanocomputers and nanomachines. Also used for products having some ability to respond to the environment.

Spinning: Fiber forming process: a multitude of fibers are spun as molten material is forced through many small orifices.

Step Coverage: If a layer of material is to be deposited on top of a thin strip of material, it is the ratio of film thickness along the walls of the step to the film thickness at the bottom of the step. LPCVD and PECVD offer excellent step coverage.

Strain: In mechanical engineering, strain is a measure of the deformation resulting from stress (that is, force per unit area); the displacement of one point with respect to another, divided by their equilibrium separation in the absence of stress. In chemistry, a molecular fragment generally has some equilibrium geometry (bond lengths, interbond angles, etc.) when the rest of the molecular structure does not impose special constraints (e.g., bending bonds to form a small ring). Deviations from this equilibrium geometry are described as strain, and increase the energy of the molecule. Strain in the mechanical engineering sense causes strain in the chemical sense.

Stress: Force per unit area applied by one part of an object to another. Pressure is an isotropic compressive stress. Suspending a mass from a fiber places it in tensile stress. Gluing a layer of rubber between two plates and then sliding one over the other (while holding their separation constant) places the rubber in shear stress.

Technocyte: A nanoscale artificial device (especially a nanite) in the human bloodstream used for repairs, cancer protection, as an artificial immune system or for other uses. [AS 1995]

Thermal Blanketing: Like insulation in a house, the covering on a satellite that regulates temperature. A material commonly used for thermal blanketing on a satellite is Mylar.

Thermal Noise: The vibration and motion of atoms and molecules caused by the fact that they have a temperature above absolute zero. Once used as an argument on why MNT could not work.

Thermocouple: A temperature-measuring device, which contains a pair of end-joined dissimilar conductors in which an electromotive force is developed by thermoelectric effects when the joined ends and the free ends of the conductors are a different temp.

Thermogenic Limit: In medical nanorobotics, the maximum amount of waste heat that may safely be released by a population of in vivo medical nanorobots that are operating within a given tissue volume.

Thermoplastic Polymer: A substance that when molded to a certain shape under appropriate conditions can later be remelted.

Thioester Bond: High-energy bond formed by a condensation reaction between an acid (acyl) group and a thiol group (-SH); seen, for example, in acetyl CoA and in many enzyme-substrate complexes.

Three-axis Stabilized: Done with an internal gyroscope and thrusters. A gyroscope's stable spin can be used as a sensor to tell the satellite when its attitude is changing. The satellite can then correct the problem using thrusters. This is one form of attitude control.

Thrombin: The enzyme that converts *fibrinogen* to *fibrin* to form a blood clot. Thrombin circulates in the inactive form, prothrombin.

Thruster: A way of controlling a satellite's attitude. Thrusters usually contain compressed gas that when sent out of the end of the thruster will move the satellite in space. The force of the compressed gas (the action) causes the satellite to move in the opposite direction (the reaction). This is Newton's third law of motion—for every action there is an equal and opposite reaction.

Tight Junction: Cell-cell junction that seals adjacent epithelial cells together, preventing the passage of most dissolved molecules from one side of the epithelial sheet to the other.

Tight-receptor Structures: A receptor structure in which a bound ligand of a particular kind is confined on all sides by repulsive interactions (note that favorable binding energies are compatible with repulsive forces). A tight-receptor structure discriminates strongly against all molecules larger than the target.

Titanium Nanopositioner Stages: If the positioning accuracy must be maintained over a prolonged time and temperatures cannot be maintained accurately, use of alternative materials may provide the necessary positioning stability. Titanium's coefficient of thermal expansion (0.9 x 10-5 mm/mm per °C) represents a significant improvement over aluminum alloys and offers the additional benefit of higher strength –improving nanopositioner performance.

Top- down Nanotechnology: Engineers taking existing devices, such as transistors, and making them smaller. Top- down or mechanical nanotechnology will have the greatest impact on our everyday lives in the near future.

Transhuman: Someone actively preparing for becoming posthuman. Someone who is informed enough to see radical future possibilities and plans ahead for them, and who takes every current option for self-enhancement.

Transhumanism: Philosophies of life (such as Extropianism) that seek the continuation and acceleration of the evolution of intelligent life beyond its currently human form and human limitations by means of science and technology, guided by life-promoting values.

Transition State: At the saddle point of a col linking two potential wells, the direction of maximum negative curvature defines the reaction coordinate; the transition state is a hypothetical system of reduced dimensionality, free to move only on a hypersurface perpendicular to the reaction coordinate at its point of maximum energy.

Triple Bond: A double bond is formed when a pi bond is superimposed on a single bond; adding a second pi bond results in a triple bond. The two pi bonds have perpendicular nodal planes, and their sum has roughly cylindrical symmetry, permitting rotation in much the same manner as a single bond.

Viscoelasticity: A type of deformation exhibiting the mechanical characteristics of viscous flow and elastic deformation.

Viscosity: The ratio of the magnitude of an applied shear stress to the velocity gradient that it produces; in other words: a measure of a noncrystalline material's resistance to permanent deformation.

Visible Spectrum: Visible light makes up only a small part of the *electromagnetic spectrum*. The visible light spectrum can be divided into different wavelenghts of light. The wavelength of the light determines the color of that light. The light spectrum goes from violet to red where red is the longest wavelength.

1

Introductory Overview of Nano Chemistry, Nanostructured Materials and Nanoparticles: Major Properties and Issues

Introduction

To analsze ultrasmall systems you need ultrasmall tools. Whether dealing with single molecules or single cells, the precision and sensitivity of the manipulation and detection schemes used are absolutely critical. We have been interested in confining single molecules in solution to perform chemistry at the lowest level possible, as well as in delivering small volumes of chemicals to single subcellular organelles. By mimicking what nature has developed over billions of years to conduct chemistry in ultrasmall packets (cells and organelles), we have arrived at what is to date the smallest test tubes known - synthetic liposomes which are manipulated and electrofused to mix their contents. Our liposomes have volumes of a few femtoliters or less, matching the probe volumes of standard single molecule detection techniques, and are made in minutes by a protocol developed in this lab.

Nanoparticles, Nanomaterials and Nanostructures

Manufactured products are made from atoms. The properties of those products depend on how those atoms are arranged. If we rearrange the atoms in coal we can make diamond. If we rearrange the atoms in sand (and add a few other trace elements) we can make computer chips. If we rearrange the atoms in dirt, water and air we can make potatoes. Todays manufacturing methods are very crude at the molecular level. Casting, grinding, milling and even lithography move atoms in great thundering statistical herds. It's like trying to make things out of LEGO blocks with boxing gloves on your hands. Yes, you can push the LEGO blocks into great heaps and pile them up, but you can't really snap them together the way you'd like. In the future, nanotechnology will let us take off the boxing gloves. We'll be able to snap together the fundamental building blocks of nature easily, inexpensively and in most of the ways permitted by the laws of physics. This will be essential if we are to continue the revolution in computer hardware beyond about the next decade, and will also let us fabricate an entire new generation of products that are cleaner, stronger, lighter, and more precise.

It's worth pointing out that the word "nanotechnology" has become very popular and is used to describe many types of research where the characteristic dimensions are less than about 1, 000 nanometers. For example, continued improvements in lithography have resulted in line widths that are less than one micron: this work is often called "nanotechnology." Sub-micron lithography is clearly very valuable (ask anyone who uses a computer!) but it is equally clear that conventional lithography will not let us build semiconductor devices in which individual dopant atoms are located at specific lattice sites. Many of the exponentially improving trends in computer hardware capability have remained steady for the last 50 years. There is fairly widespread belief that these trends are likely to continue for at least another several years, but then conventional lithography starts to reach its limits. If we are to continue these trends we will have to develop a new manufacturing technology which will let us inexpensively build computer systems with mole quantities of logic elements that are molecular in both size and precision and are interconnected in complex and highly idiosyncratic patterns. Nanotechnology will let us do this. When it's unclear from the context whether we're using the specific definition of "nanotechnology" (given here) or the broader and more inclusive definition (often used in the literature), we'll use the terms "molecular nanotechnology" or "molecular manufacturing." Whatever we call it, it should let us

- Get essentially every atom in the right place.
- Make almost any structure consistent with the laws of physics that we can specify in molecular detail.
- Have manufacturing costs not greatly exceeding the cost of the required raw materials and energy.

There are two more concepts commonly associated with nanotechnology:

- Positional assembly.
- Massive parallelism.

Clearly, we would be happy with any method that simultaneously achieved the first three objectives. However, this seems difficult without using some form of positional assembly (to get the right molecular parts in the right places) and some form of massive parallelism (to keep the costs down). The need for positional assembly implies an interest in molecular robotics, e.g., robotic devices that are molecular both in their size and precision. These molecular scale positional devices are likely to resemble very small versions of their everyday macroscopic counterparts. Positional assembly is frequently used in normal macroscopic manufacturing today, and provides tremendous advantages. Imagine trying to build a bicycle with both hands tied behind your back! The idea of manipulating and positioning individual atoms and molecules is still new and takes some getting used to. However, as Feynman said in a classic talk in 1959: "The principles of physics, as far as I can see, do not speak against the possibility of maneuvering things atom by atom." We need to apply at the molecular scale the concept that has demonstrated its effectiveness at the macroscopic scale: making parts go where we want by *putting* them where we want!

One robotic arm assembling molecular parts is going to take a long time to assemble anything large — so we need *lots* of robotic arms: this is what we mean by massive parallelism. While earlier proposals achieved massive parallelism through self replication, today's "best guess" is that future molecular manufacturing systems will use some form of convergent assembly. In this process vast numbers of small parts are assembled by vast numbers of small robotic arms into larger parts, those larger parts are assembled by larger robotic arms into still larger parts, and so forth. If the size of the parts doubles at each iteration, we can go from one nanometer parts (a few atoms in size) to one meter parts (almost as big as a person) in only 30 steps. A very important tool. Supramolecular chemistry is the chemistry beyond the molecule, and molecules are being designed to self-assemble into larger structures. In this case, biology is a place to find inspiration: cells and their pieces are made from self-assembling biopolymers such as proteins and protein complexes. One of the things being explored is synthesis of organic molecules by adding them to the ends of complementary DNA strands such as A and B, with molecules A and B attached to the end; when these are put together, the complementary DNA strands hydrogen bonds into a double helix, AB, and the DNA molecule can be removed to isolate the product AB. Natural or man-made particles or artifacts often have qualities and capabilities quite different from their macroscopic counterparts, for example, which is chemically inert at normal scales, can serve as a potent chemical catalyst at nanoscales.

Manipulation of structure in the *nanoscale* regime is an area of growing demand due to the unique physical properties observed in the *nanoscale* size regime. Well defined *nanoscale* materials provide potential applications in biotechnology, microelectronics, data storage, functional devices, communication and technology. Due to these vast technologies of importance, bottom-up assembly is currently a widely investigated methodology in meeting the material demand for the miniaturisation of electronic devices. A current limitation in nanotechnology is the lack of robust and versatile routes to template *nanoscale* building blocks into controllable structures Numeral strategies in assembling nanoparticles have been reported such as lithography templated surfaces molecular recognition processes, interfacial assembly block copolymers templates and field induced dipolar association ix into complex structures. In and co-workers demonstrated the assembly of silica particles and polystyrene beads via template assisted self assembly on surfaces. Although this method provides a wide range of one-dimensional (1D) nanostructures, there is still a need for synthetic methodologies to generate hierarchically self assembled systems in solution. Here, we investigate the feasibility of inverse ferrofluids as template for *nanoscale* building blocks because its solution morphology can be actuated by an external field. Inverse ferrofluids consist of a suspension of non magnetic particle (100-500nm) and magnetic colloids (6-10nm).

In the presence of an external magnetic field, Fe colloids phase separated into extended chains, which concurrently segregates nonmagnetic colloids into their own domains. Phase separation occurred in this system is driven by depletion demixing forces, which increases the net entropy gain in the system. De Gans et al, demonstrated the assembling of silica particles into chains that extend into micron in length in ferrofluids. It is of our interest to permanently stitch the bundles of silica shown by De Gans. It is possible to crosslink the methacrylate coated silica particles using gluteraldehyde, 2, 2-azo-bis-isobtyronitrile (AIBN) and benzoyl peroxide. Subsequently, magnetic colloids can be removed via simple centrifugation, or magnetic fractionation. We are also interested in studying the interfacial effects and/or properties of 1D structure (wire) compared to its bulk and individual nanocrystal counterpart. This template serves as an important model system in studying co-operative phenomena and complex processes. The synthetic method of preparing inverse ferrofluid template is reported. Monodisperse silica particle (100-500nm) was synthesised using the Stober method and provides a well defined model system in obtaining a wide range of uniform particle sizes. Monodispersed Co ferrofluid will be synthesised via rapid decomposition of Co 2 (CO) 8 (dicobalt octacarbonyl) in the presence of suitable surfactants such as trioctylphosphine oxide and oleic acid. Co ferrofluid and SiO 2 nanoparticles will be synthesised as reported.

The Stober process provides a well defined model system to prepare silica colloids processing a wide range of particle sizes with narrow size distribution. The surface of the silica beads will be functionalised with methacrylate groups which will served as a steric stabilizer to prevent silica particles from aggregation and also provide polymerizable moieties. In the blend of dispersion of inverse ferrofluids, 1-D chains like structure are obtained by applying an external field. The assembled silica chains will be preserved by polymerising the methacrylate group with neighboring silica particles by the addition of radical sources such as 2, 2-azo-bis- isobtyronitrile (AIBN) and benzoyl peroxide. The conventional C-shaped bar magnets are used to supply the magnetic field while observing the formation of gap spanning chains using an optical microscope. In addition to designing template for nanoscale building blocks, we are also interested in magnetic materials in the nanoscale regime. Magnetic nanostructure is worth exploring because of their application high density magnetic storage media. We are interested in investigating ferromagnetic Co nanoparticles that exhibit particle interaction stronger than thermal energy; aggregating into chains or columnar bundles depending on the concentration. Herein the mechanism of chaining behaviour in ferromagnetic Co nanoparticles stabilised with functional polymers is being studied. Preliminary observation in the retention of chaining behaviour after spin and drop casting on carbon coated grid suggested that these chains are covalently bonded. Additionally, blending of these particles with methacrylate coated silica particle exhibits unusual morphologies suggesting that magnetic nanoparticles are linked into chains.

In this blending experiment, transmission electron microscopy (TEM) images shows that the Co chains preferentially wrap around the surface of silica particles. The suggested mechanism of this interaction is the coupling of radical functional polymer with the reactive surface group on the silica particle. Free radicals generated at the particle periphery enable covalent connectivity with silica particles. This finding demonstrates that Hawker-type alkoxyamine end groups on polymer surfactants have the potential as crosslinker agents to bound nanoparticles into chains. To investigate this hypothesis, we will conduct several control experiments. Our first approach is to synthesize ferromagnetic Co nanoparticles stabilised by non reactive groups such as oleic acid and TOPO. The chaining behaviour and interactions of these Co particles with silica nanoparticles will be compared to the functionalised polymer coated Co nanopartic Nanomaterial technology is receiving a great deal of attention in the coatings industry today.

Several approaches within this technology can be used to achieve organic-inorganic nanocomposite or nanostructured coatings. These approaches include incorporation of preformed nanoparticles in organic resin systems, in-situ generation of nanoparticles or nanophases, and other nanostructuring mechanisms. Following an overview of benefits and critical issues of nanomaterial technology, this article reviews advancements in the application of nanomaterial technology to coatings. Several commercial applications of the technology are described. Conventional materials have grains sizes ranging from microns to several millimeters and contain several billion atoms each. Nanometer sized grains contain only about 900 atoms each.

As the grain size decreases, there is a significant increase in the volume fraction of grain boundaries or interfaces. This characteristic strongly influences the chemical and physical properties of the material. For example, nanostructured ceramics are tougher and stronger than the coarser grained ceramics. Nanophase metals exhibit significant increases in yield strength and elastic modulus. It has also been shown that other properties(electrical, optical, magnetic...) are influenced by the fine grained structure of these materials. Using a variety of synthesis methods, it is possible to produce nanostructured materials in the following forms: thin films, coatings, powders and as a bulk material. There is also considerable interest in the generation of carbon nanostructures, which are related to the famous Buckyball. In addition, the use of nano-sized materials as fillers for composite materials is generating interest. specifically in the case of polymer nanocomposites. There are several different types of nanostructured materials. These range from zero dimensional atom clusters to three dimensional equiaxed grain structure. Each class has at least one dimension in the nanometer range. Atom clusters and filaments are defined as zero modulation dimensionality and can have any aspect ratio from 1 to ¥2.

Any multilayered material with layer thickness in the nanometer range is classified as one-dimensionally modulated. Layers in the nanometer thickness range consisting of ultrafine grains (nanometer range diameter) are two-dimensionally modulated. This class includes coatings, buried layers and thin films. The last class is that consisting of three dimensionally modulated microstructures or nanophase materials. The methods employed to produce nanostructured materials are numerous, with each method having advantages and disadvantages depending on the desired properties or application. Atom clusters are typically synthesised via a vapor condensation route which is essentially evaporation of a solid metal followed by rapid condensation to form nano-sized clusters. These resulting powders are essentially agglomerates of the nano-sized atom clusters. The powders can be then be used as fillers for composite materials or consolidated into bulk material. The cluster assembly method can be used to produce ceramic or metal nanostructured powders. The main advantage of this method is the extremely low contamination levels and the ability to control the final cluster size by manipulation of the process parameters such as temperature, gas environment and evaporation rate. Several different methods have been developed using the vapor condensation concept. These include inert gas condensation, chemical vapor condensation, laser ablation, electron beam deposition, arc evaporation, etc. Both metals and ceramics can be produced using a variety of chemical approaches such as sol-gel or thermal decomposition. These methods provide large quantities of nano-sized agglomerates at a low cost. Chemical processes also allow effective control of the stoichiometry of the final product.

However, the precursor chemicals required may leave residual contamination on the particle surfaces, which leads to difficulties in compaction and sintering. Additionally, powders produced through wet chemical techniques often have difficulties with agglomeration. Other methods which implement chemical reactions include Examples: A common method of producing nanostructured powders is through mechanical deformation, i.e. milling or shock deformation. These processes produce nanostructured materials through severe mechanical deformation of a coarse grained precursor material. The nanometer sized grains nucleate within the dislocation cell structures located in the shear bands. The final grain size is a function of the amount of energy input during the milling as well as time, temperature during milling and milling atmosphere. The precursor materials can be crystalline or amorphous materials in the form of powders or tapes.

Through mechanical milling, it is possible to produce nanostructured powders of systems which are otherwise immicsible. A major disadvantage of this method is the possibility of contamination from the milling media due to the large forces and energies involved in the milling process. Other synthesis methods based on *mechanical deformation*: friciton/sliding wear. Examples: Three dimensional nanostructured materials are also synthesised through thermal crystallisation of an amorphous material. By controlling the nucleation and growth during annealing of an amorphous material, one can produce a bulk material with average grain sizes of less than 20 nm without the need for consolidation and sintering steps. However, this process is limited to material composition which are readily available in the form of a metallic glass, which has an amorphous microstructure. The atomic structure of nanostructured materials is unlike that seen in glasses or crystals because of the large volume fraction (greater than 50 per cent) of grain boundaries or interfaces. Coarse-grained materials have less than 3 per cent of all atoms associated with grain boundaries or interfaces.

However, in materials with nanosized grains between 5 per cent and 50 per cent of all atoms are associated with grain boundaries or interfaces. The grain boundary structure in these nanostructured materials is similar to that of the conventional coarse grained materials. The material properties of the nanostructured materials show remarkable improvement or deviation from the properties exhibited by the coarser grained material. These unique properties are attributed to the significant increase in grain boundary area due to the small grain size. In terms of the mechanical properties, nanostructured metals have shown increases in hardness values (ultimate and yield strengths).

Specifically, pure nanophase metals have shown a clear increase in hardness with decreasing grain size, following the well known Hall-Petch equation. Nanostructured ceramics have exhibited superplastic properties at low temperatures. This is significant as ceramics are conventionally brittle materials. Further deviations from trends observed in conventional materials are observed in creep of nanostructured metals. The common theme of this WTEC study is the engineering of materials with novel (i.e., improved) properties through the controlled synthesis and assembly of the material at the nanoscale level. The range of applications is extremely broad, and these will be described in further detail in subsequent chapters in this report.

The corresponding means of synthesis and assembly are similarly wide-ranging.

But however multifaceted the synthesis approaches and the ultimate applications, there are common issues and unique defining features of these nanostructured materials. First, there is the recognition of critical scale lengths that define the material structure and organisation, generally in the nanometer range, and that ultimately determine the fundamental macroscopic properties of the material. Research in nanostructured materials is motivated by the belief that ability to control the building blocks or nanostructure of the materials can result in enhanced properties at the macroscale: increased hardness, ductility, magnetic coupling, catalytic enhancement, selective absorption, or higher efficiency electronic or optical behaviour. Synthesis and assembly strategies accommodate precursors from liquid, solid, or gas phase; employ chemical or physical deposition approaches; and similarly rely on either chemical reactivity or physical compaction to integrate nanostructure building blocks within the final material structure. The variety of techniques is shown schematically. Schematic of variety of nanostructure synthesis and assembly approaches. The "bottom-up" approach first forms the nanostructured building blocks and then assembles them into the final material. An example of this approach is the formation of powder components through aerosol techniques (Wu et al. 1993) and then the compaction of the components into the final material. These techniques have been used extensively in the formation of structural composite materials. One "top-down" approach begins with a suitable starting material and then "sculpts" the functionality from the material. This technique is similar to the approach used by the semiconductor industry in forming devices out of an electronic substrate (silicon), utilising pattern formation (such as electron beam lithography) and pattern transfer processes (such as reactive ion etching) that have the requisite spatial resolution to achieve creation of structures at the nanoscale.

This particular area of nanostructure formation has tremendous scope, warranting its own separate study, is a driving issue for the electronics industry, and will not be a principal theme of this study. Another top-down approach is "ball-milling, " the formation of nanostructure building blocks through controlled, mechanical attrition of the bulk starting material (Koch 1989). Those nano building blocks are then subsequently assembled into a new bulk material. In fact, many current strategies for material synthesis integrate both synthesis and assembly into a single process, such as characterizes chemical synthesis of nanostructured materials (Murray et al. 1993; Katari et al. 1994). The degree of control required over the sizes of the nanostructure components, and the nature of their distribution and bonding within the fully formed material varies greatly, depending on the ultimate materials application.

Achieving selective optical absorption in a material (e.g., UVblocking dispersions) may allow a wide range of sizes of the component nanostructure building blocks, while quantum dot lasers or single electron transistors require a far tighter distribution of size of the nanostructure components. Compaction methods may provide excellent adhesion for nanocomposite materials of improved structural performance (e.g., ductility), but such interfaces may be unsatisfactory for electronic materials. The intention of this chapter of the report is not to recapitulate in detail the various synthesis and assembly techniques that have been and are being employed in the fabrication of nanostructured materials; that detail can be found in succeeding chapters as well as in excellent summary descriptions provided in the May 8-9, 1997 WTEC workshop proceedings (Siegel, Hu, and Roco 1998). Rather, in attempting to capture the salient features of a new impetus for and interest in a field of nanostructure science and technology, it is more useful to identify the emerging commonalities than the differences among synthesis and assembly approaches. However broad the range of synthesis approaches, the critical control points fall into two categories:

1. control of the size and composition of the nanocluster components, whether they are aerosol particles, powders, semiconductor quantum dots, or other nanocomponents

2. control of the interfaces and distributions of the nanocomponents within the fully formed materials These two aspects of nanostructure formation are inextricably linked; nevertheless, it is important to understand how to exercise separate control over the nucleation of the nanostructure building blocks and the growth (for example, minimising coagulation or agglomeration) of those components throughout the synthesis and assembly process. This latter issue is related to the importance of the following:

- the chemical, thermal, and temporal stability of such formed nanostructures
- the ability to scale-up synthesis and assembly strategies for low-cost, large-scale production of nanostructured materials, while at the same time maintaining control of critical feature size and quality of interfaces (economic viability is a compelling issue for any nanostructure technology) All researchers in this area are addressing these issues.

Nanostructure Synthesis and Assembly

There has been steady technological progress in all fields of nanostructure synthesis and assembly, in no small part because of the more general availability of characterisation tools having higher spatial, energy, and time resolution to clearly distinguish and trace the process of nanostructure formation. As transmission electron microscopy and X-ray diffraction techniques helped in an earlier period to relate the improved properties of "age-hardened" aluminum alloys to their nanostructure (Koch 1998), today's technological advances in materials characterisation are providing new insights into the role of the nanostructure in determining macroscopic properties. The tightly-coupled iteration between characterising the nanostructure, understanding the relationship between nanostructure and macroscopic material properties, and improved sophistication and control in determining nanostructure size and placement have accelerated the rate of progress and helped to define the critical components of this "new" field of nanostructure science and technology. Tightly focused (1-2 μm), high brightness synchrotron X-ray sources provide detailed structural information on colloids, polymers, alloys, and other material structures, highlighting the inhomogeneities of the material with suitable spatial resolution (Hellemans 1998).

Improved characterisation - higher spatial resolution - higher sensitivity leads t o motivat es Better control of size and placement Understanding of - structure-property link - influence of interfaces stimulates. Interactive cycle of characterisation, understanding and enhanced control in the synthesis and assembly of nanostructures. Another important enabling technology has been the now widely available scanning probe technology, including scanning tunneling microscopy and atomic force microscopy. The power of these techniques has provided impetus for developing even higher performance scanning probe tips, fabricated through microfabrication techniques. Development of different tip structures in various materials has given rise to an entire family of powerful scanning probe techniques that encompass such a wide range of characterisation capabilities that one can envision "a laboratory on a tip" (Berger et al. 1996).

The development of a tip technology also impacts the synthesis and assembly processes themselves: scanning probe technologies have been used as the basis of materials patterning and processing at nanometer scales (Held et al. 1997; Snow et al. 1997; and Wilder et al. 1997) and have provided information on the mechanical and thermal properties of materials at the nanoscale (Nakabeppu et al. 1995; Tighe et al. 1997; and Zhang et al. 1996). More sophisticated in situ monitoring strategies have provided greater understanding and control in the synthesis of nanostructured building blocks, particularly those formed in vacuum environments. Molecular beam epitaxy (MBE) represents a physical vapor (gas phase) deposition technique where sub-monolayer control can be imposed on the formation of two-dimensional and, more recently, three-dimensional nanostructured materials (Leonard et al. 1993).

A great deal of the understanding and control derives from the ability to carry out sensitive monitoring of the growth process in situ: reflection high energy electron diffraction (RHEED) details the nature of the surface and surface bonding, and oscillations of the RHEED intensity provide information on the growth rate (Neave et al. 1983). The improvements brought about by these advances in technology have been substantial, but perhaps of greater importance for this nascent field of nanostructure science and technology has been the development of strategies and technologies that have been formed across the former disciplines. More reliable means of controlling nanostructure size and placement, with an end view of being able to scale up the production of such materials while maintaining the control over the nanostructure, have given impetus to a common search for novel synthesis and assembly strategies. In that search, it is apparent that the naturally occurring synthesis and assembly of biological materials can provide us with some critical insights. Major efforts in nanoparticle synthesis can be grouped into two broad areas: gas phase synthesis and sol-gel processing. Nanoparticles with diameters ranging from 1 to 10 nm with consistent crystal structure, surface derivatisation, and a high degree of monodispersity have been processed by both gas-phase and sol-gel techniques. Typical size variances are about 20%; however, for measurable enhancement of the quantum effect, this must be reduced to less than 5% (Murray et al. 1993). Initial development of new crystalline materials was based on nanoparticles generated by evaporation and condensation (nucleation and growth) in a subatmospheric inert-gas environment (Gleiter 1989; Siegel 1991, 1994).

Various aerosol processing techniques have been reported to improve the production yield of nanoparticles (Uyeda 1991, Friedlander 1998). These include synthesis by combustion flame (Zachariah 1994, Calcote and Keil 1997, Axelbaum 1997, Pratsinis 1997); plasma (Rao et al. 1997); laser ablation (Becker et al. 1997); chemical vapor condensation (Kear et al. 1997); spray pyrolysis (Messing et al. 1994); electrospray (de la Mora et al. 1994); and plasma spray (Berndt et al. 1997). Sol-gel processing is a wet chemical synthesis approach that can be used to generate nanoparticles by gelation, precipitation, and hydrothermal treatment (Kung and Ko 1996). Size distribution of semiconductor, metal, and metal oxide nanoparticles can be manipulated by either dopant introduction (Kyprianidou-Leodidou et al. 1994) or heat treatment (Wang et al. 1997). Better size and stability control of quantum-confined semiconductor nanoparticles can be achieved through the use of inverted micelles (Gacoin 1997), polymer matrix architecture based on block copolymers (Sankaran et al. 1993) or polymer blends (Yuan et al. 1992), porous glasses (Justus et al. 1992), and ex-situ particle-capping techniques (Majetich and Canter 1993; Olshavsky and Allcock 1997). Additional nanoparticle synthesis techniques include sonochemical processing, cavitation processing, microemulsion processing, and highenergy ball milling. In sonochemistry, an acoustic cavitation process can generate a transient localised hot zone with extremely high temperature gradient and pressure (Suslick et al. 1996).

Such sudden changes in temperature and pressure assist the destruction of the sonochemical precursor (e.g., organometallic solution) and the formation of nanoparticles. The technique can be used to produce a large volume of material for industrial applications. In hydrodynamic cavitation, nanoparticles are generated through creation and release of gas bubbles inside the sol-gel solution (Sunstrom et al. 1996). By rapidly pressurising in a supercritical drying chamber and exposing to cavitational disturbance and high temperature heating, the sol-gel solution is mixed. The erupted hydrodynamic bubbles are responsible for nucleation, growth, and quenching of the nanoparticles. Particle size can be controlled by adjusting the pressure and the solution retention time in the cavitation chamber.

Microemulsions have been used for synthesis of metallic (Kishida et al. 1995), semiconductor (Kortan et al. 1990; Pileni et al. 1992), silica (Arriagada and Osseo-Assave 1995), barium sulfate (Hopwood and Mann 1997), magnetic, and superconductor (Pillai et al. 1995) nanoparticles. By controlling the very low interfacial tension (~10-3 mN/m) through the addition of a cosurfactant (e.g., an alcohol of intermediate chain length), these microemulsions are produced spontaneously without the need for significant mechanical agitation. The technique is useful for large-scale production of nanoparticles using relatively simple and inexpensive hardware (Higgins 1997). Finally, high energy ball milling, the only top-down approach for nanoparticle synthesis, has been used for the generation of magnetic (Leslie- Pelecky and Reike 1996), catalytic (Ying and Sun 1997), and structural (Koch 1989) nanoparticles. The technique, which is already a commercial technology, has been considered dirty because of contamination problems from ball-milling processes. However, the availability of tungsten carbide components and the use of inert atmosphere and/or high vacuum processes have reduced impurities to acceptable levels for many industrial applications. Common drawbacks include the low surface area, the highly polydisperse size distributions, and the partially amorphous state of the asprepared powders.

Field-Controlled Molecular Switching Devices

A second category of molecular electronics device that appears promising for the development of molecular nanoelectronic digital computers is the electric-field controlled molecular switching devices, including molecular quantum-effect devices. These are most closely descended from the solid-state microelectronics and nanoelectronic devices described earlier, and promise to be the fastest and most densely integrated among all of the current alternatives. A group at Purdue University has used self-assembly to fabricate and demonstrate functioning arrays of molecular electronic quantum confinement structures connected by molecular wires.

As another example, J.M. Tour proposes embedding a quantum well in a molecular wire by inserting pairs of barrier groups that break the sequence of conjugated porbitals, forming a two-terminal molecular RTD; other structures for three-terminal molecules have been suggested. (In February 1999, Tour reported the demonstration of RTD effects in the manner proposed above; at press time, the paper was still in review [J.M. Tour, personal communication, 1999].) Nondissipative (adiabatic) Thouless electron pumping has also been described at 0.33 K.

Merkle and Drexler consider a hypothetical "helical logic" device in which a single electron can switch another single electron, with 1 or 0 represented as the presence or absence of individual electrons. The electrons are constrained to move along helical paths, driven by a rotating immersive electric field (normal to the helical axis) that confines each charge carrier to a fraction of a turn of a single helical loop, moving it like water in an Archimedean screw. A logic operation involves two helices, one of which splits into two "descendant" helices. At the point of divergence, differences in the electrostatic potential resulting from the presence or absence of a carrier in the adjacent helix control the direction taken by a carrier in the splitting helix; the sequence is reversible, allowing two initially distinct helical paths to merge into a single outgoing helical path without forcing a dissipative transition. Energy dissipation at ~10 GHz is ~10^{-6} zJ/gate-cycle at an operating temperature of 1 K.

Other classes of molecular electronic devices have been studied, including picosecond photoactive or photochromic molecular switching devices using light, and electrochemical molecular devices using electrochemical reactions, to change the shape, orientation, or electron configuration of a molecule in order to switch a current. However, photoactive devices in a dense network would require an NSOM-like emitter to switch individually, since optical photon pathways are not easily confined at length scales very much below optical wavelengths of ~500-1000 nm; electrochemical molecular devices might require immersion in solvent to operate.

Fullerenes exhibit a wide range of quantised conductivities that may be useful in nanoelectronic computers. For example, chemisorption alters properties: 3 alkali dopant atoms per C_{60} buckyball gives high conductivity, while none or 6 dopant atoms gives an insulator. Buckytube conductivity also varies with mechanical stress and torsion, tube diameter, and other geometrical parameters. Depending upon the direction of the nanotube chiral vector, carbon nanotubes may be either metallic or semiconducting. Suitably-sized, doped, flexed, stressed, or chiralised fullerenes could be nested to make insulated wires or joined to make nanoelectronic computer components including resistors with quantised "staircase" resistance of $nh/2e^2$, n = 1, 2, ... (e is electronic charge, h is Planck's constant), two-terminal devices such as diodes, three- or four-terminal devices such as heterojunction, TUBEFET, or thin-film transistors, and gates that introduce electric fields. A single-molecule transistor using semiconducting buckytubes has been demonstrated experimentally at room temperature.

Carbon nanotubes have been cut to length, size-sorted by field flow fractionation, their open ends derivatised with thiol groups and then tethered to 10-nm gold particle "junction boxes" which AFM images show can connect together at least two separate tubes. If the tethering process can be spatially controlled using DNA complementarity or by other means, then fullerene nanoelectronic circuit elements could be assembled into three-dimensional CPU-like structures. P. Collins expects an all-carbon 8-bit adder with nanometer dimensions by the year 2002, and J.C.

Ellenbogen of MITRE Corp. has designed a molecular half-adder that measures ~10 nm x 10 nm in size. This research area was extremely active in 1998.

Fullerenes may also allow compact memory devices. For example, it has long been speculated that a ~1 nm buckyball encapsulated in a ~1 nm diameter nanotube segment may glide freely back and forth, trapped weakly in each end cap by van der Waals forces. An external voltage could nudge a charged C_{60} molecule to one end of the other, creating a two-state "buckyshuttle RAM" storage device. Such a device could store one bit of information in a ~10 nm^3 volume, a storage density of ~10^8 bits/micron^3; the C_{60} thermal velocity of ~100 m/sec at 310 K inside a ~10 nm memory element gives a typical read/write time of ~0.1 nanosec. In early 1999, Kwon and colleagues reported that finely dispersed 4-6 nm diamond powder thermally annealed by heating in a graphite crucible under argon at 1800°C for 1 hour produced oblong multiwall carbon nanostructures which in some cases could move around inside each other, like C_{60} trapped within C_{480}. Kwon noted that there are potential energy minima at either end, so a charged buckyball (e.g., an enclosed ion) could be shuttled with an applied electric field, which could be used for memory storage. Computational simulation of writing a bit showed that the interior buckyball neatly shuttled to the correct end, then gently bounced to rest in ~0.01 nanosec (i.e., ~100 GHz operation) dissipating only ~10 kT in the C_{480}. Kwon proposed a high density memory board using aligned buckyshuttles in a hexagonal lattice with addressing wires above and below, not unlike a ferrite core memory; the memory would likely be nonvolatile at room temperature, with mass production based on self-assembly.

Possibilities of Molecular Manufacturing

Most contemporary industrial fabrication processes are based on "top-down" technologies, wherein small objects are sawn or machined from larger objects, or small features are imposed on larger objects, in either case by removing unwanted matter. The results of such processes may be small, such as micron-featured integrated circuits, or very large, such as jet aircraft, but in most cases the material is being processed in chunks far larger than molecular scale. Molecular manufacturing, on the other hand, represents a "bottom-up" technology. Desired products will be built directly by "assembler" machines, molecule by molecule, making larger and larger objects with atomic precision. The results of such processes may also be very small or very large, much as biology builds both micron-sized bacteria and 100-meter tall sequoia trees. However, since assemblers add matter only where it is intended, little need be removed and hence there may be minimal waste during the process. By guiding with precision the assembly of molecules and supra-molecular structures, such a manufacturing system could construct an extraordinarily wide range of products of unprecedented quality and performance.

In 1959, the Nobel physicist Richard P. Feynman observed that:

> "The principles of physics, as far as I can see, do not speak against the possibility of maneuvering things atom by atom. It is not an attempt to violate any laws; it is something, in principle, that can be done; but, in practice, it has not been done because we are too big . . . Ultimately, we can do chemical synthesis."
>
> "A chemist comes to us and says, 'Look, I want a molecule that has the atoms arranged thus and so; make me that molecule. The chemist does a mysterious thing when he wants to make a molecule. He sees that it has got that ring, so he mixes this and that, and he shakes it, and he fiddles around. And, at the end of a difficult process, he usually does succeed in synthesising what he wants.
>
> "But it is interesting that it would be, in principle, possible (I think) for a physicist to synthesize any chemical substance that the chemist writes down. Give the orders and the physicist synthesizes it. How? Put the atoms down where the chemist says, and so you make the substance. The problems of chemistry and biology can be greatly helped if our ability to see what we are doing, and to do things on an atomic level, is ultimately developed a development which I think cannot be avoided."

Nearly four decades after Feynman's famous "Plenty of Room at the Bottom" speech, and a decade after Drexler's original proposal for a bottom-up approach to machinebuilding using molecular assemblers, Nobel chemist Richard Smalley also largely agreed that this objective should prove feasible. Noted Smalley: "On a length scale of more than one nanometer, the mechanical robot assembler metaphor envisioned by Drexler almost certainly will work..." Many skeptical questions arise when one first encounters the ideas of molecular nanotechnology and molecular assemblers. It is useful to keep in mind the proven feasibility of such systems in the biological tradition over billions of years of natural evolution. In one sense, molecular nanotechnology will be a refinement and expansion upon how nature works at the molecular scale. Nature's examples, such as human beings, certainly answer the most basic skeptical questions, such as: Can macroscopic objects be built from molecular scale processes? (Yes, thanks to cellular replication.) Are molecular objects stable? (Of course; the human population alone contains $>10^{29}$ reasonably stable and quite functional ribosomal "nanomachines"). Observes Nobel chemist Jean-Marie Lehn: "The chemist finds illustration, inspiration and stimulation in natural processes, as well as confidence and reassurance since they are proof that such highly complex systems can indeed be achieved on the basis of molecular components."

What about quantum effects? The uncertainty principle makes electron positions somewhat fuzzy, but the atom as a whole has a comparatively definite position set by the relatively great mass of the atomic nucleus. The quantum probability function of electrons in atoms tends to drop off exponentially with distance outside the atom, giving atoms a moderately sharp "edge". Mathematically, the positional uncertainty of a single carbon atom of mass $m_C = 2 \times 10^{26}$ kg bound in a single C-C bond of stiffness $k_C = 440$ N/m may be crudely estimated from the classical vibrational frequency $n_C = (k_C/m_C)^{1/2} = 1.5 \times 10^{14}$ Hz. This sets the zero-point vibrational bond energy $E_C = h\ n_C / 2 = 4.9 \times 10^{20}$ J $= k_C\ DX_C{}^2 / 2$ where $h = 6.63 \times 10^{34}$ J-sec (Planck's constant) and $DX_C \sim 0.015$ nm is the maximum classical amplitude of the bound carbon atom (roughly the same as the 3 dB point for the gaussian wavefunction, notes J. Soreff). Thus DX_C is just ~5% of the typical atomic electron cloud diameter of ~0.3 nm, imposing only a modest additional constraint on the fabrication and stability of nanomechanical structures. (Even in most liquids at their boiling points, each molecule is free to move only ~0.07 nm from its average position.)

How about the effects of Brownian bombardment on nano-machines? Describing a nanomechanical component as a harmonic oscillator embedded in a gas, Drexler notes: "At equilibrium, an impinging gas molecule is as likely to absorb energy as to deliver it, and so molecular bombardment has no net effect on the amplitude of vibration. How a system is coupled to a thermal bath can affect its detailed dynamics, but not the statistical distribution of dynamical quantities." Nanomachines must also obey the laws of thermodynamics. Nanorobots cannot be used as a "Maxwell's demon" — energy must be expended to do useful work; there can be no free lunch.

Will high-energy radiation damage nanomachines? Radiation can break chemical bonds and disrupt molecular machines. The annual failure rate for a properly designed ~1 micron^3 molecular machine (containing many billions of atoms) can be made as low as several percent. This estimate is derived from a consideration of all known thermal, photochemical, radiation, and other damage mechanisms, but the dominant error mechanism that appears difficult to substantially reduce is damage caused by background radiation. Failure rates are not zero, but are nonetheless remarkable by today's standards, especially given error detection and correction at the module level. During nanodevice fabrication, the problems of crossbonding and reactive intermediates pose additional constraints that must be dealt with using appropriate assembler designs. Specifically, assemblers with rigid positioning components operating in vacuo can restrict reactive intermediates and molecular tool tips to desired locations with a precision of at least ~0.1 nm, or slightly less than an atomic diameter. Within the confines of an evacuated or "eutactic" assembler workspace, the location of every structural atom may be known to within the uncertainty created by thermal noise.

Internal structural components may possess relatively high stiffness, so positional uncertainty caused by thermal noise may be small. In the classical analysis, positional variance $DX^2 = kT / k_s$, where k = 1.381 x 10^{21} J/kelvin (Boltzmann's constant), T is temperature in kelvins, and k_s is component stiffness, typically 10 N/m for nanometer-scale diamondoid components, hence the standard deviation of component position is ~0.02 nm at room temperature, or ~0.01 nm at cryogenic (e.g., LN_2) temperatures. By employing sufficiently stiff mechanisms the position of every structural atom in the system can be known to within a fraction of an atomic diameter with high reliability and without the need for explicit positional sensing. Steric tool hindrance — in molecular manufacturing, the conical volume occupied by a tool tip or manipulator platform — is a technical design problem that may be overcome by using sufficiently stiff tools or multi-tip tools, working in vacuo. Additionally, while some synthetic methods might involve the manipulation of individual atoms by appropriate tools, such as the hydrogen abstraction tool which removes a single selected hydrogen atom from a diamondoid surface, other reactions will involve small clusters of atoms or larger molecular components. What about friction and wear among nanomechanical components? Properly designed molecular machines lack wear mechanisms, although other damage mechanisms remain. Dry bearings with atomically precise surfaces can have negligible static friction and wear, and very low dynamic friction or drag.

Nanomechanical components are better viewed as moving smoothly in a force field than as sliding subject to friction. For example, total drag power dissipated by an isolated molecular sleeve bearing of radius 2 nm spinning at 1 MHz is dominated by band-stiffness scattering, amounting to 0.000004 picowatts (pW), very small compared to the 1-1000 pW power draw that will be typical of micron-size medical nanorobots. The strong precise surfaces of nanomachines experience no change during a typical operational cycle, hence zero wear. Within the single-point failure model, the first step in a wear process (e.g., a dislocated atom) is regarded as fatal, hence cumulative wear plays no role in determining device lifetimes. In eutactic nanomachines, contaminants of all kinds are rigorously excluded. The equivalent of wear particles cannot appear until the device has failed. On this scale, repulsive fields provide "lubrication" and an oil molecule would be a contaminant object, not a lubricant. The general conclusion is that molecular nanotechnology violates no physical laws. In 1998, it was generally accepted that molecular nanotechnology will be developed, although there remains some disagreement about how long it will take. To progress from today's limited capabilities to the complex nanorobots described later in this trilogy will clearly require a great deal of research and development effort. Still, the proper question is no longer "if" but "when."

This study presents a few of the many possible technical paths leading to a working molecular assembler that were being discussed in 1998. The reader is strongly urged to consult the most current literature for the latest results. The conventional top-down approaches to nanotechnology. Several possible direct bottom-up pathways to molecular manufacturing including biotechnology, supramolecular chemistry, and scanning probes. Gimzewski and Joachim note that "bottom-up approaches to nanofabrication may one day compete with conventional top-down approaches in providing nanotechnologies for the next millenium. Top-down refers to increasing miniaturisation through extension of existing microfabrication schemes, [whereas] the bottom-up scenario is one of ever-increasing complexity at the molecular level while maintaining control on an atom-by-atom basis". Surveys molecular mechanical components and design work on molecular assemblers and molecular manufacturing systems from the perspective of 1998.

Other Synthesis Issues

One of the most challenging problems in synthesis is the controlled generation of monodispersed nanoparticles with size variance so small that size selection by centrifugal precipitation or mobility classification is not necessary. Among all the synthesis techniques discussed above, gas-phase synthesis is one of the best techniques with respect to size monodispersity, typically achieved by using a combination of rigorous control of nucleationcondensation growth and avoidance of coagulation by diffusion and turbulence as well as by the effective collection of nanoparticles and their handling afterwards. The stability of the collected nanoparticle powders against agglomeration, sintering, and compositional changes can be ensured by collecting the nanoparticles in liquid suspension. For semiconducting particles, stabilisation of the liquid suspension has been demonstrated by the addition of polar solvent (Murray et al. 1993); surfactant molecules have been used to stabilize the liquid suspension of metallic nanoparticles. Alternatively, inert silica encapsulation of nanoparticles by gas-phase reaction and by oxidation in colloidal solution has been shown to be effective for metallic nanoparticles (Andres et al. 1998).

New approaches need to be developed for the generation of monodisperse nanoparticles that do not require the use of a size classification procedure. An example of this is a process developed in Japan where very monodispersed gold colloidal nanoparticles with diameters of about 1 nm have been prepared by reduction of metallic salt with UV irradiation in the presence of dendrimers (Esumi et al. 1998). Poly(amidoamine) dendrimers with surface amino groups of higher generations have spherical 3-D structures, which may have an effective protective action for the formation of gold nanoparticles. Although the specific role of dendrimers for the formation of monodispersed nanoparticles has yet to be defined, good monodispersity is thought to come from the complex reaction accompanying the decomposition of dendrimers, which eventually leads to the conversion of solution ions to gold nanoparticles. Scaleup production is of great interest for nanoparticle synthesis. High energy ball milling, already a commercial high-volume process, as mentioned above, has been instrumental in generating nanoparticles for the preparation of magnetic, structural, and catalytic materials. However, the process produces polydispersed amorphous powder, which requires subsequent partial recrystallisation before the powder is consolidated into nanostructured materials. Although gas-phase synthesis is generally a low production rate process (typically in the 100 milligrams per hour range) in research laboratories, higher rates of production (about 20 grams per hour) are being demonstrated at Ångström Laboratory at Uppsala University in Sweden.

Even higher production rates (about 1 kg per hour) are now being achieved commercially. For sol-gel processing, the development of continuous processing techniques based on present knowledge of batch processing has yet to be addressed for economical scaleup production of nanoparticles. Other related sol-gel issues concern the cost of precursors and the recycling of solvent. Overall, sol-gel processing is attractive for commercial scale-up production. A recent paradigm shift envisioned for optoelectronics and computational devices involves the assembly of molecular or quantum wires (Chidsey and Murray 1986). Large polymeric molecules have been used as nano building blocks for nanoporous molecular sieves, biocompatible materials, optical switching, data processing, and other nonlinear optical components. Chain aggregates of nanoparticles can be considered as polymer-like units with their primary particles composed of a few hundred to a few thousand molecules.

Thus, these chain aggregates can be considered "heavy" quantum wires. In fact, nanoparticle chain aggregates have been studied extensively as magnetic materials (Zhang and Manthiram 1997), as reinforced elastomers (Pu et al. 1997), and as additives in concrete (Sabir 1997). These aggregates have been shown to have chemical and mechanical properties different from those of individual primary particles (Friedlander et al. 1998). Depending on the particle size and its compositional material, the bonding force responsible for holding the aggregates together varies from weak van der Waals force for micrometer particles to strong chemical bonds for nanometer particles to very strong magnetic dipolar bonds for nanosized magnetic particles. The mechanical, optical, and electronic transport properties can be varied by controlling the diameter and the monodispersity of the primary particles, the crystalline structure and morphology, aggregate length, interfacial properties, and material purity. These chain aggregates can be formed by allowing agglomeration of nanoparticles generated by any of the synthesis techniques discussed above, with the exception of high energy ball milling, which generates particles with low surface area and high anisotropic morphologies, both of which are detrimental for the formation of chain aggregates. Depending on the magnetic and electric charging properties of the nanoparticles, an external applied magnetic or electric field can be used to control the fractal dimension of aggregates. For optical applications of chain aggregates, lower fractal dimensions (i.e., relatively straight chain aggregates with few branches) are desirable. Recent advances in the fabrication of nanometer fibers or tubes offer another form of building blocks for nanostructured materials.

An effective way to generate nanometer fibers (or tubes) is based on the use of membrane-template techniques (Martin 1994). Membranes, with nanochannels generated by fission-fragment tracks or by electrochemical etching of aluminum metal, are used as templates for either chemical or electrochemical deposition of conductive polymers (Pathasarathy and Martin 1994), metal (van de Zande et al. 1997), semiconductor (Klein et al. 1993), and other materials for the generation of nanofibers or tubes. Since the nanochannels on membranes are very uniform in size, the diameter and the aspect ratio of the nanofibers (or tubes) synthesised by the membranetemplate technique can be precisely controlled.

This greatly facilitates the interpretation of optical data and the processing of these fibers (or tubes) into 2-D nanostructured materials (de Heer et al. 1995). Single-crystal semiconductor nanofibers can also be grown catalytically by metalorganic vapor phase epitaxy and laser ablation vapor-liquid-solid techniques (Hiruma et al. 1995; Morales and Lieber 1998). The synthesis of these onedimensional structures with diameters in the range of 3 to 15 nm holds considerable technological promise for optoelectronic device applications, such as the p-n junctions for light emission at Hitachi Central Research Laboratory in Japan. The advent of carbon-based nanotubes has created yet another way to fabricate nanometer fibers and tubes. These nanotubes have been used as templates for the fabrication of carbide and oxide nanotubes (Dai et al. 1995; Kasuga et al. 1998).

Synthesis of nanotubes based on BN, BC3 and BC2N have also been reported (Chopra et al. 1995; Miyamoto et al. 1994). These nanotubes potentially possess large third-order optical non-linearity and other unusual properties (Xie and Jiang 1998). Metallic nanofibers synthesised by carbon-nanotube-template techniques are useful in the design of infrared absorption materials. The carbon nanotubes can now be catalytically produced in large quantities and have been used for reinforcement of nanostructural composite materials and concrete (Peigney et al. 1997). The elegant complexity of biological materials represents the achievement of structural order over many length scales, with the full structure developed from the "nested levels of structural hierarchy" (Aksay et al. 1992), in which self-assembled organic materials can form templates or scaffolding for inorganic components. These notions of a multilevel material structure with strong interactions among levels and an interplay of perfection and imperfection forming the final material was discussed earlier by Cyril Stanley Smith (1981). Along with characteristic length scales, there are characteristic relaxation times of the material, bringing in the consideration of the temporal stability of the structured materials (Zener 1948). A more detailed discussion of the role of biological materials as both paradigm and tool for the fabrication of nanostructured materials is given. It is interesting how many of the synthesis and assembly approaches seek out and adapt two key features of biogenic fabrication: that of "self-assembly, " and the use of natural nanoscaled templates.

Self-Assembly

"Self-assembly" is a term that by now figures prominently in the literature of nanostructured materials and nanofabrication. The term therefore carries a variety of implicit or explicit meanings, and we cite the definition given by Kuhn and Ulman: "Self-assembly is a process in which supermolecular hierarchical organisation is established in a complex system of interlocking components." The mechanisms that produce the hierarchical organisation are determined by competing molecular interactions (e.g., interactions between hydrophobic versus hydrophilic components, van der Waals, Coulombic, or hydrogen bonding), resulting in particular microphase separation or surface segregation of the component materials. Thus, the use of a hierarchy of bond strengths and/or chemical specificity can produce a hierarchy of lengths in the final nanostructured material (Muthukumar et al. 1998; Stupp and Braun 1998). As one example of such self-assembled or self-organised materials, McGehee et al. (1994) have mixed silica precursors with surfactants that have self-ordered to form various surfactant-water liquid crystals, producing various structures built from walls of amorphous silica, organised about a repetitive arrangement of pores up to a hundred angstroms in diameter.

In the "natural" formation of inorganic nanostructures, the addition of organic molecules can strongly influence the resulting structure of inorganic components. Such strategies have been adopted in synthetic formation of nanostructures, such as in the formation of networks of gold clusters (Andres et al. 1998). Gold clusters 3.7 nm in diameter, formed in the vapor phase, are encapsulated in organic surfactants, such as dodecanethiol, forming a colloidal suspension. The surfactants prevent the agglomeration of the gold clusters. Addition of small amount of dithiol precipitates out a 3-D cluster network, which can in turn be deposited onto another solid substrate.

A transmission electron microscope (TEM) image of a cluster array spin-cast onto MoS2. The methodology of self-assembly has even been extended to physical vapor deposition processes where it would seem more difficult to control the nucleation and growth of three-dimensional nanostructures. Utilising the strain inherent in the epitaxial growth of lattice-mismatched materials, and the expected strain-induced transition from two-dimensional (layered) to three-dimensional (islanded) growth, together with careful monitoring of the growth process through RHEED analysis, researchers have been able to form arrays of semiconductor quantum dots, ~ 200-300 Å in diameter, ~ 1011 cm-2 in density, and with a size variation of about ±7% (Leonard et al. 1993). An example of such "self-assembled" semiconductor quantum dots. The achievement of arrays of several billions of quantum dots of these dimensions with such a size variation is beyond the capability of standard high resolution lithographic and pattern transfer processes. Moreover, the controlled formation of critical surfaces and interfaces without the intercession of ion-assisted processing that can introduce potential defects into the materials has produced a rich source of optically and electronically efficient quantum structures.

A number of researchers have already incorporated such self-assembled dots into laser structures (Bimberg et al. 1998). Chemical specificity may provide the most robust means of ensuring control of size and placement of nanostructured building blocks, and recent work in the synthesis of compound semiconductor quantum dots from chemical precursors have provided even tighter distributions of size variation (±5%) than those shown in the strain-induced self-assembled dots (Katari et al. 1994; Murray et al. 1993).

Natural Templates or "Scaffolds"

Biogenic systems employ natural templates or scaffolding in the construction of nanoscaled materials, where the templates and scaffolding can help set the proper, critical dimensions that are characteristic of the final material. The templates can be formed "artificially, " through the use of lithographically defined patterning and processing of the substrate material in order to achieve selective nucleation and growth of the nanostructured materials. However, there are also a host of "natural" templating materials that researchers can avail themselves of. The zeolites have for a long time been utilised as the basis of high surface area materials that enable catalysis. Recent work has taken these natural cage structures and formed large cage structures with varying pore sizes (Estermann et al. 1991; Ying and Sun 1997; Bu et al. 1998). When zeolite cages are "loaded" with various materials, the controlled proximity enforced by the cage structure can result in various magnetically active or nonlinear optical materials. More recently, the C-based fullerenes and C-nanotubes have provided a tremendously rich basis for nanostructure research; the impact of such materials is further described. The by now "easy" formation of such structures and ready availability of such materials provide a rich source of nanostructures with nearly immediate applications for electronic devices, storage, and enhancement of structural materials. Further, these materials can serve as the nanostructured precursors of nanostructure synthesis approaches (such as the starting material for cluster or aerosol deposition), or can provide an avenue for the synthesis of templates of varying sizes. Such templates can be electronically "doped" and filled with other materials. Recent work reported the synthesis of "nanorods" of GaN through confined reactions in carbon nanotubes (Han et al. 1998).

Opportunities and Challenges

The broad explorations of this WTEC study have shown a great diversity of impressive work on nanostructured materials. One could argue that many aspects of the work on nanostructured materials have been long-established efforts, with well-developed techniques that have been brought forth into the manufacturing arena.

How do we then explain the current excitement and interest in nanostructured materials as a promising new endeavor? Part of the answer lies in the recognition of common scientific issues and common enabling technologies that link this group of researchers together. Ready availability of ever more sophisticated characterisation methods that allow us to visualize and probe materials at the nanoscale has accelerated the pace of activities in this field; at the same time, recognition of the common critical issues of control over nanostructure size and placement motivates sharing of solutions over the boundaries of conventional disciplines.

Thus, some of the most exciting findings of this study manifest the crossfertilisation of techniques and ideas; for example, aerosol particles are being integrated with more traditionally fabricated electronic nanostructures, with placement achieved through the manipulation of STM tips, in order to explore ideas of electronic device enhancement at the nanostructure level (Junno et al. 1998). The ordered assembly of diblock copolymers define three-dimensional nanostructures, and those structures are transferred into electronic (semiconductor) substrates through high resolution pattern transfer processes (Möller 1998). Recognising the need for effective utilisation of ideas across disciplines brings the responsibility of establishing the educational infrastructure that will adequately train young scientists to more fully develop the concepts and applications of nanostructured materials in the future. Such an infrastructure will require researchers and educators who are familiar with the properties of a broad range of materials, including polymers, biomolecular materials, metals, ceramics, and semiconductors. It will benefit from an informed perspective on critical applications, and must provide access to a wide range of synthesis and assembly techniques and characterisation methods that currently reside separately in the disciplines of physics, chemistry, electrical engineering, biology, etc.

Finally, a critical enabler for the future of this field and for its educational infrastructure is further development of computational tools that encompass the full range of atomistic calculations to macroscopic materials properties. This will require a "systems approach" (Olson 1997) that will span a variety of different computational models, addressing the different critical length scales, starting with solutions of the Schrodinger and Poisson equations, solving interatomic forces, and scaling all the way up to simulations of macroscopic properties and behaviour (Goddard 1998). Increased appreciation of and access to the diverse means of nanoparticle synthesis and assembly that have been developed within many different disciplines, and a common development of enabling tools and technologies, will enhance the pace of accomplishments in this new area of nanoscale synthesis and assembly.

Dispersions and Coatings

The object of the controlled assembly of nanoparticulates is to make materials with new properties and assemble them with practical applications. A unique value of nanoparticulates is their extremely high particle surface area; having many more sites for achieving property enhancements makes them ideal for a wide variety of applications as dispersions and coatings. Dispersive and coating applications of nanoparticles include optical, thermal, and diffusion barriers. Significant work on nanoscale dispersions and coatings is underway worldwide in the areas of ceramics, cosmetics, biosensors, colorants, and abrasion-resistant polymers.

Other applications include imaging ink jet materials, electrophotography, pharmaceuticals, flavor enhancers, pesticides, lubricants, and other proprietary applications specific to industry. Still another application is in a new, post-silicon generation of electronic devices that includes nanotubes and fullerenes as constituent units of carbon nanoelectronic devices; here, dispersion takes on a more quantum consideration in which the number of atoms in a cluster is compared to the number of surface atoms to determine its dispersion function. Also in the semiconductor industry, a monolayer or thin film coating of atoms or molecules is deposited on foils, metal sheets, or glass to enhance storage capacity and accelerate responses from the electronic component. All these applications deal with dispersions or coatings of particles that enhance specific features. Also, ability to manufacture smaller functional systems enhances performance, cost, and efficiency. These considerations power the drive towards miniaturisation and precision finishing.

Property Amplification

What's particularly exciting in the study of nanoparticles, nanostructured materials, and nanodevices is their ability to also add value to materials and products through enhancement of specific properties, such as the following:

- Mechanical strength. Nanostructured powders have been produced by plasma processing where the reactor vaporizes coarse metal particles (Froes 1998, 105-6); by combustion synthesis where redox reactions take place at elevated temperatures, followed by quenching; and by mechanical alloying with gas atomisation. When nanocrystalline powders achieved by these means are compacted and applied as a coating, they lend significant strength and ductility to a variety of conventional materials such as ceramic, composites, and metal alloys.
- Superconductivity. Brus (1996) describes the nature of the superconductivity effect in detail. It involves a deposition of nanocrystals on substrates, leading to improved optical and electrical properties.
- Covering power. Because nanostructuring increases the number of active sites—there are many more atoms per grain boundary—the enhanced surface area leads to a reduced material requirement, which in turn can lower cost (Solomaon and Hawthorne 1983).

- **Ability to incorporate high cost materials. Expensive materials such as colorants and drugs may be effectively dispersed in small and controlled quantities through nanostructuring (Schnur 1994).**
- **Environmental value. Improvements in environmental impact are achieved by utilising nanostructure particulates in coatings and thus eliminating the requirement for toxic solvents. By eliminating hazardous wastes, nanocoatings can both reduce a company's disposal costs and improve its environmental position. Thus, nanostructured dispersions and coatings can significantly reduce material costs and improve performance and functionality in a large variety of applications. In all of the sites the WTEC panel visited in Europe, Russia, and Japan, research groups were interested in achieving one or more of the properties listed above.**

Enablers

For nanoparticulate dispersions and coatings, certain enablers must be present to achieve success: (1) effective particle preparation, (2) stabilisation of the dispersed phase, (3) scaleup and control of the process, and (4) the existence of excellent analytical capabilities. The main methods or issues for each are described below.

1. **Particle preparation. Wet chemical methods such as liquid phase precipitation or sol-gel methods are of high interest in particle preparation (Friedlander 1993). In the area of hybrid methods, both spray pyrolysis and flame hydrolysis are utilised. Numerous physical methods such as mechanical size reduction are also often employed.**
2. **Stabilisation of the dispersed phase. For stabilisation of the dispersed phase, it is necessary to understand how particles can be kept as distinct entities. Either a charged stabilised system or a sterically stabilised approach is required. A successful preparation for use as a dispersion should be free from agglomeration in the liquid state so as to maintain particle integrity. The dry-coated format of nanoparticles should also minimize any presence of particle aggregation**
3. **Scaleup and control of the process. The scaleup and control of nanoparticle processing is well described by Kear (1998). Here, the issue in achieving a high rate of production of powder is the effective pyrolysis of the gas stream containing the precursor. Issues such as scaling and reproducibility from one run of nanoparticle material or dispersions to another are all inherent in the concept of precision engineering or invariant process control.**
4. **Analytical capabilities. Analytical capabilities are absolutely essential for characterising dispersions and coatings (Angstrom 1995). Particle size**

determinations, assay analysis, and interfacial properties are all important. Transmission electron microscopes, atomic force microscopy, nuclear magnetic resonance, and scanning tunneling microscopy are just some of the tools utilised in characterising nanoparticle dispersions and coatings, particularly at the very small end of the nanoscale.

Applications

There are some general examples of the vast array of applications of nanoparticulate dispersions and coatings. There are a number of applications in all of the categories below that are already in the public domain; however, most are still highly proprietary, especially drug delivery applications. Cosmetics. An area of nanoparticle technology that has incredible commercial potential is the cosmetic industry (Crandall 1996, 251-267). Here there is a great demonstrated demand, and the technology can be made simple, since properties of color and light fastness are achieved by component mixing in the cosmetic preparation. A survey in 1990 indicated a worldwide gross volume of $14-18 billion for toiletries (Crandall 1996, 61), i.e., traditional hygiene products such as powders, sprays, perfumes, and deodorants. The large markets for sunscreens and skin rejuvenation preparations promise additional revenues. The diet industry is said to gross $33 billion annually (Crandall 1996, 61).

One way that nanoparticle technology is addressing this market is through introducing nanoparticulate taste enhancers into low-calorie substrates. Medicine/ Pharmacology. In the area of medical applications, finely dispersed pharmaceuticals offer rapid drug delivery and reduced dosages for patients (POST 1995). Dispersions of strong and resilient biocompatible materials suggest opportunities for artificial joints. These generally are ceramic materials containing nanoparticulates. Overall, much of the demand for nanoparticulate dispersions and coatings comes from the cosmetic and pharmaceutical industries; in particular, liquid dispersion preparations will be widely used to apply topical coatings to the human epidermis because they can be absorbed faster and more completely than conventional coatings.

Microelectromechanical systems (MEMS). Although MEMS technologies will support the semiconductor industry in particular, there are many other applications being explored, such as in medicine, ceramics, thin films, metal alloys, and other proprietary applications. In the United States a particular focus is applying sputtering coatings to achieve MEMS technology in concert with these applications. Printing. In the areas of image capture/image output addressed by ink jet technology, nanoscience can help control the properties of the inks themselves. The production and use of nanoengineered ink products benefits from such complimentary technology as laser-assist delivery of the ink jet droplet to maintain an accurate deposit of the ink on its target (POST 1996). Another application in this field is using nanoscale properties to tailor the inks to achieve ideal absorption and drying times for desired color properties and permanency.

Here single atoms or molecules are deposited by physical vapor deposition, which could be achieved through sputtering, molecular beam epitaxy, or chemical vapor deposition. Sputtering is used on a large scale to coat metal sheets, glass, polymer substrates and other receptive materials in order to produce enhanced electronic properties for information storage and processing speed. Sensors. Chemical or physical sensors often use nanoparticles because they provide high surface area for detecting the state of chemical reactions, because the quality of detection signals is improved, and because earlier and more accurate determination of leakage reduces waste. Some commercial sensors and actuators composed of thin films are already used for environmental vapor monitoring in reactors, for example. Other likely applications of nanotechnology involving dispersions and coatings include nanofabricated surface coatings for keeping windows and surfaces clean (POST 1996). Here, the transparent nanocoating on a surface prevents fog and dirt particles from depositing on the substrate. Commercial products (achieved through gas phase condensation) also include aluminum oxide/epoxy dispersions yielding 19 times more wear resistance than conventional products.

Molecular Technology Today

One dictionary definition of a machine is “any system, usually of rigid bodies, formed and connected to alter, transmit, and direct applied forces in a predetermined manner to accomplish a specific objective, such as the performance of useful work.” Molecular machines fit this definition quite well. To imagine these machines, one must first picture molecules. We can picture atoms as beads and molecules as clumps of beads, like a child’s beads linked by snaps. In fact, chemists do sometimes visualize molecules by building models from plastic beads (some of which link in several directions, like the hubs in a Tinkertoy set). Atoms are rounded like beads, and although molecular bonds are not snaps, our picture at least captures the essential notion that bonds can be broken and reformed. If an atom were the size of a small marble, a fairly complex *molecule* would be the size of your fist. This makes a useful mental image, but atoms are really about 1/10, 000 the size of *bacteria*, and bacteria are about 1/10, 000 the size of mosquitoes. (An atomic *nucleus*, however, is about 1/100, 000 the size of the atom itself; the difference between an atom and its nucleus is the difference between a fire and a nuclear reaction.) The things around us act as they do because of the way their molecules behave. Air holds neither its shape nor its volume because its molecules move freely, bumping and ricocheting through open space.

Water molecules stick together as they move about, so water holds a constant volume as it changes shape. Copper holds its shape because its atoms stick together in regular patterns; we can bend it and hammer it because its atoms can slip over one another while remaining bound together. Glass shatters when we hammer it because its atoms separate before they slip. Rubber consists of networks of kinked molecules, like a tangle of springs. When stretched and released, its molecules straighten and then coil again. These simple molecular patterns make up passive substances. More complex patterns make up the active nanomachines of living *cells*.

Biochemists already work with these machines, which are chiefly made of protein, the main engineering material of living cells. These molecular machines have relatively few atoms, and so they have lumpy surfaces, like objects made by gluing together a handful of small marbles. Also, many pairs of atoms are linked by bonds that can bend or rotate, and so protein machines are unusually flexible. But like all machines, they have parts of different shapes and sizes that do useful work. All machines use clumps of atoms as parts. Protein machines simply use very small clumps.

Biochemists dream of designing and building such devices, but there are difficulties to be overcome. Engineers use beams of light to project patterns onto silicon chips, but chemists must build much more indirectly than that. When they combine molecules in various sequences, they have only limited control over how the molecules join. When biochemists need complex molecular machines, they still have to borrow them from cells. Nevertheless, advanced molecular machines will eventually let them build nanocircuits and nanomachines as easily and directly as engineers now build microcircuits or washing machines. Then progress will become swift and dramatic.

Genetic engineers are already showing the way. Ordinarily, when chemists make molecular chains—called "polymers"—they dump molecules into a vessel where they bump and snap together haphazardly in a liquid. The resulting chains have varying lengths, and the molecules are strung together in no particular order. But in *modern gene synthesis machines*, genetic engineers build more orderly polymers—specific *DNA* molecules—by combining molecules in a particular order. These molecules are the nucleotides of DNA (the letters of the genetic alphabet) and genetic engineers don't dump them all in together. Instead, they direct the machine to add different nucleotides in a particular sequence to spell out a particular message.

They first bond one kind of *nucleotide* to the chain ends, then wash away the leftover material and add chemicals to prepare the chain ends to bond the next nucleotide. They grow chains as they bond on nucleotides, one at a time, in a programmed sequence. They anchor the very first nucleotide in each chain to a solid surface to keep the chain from washing away with its chemical bathwater. In this way, they have a big clumsy machine in a cabinet assemble specific molecular structures from parts a hundred million times smaller than itself.

But this blind assembly process accidentally omits nucleotides from some chains. The likelihood of mistakes grows as chains grow longer. Like workers discarding bad parts before assembling a car, genetic engineers reduce errors by discarding bad chains. Then, to join these short chains into working genes (typically thousands of nucleotides long), they turn to molecular machines found in bacteria. These protein machines, called restriction enzymes, "read" certain DNA sequences as "cut here." They read these genetic patterns by touch, by sticking to them, and they cut the chain by rearranging a few atoms. Other *enzymes* splice pieces together, reading matching parts as "glue here"—likewise "reading" chains by selective stickiness and splicing chains by rearranging a few atoms. By using gene machines to write, and restriction enzymes to cut and.paste, genetic engineers can write and edit whatever DNA messages they choose.

But by itself, DNA is a fairly worthless molecule. It is neither strong like Kevlar, nor colorful like a dye, nor active like an enzyme, yet it has something that industry is prepared to spend millions of dollars to use: the ability to direct molecular machines called ribosomes. In cells, molecular machines first transcribe DNA, copying its information to make *RNA* "tapes." Then, much as old numerically controlled machines shape metal based on instructions stored on tape, ribosomes build proteins based on instructions stored on RNA strands. And proteins are useful.

Proteins, like DNA, resemble strings of lumpy beads. But unlike DNA, protein molecules fold up to form small objects able to do things. Some are enzymes, machines that build up and tear down molecules (and copy DNA, transcribe it, and build other proteins in the cycle of life). Other proteins are hormones, binding to yet other proteins to signal cells to change their behaviour. Genetic engineers can produce these objects cheaply by directing the cheap and efficient molecular machinery inside living organisms to do the work. Whereas engineers running a chemical plant must work with vats of reacting chemicals (which often misarrange atoms and make noxious byproducts), engineers working with bacteria can make them absorb chemicals, carefully rearrange the atoms, and store a product or release it into the fluid around them.

Genetic engineers have now programmed bacteria to make proteins ranging from human growth hormone to rennin, an enzyme used in making cheese. The pharmaceutical company *Eli Lilly* (*Indianapolis*) is now marketing Humulin, human insulin molecules made by bacteria.

Existing Protein Machines

These protein hormones and enzymes selectively stick to other molecules. An enzyme changes its target's structure, then moves on; a hormone affects its target's behaviour only so long as both remain stuck together. Enzymes and hormones can be described in mechanical terms, but their behaviour is more often described in chemical terms.

But *other proteins serve basic mechanical functions*. Some push and pull, some act as cords or struts, and parts of some molecules make excellent bearings. The machinery of muscle, for instance, has gangs of proteins that reach, grab a "rope" (also made of protein), pull it, then reach out again for a fresh grip; whenever you move, you use these machines. Amoebas and human cells move and change shape by using fibers and rods that act as molecular muscles and bones. A reversible, variable-speed motor drives bacteria through water by turning a corkscrew-shaped propeller. If a hobbyist could build tiny cars around such motors, several billions of billions would fit in a pocket, and 150-lane freeways could be built through your finest *capillaries*.

Simple molecular devices combine to form systems resembling industrial machines. In the 1950s engineers developed machine tools that cut metal under the control of a punched paper tape. A century and a half earlier, Joseph-Marie Jacquard had built a loom that wove complex patterns under the control of a chain of punched cards. Yet over three billion years before Jacquard, cells had developed the machinery of the *ribosome*. Ribosomes are proof that nanomachines built of protein and RNA can be programmed to build complex molecules.

Then consider viruses. One kind, the *T4 phage*, acts like a spring-loaded syringe and looks like something out of an industrial parts catalog. It can stick to a bacterium, punch a hole, and inject viral DNA (yes, even bacteria suffer infections). Like a conqueror seizing factories to build more tanks, this DNA then directs the cell's machines to build more viral DNA and syringes. Like all organisms, these viruses exist because they are fairly stable and are good at getting copies of themselves made.

Whether in cells or not, nanomachines obey the universal laws of nature. Ordinary chemical bonds hold their atoms together, and ordinary chemical reactions (guided by other nanomachines) assemble them. Protein molecules can even join to form machines without special help, driven only by thermal agitation and chemical forces. By mixing viral proteins (and the DNA they serve) in a test tube, molecular biologists have assembled working T_4 viruses. This ability is surprising: imagine putting automotive parts in a large box, shaking it, and finding an assembled car when you look inside! Yet the T_4 *virus* is but one of many *self-assembling structures*. Molecular biologists have taken the machinery of the ribosome apart into over fifty separate protein and RNA molecules, and then combined them in test tubes to form working ribosomes again.

To see how this happens, imagine different T_4 protein chains floating around in water. Each kind folds up to form a lump with distinctive bumps and hollows, covered by distinctive patterns of oiliness, wetness, and electric charge. Picture them wandering and tumbling, jostled by the thermal vibrations of the surrounding water molecules. From time to time two bounce together, then bounce apart. Sometimes, though, two bounce together and fit, bumps in hollows, with sticky patches matching; they then pull together and stick. In this way protein adds to protein to make sections of the virus, and sections assemble to form the whole.

Protein engineers will not need nanoarms and nanohands to assemble complex nanomachines. Still, tiny manipulators will be useful and they will be built. Just as today's engineers build machinery as complex as player pianos and robot arms from ordinary motors, bearings, and moving parts, so tomorrow's biochemists will be able to use protein molecules as motors, bearings, and moving parts to build robot arms which will themselves be able to handle individual molecules.

Designing with Protein

How far off is such an ability? Steps have been taken, but much work remains to be done. Biochemists have already mapped the structures of many proteins. With gene machines to help write DNA tapes, they can direct cells to build *any protein they can design*. But they still don't know how to design chains that will fold up to make proteins of the right shape and function. The forces that fold proteins are weak, and the number of plausible ways a protein might fold is astronomical, so designing a large protein from scratch isn't easy.

The forces that stick proteins together to form complex machines are the same ones that fold the protein chains in the first place. The differing shapes and kinds of stickiness of *amino acids*—the lumpy molecular "beads" forming protein chains—make each protein chain fold up in a specific way to form an object of a particular shape. Biochemists have learned rules that suggest how an amino acid chain might fold, but the rules aren't very firm. Trying to predict how a chain will fold is like trying to work a jigsaw puzzle, but a puzzle with no pattern printed on its pieces to show when the fit is correct, and with pieces that seem to fit together about as well (or as badly) in many different ways, all but one of them wrong. False starts could consume many lifetimes, and a correct answer might not even be recognised. Biochemists using the best computer programmes now available still cannot predict how a long, natural protein chain will actually fold, and some of them have despaired of designing protein molecules soon.

Yet most biochemists work as scientists, not as engineers. They work at predicting how natural proteins will fold, not at *designing* proteins that will fold predictably. *These tasks may sound similar*, but they differ greatly: the first is a scientific challenge, the second is an *engineering* challenge. Why should natural proteins fold in a way that scientists will find easy to predict? All that nature requires is that they in fact fold correctly, not that they fold in a way obvious to people.

Proteins *could* be designed from the start with the goal of making their folding more predictable. Carl Pabo, *writing in the journal Nature*, has suggested a design strategy based on this insight, and some biochemical engineers have designed and built *short chains of a few dozen pieces* that fold and nestle onto the surfaces of other molecules as planned. *They have designed from scratch a protein* with properties like those of melittin, a toxin in bee venom. They have modified existing enzymes, *changing their behaviours in predictable ways*. Our understanding of proteins is growing daily.

In 1959, *according to biologist Garrett Hardin*, some geneticists called genetic engineering impossible; today, it is an industry. Biochemistry and computer-aided design are now exploding fields, and as *Frederick Blattner wrote in the journal Science*, "computer chess programmes have already reached the level below the grand master. Perhaps the solution to the protein-folding problem is nearer than we think." *William Rastetter* of *Genentech*, writing *in Applied Biochemistry and Biotechnology* asks, "How far off is *de novo* enzyme design and synthesis? Ten, fifteen years?" He answers, "Perhaps not that long."

Forrest Carter of the *U.S. Naval Research Laboratory*, Ari Aviram and Philip Seiden of *IBM*, Kevin Ulmer of Genex Corporation, and other researchers in university and industrial laboratories around the globe have already begun theoretical work and experiments aimed at developing molecular switches, memory devices, and other structures that could be incorporated into a protein-based computer. The U.S. Naval Research Laboratory has held *two international workshops on molecular electronic devices*, and a meeting sponsored by the *U.S. National Science Foundation* has *recommended support for basic research* aimed at developing molecular computers. Japan has reportedly begun a multimillion-dollar programme aimed at developing self-assembling molecular motors and computers, and *VLSI Research Inc.*, of San Jose, *reports* that "It looks like the race to bio-chips [another term for molecular electronic systems] has already started. *NEC*, *Hitachi*, *Toshiba*, *Matsushita*, *Fujitsu*, *Sanyo-Denki* and *Sharp* have commenced full-scale research efforts on bio-chips for bio-computers."

Biochemists have other reasons to want to learn the art of protein design. New enzymes promise to perform dirty, expensive chemical processes more cheaply and cleanly, and novel proteins will offer a whole new spectrum of tools to biotechnologists. We are already on the road to protein engineering, and as Kevin Ulmer notes in the quote from *Science* that heads this chapter, this road leads "toward a more general capability for molecular engineering which would allow us to structure matter atom by atom."

New Replicators

History shows us that hardware evolves. Test tube RNA, viruses, and dogs all show how evolution proceeds by the modification and testing of replicators. But hardware (today) cannot reproduce itself—so where are the replicators behind the evolution of technology? What are the machine genes?

Of course, we need not actually identify replicators in order to recognize evolution. Darwin described evolution before Mendel discovered genes, and geneticists learned much about heredity before *Watson* and *Crick* discovered the structure of DNA. Darwin needed no knowledge of molecular genetics to see that organisms varied and that some left more descendants. A replicator is a pattern that can get copies of itself made. It may need help; without protein machines to copy it, DNA could not replicate. But by this standard, some *machines* are replicators! Companies often make machines that fall into the hands of a competitor; the competitor then learns their secrets and builds copies. Just as genes "use" protein machines to replicate, so such machines "use" human minds and hands to replicate. With nanocomputers directing assemblers and *disassemblers*, the replication of hardware could even be automated.

The human mind, though, is a far subtler engine of imitation than any mere protein machine or assembler. Voice, writing, and drawing can transmit designs from mind to mind before they take form as hardware. The ideas behind methods of design are subtler yet: more abstract than hardware, they replicate and function exclusively in the world of minds and symbol systems.

Where genes have evolved over generations and eons, mental replicators now evolve over days and decades. Like genes, ideas split, combine, and take multiple forms (genes can be transcribed from DNA to RNA and back again; ideas can be translated from language to language). *Science* cannot yet describe the neural patterns that embody ideas in brains, but anyone can see that ideas mutate, replicate, and compete. Ideas evolve.

Richard Dawkins calls bits of replicating mental patterns "memes" (*meme* rhymes with *cream*). He says "examples of memes are tunes, ideas, catch-phrases, clothes fashions, ways of making pots or of building arches. Just as genes propagate themselves in the gene pool by leaping from body to body [generation to generation] via sperms or eggs, so memes propagate themselves in the meme pool by leaping from brain to brain via a process which, in the broad sense, can be called imitation."

Creatures of the Mind

Memes replicate because people both learn and teach. They vary because people create the new and misunderstand the old. They are selected (in part) because people don't believe or repeat everything they hear. As test tube RNA molecules compete for scarce copying machines and subunits, so memes must compete for a scarce resource—human attention and effort. Since memes shape behaviour, their success or failure is a deadly serious matter.

Since ancient times, mental models and patterns of behaviour have passed from parent to child. Meme patterns that aid survival and reproduction have tended to spread. (Eat this root only after cooking; don't eat those berries, their evil spirits will twist your guts.) Year by year, people varied their actions with varying results. Year by year, some died while others found new tricks of survival and passed them on. Genes built brains skilled at imitation because the patterns imitated were, on the whole, of value—their bearers, after all, had survived to spread them.

Memes themselves, though, face their own matters of "life" and "death": as replicators, they evolve solely to survive and spread. Like viruses, they can replicate without aiding their host's survival or well-being. Indeed, the meme for martyrdom-in-a-cause can spread itself through the very act of killing its host. Genes, like memes, survive by many strategies. Some duck genes have spread themselves by encouraging ducks to pair off to care for their gene-bearing eggs and young. Some duck genes have spread themselves (when in male ducks) by encouraging rape, and some (when in female ducks) by encouraging the planting of eggs in other ducks' nests.

Still other genes found in ducks are virus genes, able to spread without making more ducks. Protecting eggs helps the duck species (and the individual duck genes) survive; rape helps one set of duck genes at the expense of others; infection helps viral genes at the expense of duck genes in general. As Richard Dawkins points out, genes "care" only about their own replication: they appear selfish. But *selfish motives can encourage cooperation*. People seeking money and recognition for themselves cooperate to build corporations that serve other people's wants. Selfish genes cooperate to build organisms that themselves often cooperate. Even so, to imagine that genes *automatically* serve some greater good (—of their chromosome?—their cell?—their body?—their species?) is to mistake a common effect for an underlying cause. To ignore the selfishness of replicators is to be lulled by a dangerous illusion. Some genes in cells are out-and-out parasites. Like herpes genes inserted in human chromosomes, they exploit cells and harm their hosts. Yet if genes can be parasites, why not memes as well?

In *The Extended Phenotype*, Richard Dawkins describes a worm that parasitizes bees and completes its life cycle in water. It gets from bee to water by making the host bee dive to its death. Similarly, ant brainworms must enter a sheep to complete their life cycle. To accomplish this, they burrow into the host ant's brain, somehow causing changes that make the ant "want" to climb to the top of a grass stem and wait, eventually to be eaten by a sheep. As worms enter other organisms and use them to survive and replicate, so do memes. Indeed, the *absence* of memes exploiting people for their own selfish ends would be amazing, a sign of some powerful—indeed, nearly perfect—mental immune system. But parasitic memes clearly do exist. Just as viruses evolve to stimulate cells to make viruses, so rumors evolve to sound plausible and juicy, stimulating repetition. Ask not whether a rumor is true, ask instead how it spreads. Experience shows that ideas evolved to be successful replicators need have little to do with the truth.

At best, chain letters, spurious rumors, fashionable lunacies, and other mental parasites harm people by wasting their time. At worst, they implant deadly misconceptions. These meme systems exploit human ignorance and vulnerability. Spreading them is like having a cold and sneezing on a friend. Though some memes act much like viruses, infectiousness isn't necessarily bad (think of an infectious grin, or infectious good nature). If a package of ideas has merit, then its infectiousness simply increases its merit—and indeed, the best ethical teachings also teach us to teach ethics. Good publications may entertain, enrich understanding, aid judgment—and advertise gift subscriptions. Spreading useful meme systems is like offering useful seeds to a friend with a garden.

Main Ideas

Parasites have forced organisms to evolve immune systems, such as the enzymes that bacteria use to cut up invading viruses, or the roving white blood cells our bodies use to destroy bacteria. Parasitic memes have forced minds down a similar path, evolving meme systems that serve as mental immune systems. The oldest and simplest mental immune system simply commands "believe the old, reject the new." Something like this system generally kept tribes from abandoning old, tested ways in favour of wild new notions—such as the notion that obeying alleged ghostly orders to destroy all the tribe's cattle and grain would somehow bring forth a miraculous abundance of food and armies of ancestors to drive out foreigners. (*This meme package infected the Xhosa people* of southern Africa in 1856; by the next year 68, 000 had died, chiefly of starvation.) Your body's immune system follows a similar rule: it generally accepts all the cell types present in early life and rejects new types such as potential cancer cells and invading bacteria, as foreign and dangerous. This simple reject-the-new system once worked well, yet in this era of organ transplantation it can kill.

Similarly, in an era when science and technology regularly present facts that are both new and trustworthy, a rigid mental immune system becomes a dangerous handicap. For all its shortcomings, though, the reject-the-new principle is simple and offers real advantages. Tradition holds much that is tried and true (or if not true, then at least workable). Change is risky: just as most mutations are bad, so most new ideas are wrong. Even reason can be dangerous: if a tradition links sound practices to a fear of ghosts, then overconfident rational thought may throw out the good with the bogus. Unfortunately, traditions evolved to *be* good may have less appeal than ideas evolved to *sound* good—when first questioned, the soundest tradition may be displaced by worse ideas that better appeal to the rational mind. Yet memes that seal the mind against new ideas protect themselves in a suspiciously self-serving way. While protecting valuable traditions from clumsy editing, they may also shield parasitic claptrap from the test of truth. In times of swift change they can make minds dangerously rigid. Much of the history of philosophy and science may be seen as a search for better mental immune systems, for better ways to reject the false, the worthless, and the damaging.

The best systems respect tradition, yet encourage experiment. They suggest standards for judging memes, helping the mind distinguish between parasites and tools. The principles of evolution provide a way to view change, whether in molecules, organisms, technologies, minds, or cultures. The same basic questions keep arising: What are the replicators? How do they vary? What determines their success? How do they defend against invaders? These questions will arise again when we consider the consequences of the assembler revolution, and yet again when we consider how society might deal with those consequences. The deep-rooted principles of evolutionary change will shape the development of nanotechnology, even as the distinction between hardware and life begins to blur.

These principles show much about what we can and cannot hope to achieve, and they can help us focus our efforts to shape the future. They also tell us much about what we can and cannot foresee, because they guide the evolution not only of hardware, but of knowledge itself.

Molecular Replicators

Cells replicate. Their machines copy their DNA, which directs their ribosomal machinery to build other machines from simpler molecules. These machines and molecules are held in a fluid-filled bag. Its membrane lets in fuel molecules and parts for more nanomachines, DNA, membrane, and so forth; it lets out spent fuel and scrapped components. A cell replicates by copying the parts inside its membrane bag, sorting them into two clumps, and then pinching the bag in two. Artificial replicators could be built to work in a similar way, but using assemblers instead of ribosomes. In this way, we could build cell-like replicators that are not limited to molecular machinery made from the soft, moist folds of protein molecules.

But engineers seem more likely to develop other approaches to replication. *Evolution* had no easy way to alter the fundamental pattern of the cell, and this pattern has shortcomings. In synapses, for example, the cells of the brain signal their neighbors by emptying bladders of chemical molecules. The molecules then jostle around until they bind to sensor molecules on the neighboring cell, sometimes triggering a neural impulse. A chemical *synapse* makes a slow switch, and neural impulses move slower than sound. With assemblers, molecular engineers will build entire computers smaller than a synapse and a millionfold faster.

Mutation and selection could no more make a synapse into a mechanical *nanocomputer* than a breeder could make a horse into a car. Nonetheless, engineers have built cars, and will also learn to build computers faster than brains, and replicators more capable than existing cells.

Some of these replicators will not resemble cells at all, but *will instead resemble factories* shrunk to cellular size. They will contain nanomachines mounted on a molecular framework and conveyor belts to move parts from machine to machine. Outside, they will have a set of assembler arms for building replicas of themselves, an atom or a section at a time.

How fast these replicators can replicate will depend on their assembly speed and their size. Imagine an advanced assembler that contains a million atoms: it can have as many as ten thousand moving parts, each containing an average of one hundred atoms—enough parts to make up a rather complex machine. In fact, the assembler itself looks like a box supporting a stubby robot arm a hundred atoms long. The box and arm contain devices that move the arm from position to position, and others that change the molecular tools at its tip.

Behind the box sits a device that reads a tape and provides mechanical signals that trigger arm motions and tool changes. In front of the arm sits an unfinished structure. Conveyors bring molecules to the assembler system. Some supply energy to motors that drive the tape reader and arm, and others supply groups of atoms for assembly. Atom by atom (or group by group), the arm moves pieces into place as directed by the tape; chemical reactions bond them to the structure on contact.

These assemblers will work fast. *A fast enzyme*, such as carbonic anhydrase or ketosteroid isomerase, can process almost a million molecules per second, even without conveyors and power-driven mechanisms to slap a new *molecule* into place as soon as an old one is released. It might seem too much to expect an assembler to grab a molecule, move it, and jam it into place in a mere millionth of a second. But small appendages can move to and fro very swiftly. A human arm can flap up and down several times per second, fingers can tap more rapidly, a fly can wave its wings fast enough to buzz, and a mosquito makes a maddening whine. Insects can wave their wings at about a thousand times the frequency of a human arm because an insect's wing is about a thousand times shorter. An assembler arm will be about fifty million times shorter than a human arm, and so (as it turns out) it will be able to move back and forth *about fifty million times more rapidly*. For an assembler arm to move a mere million times per second would be like a human arm moving about once per minute: sluggish. So it seems a very reasonable goal. The speed of replication will depend also on the total size of the system to be built. Assemblers will not replicate by themselves; they will need materials and energy, and instructions on how to use them. Ordinary chemicals can supply materials and energy, but nanomachinery must be available to process them.

Bumpy *polymer* molecules can code information like a punched paper tape, but a reader must be available to translate the patterns of bumps into patterns of arm motion. Together, these parts form the essentials of a replicator: the tape supplies instructions for assembling *a copy of the assembler, of the reader, of the other nanomachines, and of the tape itself*. A reasonable design for this sort of replicator will likely include several assembler arms and several more arms to hold and move workpieces. Each of these arms will add another million atoms or so. The other parts—tape readers, chemical processors, and so forth-may also be as complicated as assemblers. Finally, a flexible replicator system will probably include a simple computer; following the mechanical approach that I mentioned in this encyclopaedia, this will add roughly 100 million atoms.

Altogether, these parts will total less than 150 million atoms. Assume instead a total of one billion, to leave a wide margin for error. Ignore the added capability of the additional assembler arms, leaving a still wider margin. Working at one million atoms per second, the system will still copy itself in one thousand seconds, or a bit over fifteen minutes—about the time a bacterium takes to replicate under good conditions. Imagine such a replicator floating in a bottle of chemicals, making copies of itself. It builds one copy in one thousand seconds, thirty-six in ten hours. In a week, it stacks up enough copies to fill the volume of a human cell. In a century, it stacks up enough to make a respectable speck. If this were all that replicators could do, we could perhaps ignore them in safety.

Each copy, though, will build yet more copies. Thus the first replicator assembles a copy in one thousand seconds, the two replicators then build two more in the next thousand seconds, the four build another four, and the eight build another eight. At the end of ten hours, there are not thirty-six new replicators, but over 68 billion. In less than a day, they would weigh a ton; in less than two days, they would outweigh the Earth; in another four hours, they would exceed the mass of the Sun and all the planets combined—if the bottle of chemicals hadn't run dry long before. Regular doubling means *exponential growth*. Replicators multiply exponentially unless restrained, as by lack of room or resources. Bacteria do it, and at about the same rate as the replicators just described. People replicate far more slowly, yet given time enough they, too, could overshoot any finite resource supply. Concern about population growth will never lose its importance. Concern about controlling rapid new replicators will soon become important indeed.

Molecules & Skyscrapers

Machines able to grasp and position individual atoms will be able to build almost anything by bonding the right atoms together in the right patterns. To be sure, building large objects one atom at a time will be slow. A fly, after all, contains about a million atoms for every second since the dinosaurs were young. Molecular machines can nonetheless build objects of substantial size—they build whales, after all.

To make large objects rapidly, a vast number of assemblers must cooperate, but replicators will produce assemblers by the ton. Indeed, with correct design, the difference between an assembler system and a replicator will lie entirely in the assembler's programming.

If a replicating assembler can copy itself in one thousand seconds, then it can be programmed to build something else its own size just as fast. Similarly, a ton of replicators can swiftly build a ton of something else—and the product will have all its billions of billions of billions of atoms in the right place, with *only a minute fraction misplaced*. To see the abilities and limits of one method for assembling large objects, imagine a flat sheet covered with small assembly arms-perhaps an army of replicators reprogrammed for construction work and arrayed in orderly ranks. Conveyors and communication channels behind them supply reactive molecules, energy, and assembly instructions. If each arm occupies an area 100 atomic diameters wide, then behind each assembler will be room for conveyors and channels totaling about 10, 000 atoms in cross sectional area.

This seems room enough. A space ten or twenty atoms wide can hold a conveyor (perhaps based on molecular belts and pulleys). A channel a few atoms wide can hold a molecular rod which, like those in the mechanical computer mentioned in this encyclopaedia, will be pushed and pulled to transmit signals. All the arms will work together to build a broad, solid structure layer by layer. Each arm will be responsible for its own area, handling about 10, 000 atoms per layer. A sheet of assemblers handling 1, 000, 000 atoms per second per arm will complete about one hundred atomic layers per second. This may sound fast, but at this rate piling up a paper-sheet thickness will take about an hour, and making a meter-thick slab will take over a year.

Faster arms might raise the assembly speed to over a meter per day, but they would produce more waste heat. If they could build a meter-thick layer in a day, the heat from one square meter could cook hundreds of steaks simultaneously, and might fry the machinery. At some size and speed, cooling problems will become a limiting factor, but there are other ways of assembling objects faster without overheating the machinery.

Imagine trying to build a house by gluing together individual grains of sand. Adding a layer of grains might take grain-gluing machines so long that raising the walls would take decades. Now imagine that machines in a factory first glue the grains together to make bricks. The factory can work on many bricks at once. With enough grain-gluing machines, bricks would pour out fast; wall assemblers could then build walls swiftly by stacking the preassembled bricks. Similarly, molecular assemblers will team up with larger assemblers to build big things quickly—machines can be any size from molecular to gigantic. With this approach, most of the assembly heat will be dissipated far from the work site, in making the parts.

Skyscraper construction and the architecture of life suggest a related way to construct large objects. Large plants and animals have vascular systems, intricate channels that carry materials to molecular machinery working throughout their tissues. Similarly, after riggers and riveters finish the frame of a skyscraper, the building's "vascular system"—its elevators and corridors, aided by cranes—carry construction materials to workers throughout the interior. Assembly systems could also employ this strategy, first putting up a scaffold and then working throughout its volume, incorporating materials brought through channels from the outside.

Imagine this approach being used to "grow" a large rocket engine, working inside a vat in an industrial plant. The vat—made of shiny steel, with a glass window for the benefit of visitors—stands taller than a person, since it must hold the completed engine. Pipes and pumps link it to other equipment and to water-cooled heat exchangers. This arrangement lets the operator circulate various fluids through the vat.

To begin the process, the operator swings back the top of the vat and lowers into it a base plate on which the engine will be built. The top is then resealed. At the touch of a button, pumps flood the chamber with a thick, milky fluid which submerges the plate and then obscures the window. This fluid flows from another vat in which replicating assemblers have been raised and then reprogrammed by making them copy and spread a new instruction tape (a bit like infecting bacteria with a *virus*). These new assembler systems, smaller than bacteria, scatter light and make the fluid look milky. Their sheer abundance makes it viscous.

At the center of the base plate, deep in the swirling, assembler-laden fluid, sits a "seed." It contains a nanocomputer with stored engine plans, and its surface sports patches to which assemblers stick. When an assembler sticks to it, they plug themselves together and the seed computer transfers instructions to the assembler computer. This new programming tells it where it is in relation to the seed, and directs it to extend its manipulator arms to snag more assemblers. These then plug in and are similarly programmed. Obeying these instructions from the seed (which spread through the expanding network of communicating assemblers) a sort of assembler-*crystal* grows from the chaos of the liquid. Since each assembler knows its location in the plan, it snags more assemblers only where more are needed. This forms a pattern less regular and more complex than that of any natural crystal. In the course of a few hours, the assembler scaffolding grows to match the final shape of the planned rocket engine.

Then the vat's pumps return to life, replacing the milky fluid of unattached assemblers with a clear mixture of organic solvents and dissolved substances—including aluminum compounds, oxygen-rich compounds, and compounds to serve as assembler fuel. As the fluid clears, the shape of the rocket engine grows visible through the window, looking like a full-scale model sculpted in translucent white plastic. Next, a message spreading from the seed directs designated assemblers to release their neighbors and fold their arms. They wash out of the structure in sudden streamers of white, leaving a spongy lattice of attached assemblers, now with room enough to work. The engine shape in the vat grows almost transparent, with a hint of iridescence.

Each remaining assembler, though still linked to its neighbors, is now surrounded by tiny fluid-filled channels. Special arms on the assemblers work like flagella, whipping the fluid along to circulate it through the channels. These motions, like all the others performed by the assemblers, are powered by molecular engines fueled by molecules in the fluid. As dissolved sugar powers yeast, so these dissolved chemicals power assemblers. The flowing fluid brings fresh fuel and dissolved raw materials for construction; as it flows out it carries off waste heat. The communications network spreads instructions to each assembler.

The assemblers are now ready to start construction. They are to build a rocket engine, consisting mostly of pipes and pumps. This means building strong, light structures in intricate shapes, some able to stand intense heat, some full of tubes to carry cooling fluid. Where great strength is needed, the assemblers set to work constructing rods of interlocked fibers of carbon, in its diamond form. From these, they build a lattice tailored to stand up to the expected pattern of stress. Where resistance to heat and corrosion is essential (as on many surfaces), they build similar structures of aluminum oxide, in its sapphire form. In places where stress will be low, the assemblers save mass by leaving wider spaces in the lattice. In places where stress will be high, the assemblers reinforce the structure until the remaining passages are barely wide enough for the assemblers to move. Elsewhere the assemblers lay down other materials to make sensors, computers, motors, solenoids, and whatever else is needed.

To finish their jobs, they build walls to divide the remaining channel spaces into almost sealed cells, then withdraw to the last openings and pump out the fluid inside. Sealing the empty cells, they withdraw completely and float away in the circulating fluid. Finally, the vat drains, a spray rinses the engine, the lid lifts, and the finished engine is hoisted out to dry. Its creation has required less than a day and almost no human attention. What is the engine like? Rather than being a massive piece of welded and bolted metal, it is a seamless thing, gemlike. Its empty internal cells, patterned in arrays about a wavelength of light apart, have a side effect: like the pits on a laser disk they diffract light, producing a varied iridescence like that of a fire opal. These empty spaces lighten a structure already made from some of the lightest, strongest materials known. Compared to a modern metal engine, this advanced engine has over 90 percent less mass.

Tap it, and it rings like a bell of surprisingly high pitch for its size. Mounted in a spacecraft of similar construction, it flies from a runway to space and back again with ease. It stands long, hard use because its strong materials have let designers include large safety margins. Because assemblers have let designers pattern its structure to yield before breaking (blunting cracks and halting their spread), the engine is not only strong but tough.

For all its excellence, this engine is fundamentally quite conventional. It has merely replaced dense metal with carefully tailored structures of light, tightly bonded atoms. The final product contains no nanomachinery.

More advanced designs will exploit *nanotechnology* more deeply. They could leave a vascular system in place to supply assembler and *disassembler systems*; these can be programmed to mend worn parts. So long as users supply such an engine with energy and raw materials, it will renew its own structure. More advanced engines can also be literally more flexible. Rocket engines work best if they can take different shapes under different operating conditions, but engineers cannot make bulk metal strong, light, and limber. With nanotechnology, though, a structure stronger than steel and lighter than wood could change shape like muscle (*working, like muscle*, on the sliding fiber principle), An engine could then expand, contract, and bend at the base to provide the desired thrust in the desired direction under varying conditions. With properly programmed assemblers and disassemblers, it could even remodel its fundamental structure long after leaving the vat.

In short, replicating assemblers will copy themselves by the ton, then make other products such as computers, rocket engines, chairs, and so forth. They will make disassemblers able to break down rock to supply raw material. They will make solar collectors to supply energy. Though tiny, they will build big. Teams of nanomachines in nature build whales, and seeds replicate machinery and organise atoms into vast structures of cellulose, building redwood trees. There is nothing too startling about growing a rocket engine in a specially prepared vat. Indeed, foresters given suitable assembler "seeds" could grow spaceships from soil, air, and sunlight.

Assemblers will be able to make virtually anything from common materials without labor, replacing smoking factories with systems as clean as forests. They will transform technology and the economy at their roots, opening a new world of possibilities. They will indeed be engines of abundance.

Manufactured products are made from atoms. The properties of those products depend on how those atoms are arranged. If we rearrange the atoms in coal, we get diamonds. If we rearrange the atoms in sand (and add a pinch of impurities) we get computer chips. If we rearrange the atoms in dirt, water and air we get grass.

Since we first made stone tools and flint knives we have been arranging atoms in great thundering statistical heards by casting, milling, grinding, chipping and the like. We've gotten better at it: we can make more things at lower cost and greater precision than ever before. But at the molecular scale we're still making great ungainly heaps and untidy piles of atoms.

That's changing. In special cases we can already arrange *atoms* and *molecules* exactly as we want. *Theoretical analyses make it clear we can do a lot more.* Eventually, we should be able to arrange and rearrange atoms and molecules much as we might arrange LEGO blocks. In not too many decades we should have a manufacturing technology able to:

Build products with almost every atom in the right place. Do so inexpensively. Make most arrangements of atoms consistent with physical law. Often called *nanotechnology, molecular nanotechnology or molecular manufacturing*, it will let us make most products lighter, stronger, smarter, cheaper, cleaner and more precise. One warning: in contrast to the useage in this article *some researchers use the word "nanotechnology" to refer to high resolution lithographic technology while others use it to refer to almost any research where some critical size is less than a micron (1, 000 nanometers)*. When there is risk of confusion, the more specific terms "molecular nanotechnology" or "molecular manufacturing" should be used. There are two main issues in nanotechnology:

- What might molecular manufacturing systems look like?
- How could we build such systems given our current technology?

As molecular manufacturing systems do not yet exist, and as it will likely be a few decades before we can build them, the answer to the first question must be based on theoretical and computational models. Such models serve several purposes. *First* and most obvious, some might argue that the goal is itself inherently impossible, e.g., that such systems cannot be made within the framework of existing physical law. Theoretical and computational models provide an inexpensive way to examine this question and provide assurances that proposed systems are possible. *Second,* such models give us a feel for what molecular manufacturing systems might look like. The better you understand the goal, the better your chances of actually achieving it. To illustrate this point, we consider a historical example based on the single most important invention of the 20th century: the computer. The basic idea of the relay was known in the 1820's, and *the concept of a mechanical stored programme computer was understood by Babbage by the mid 1800's.* Practical computers could have been built using relays by the 1860's if there had been an adequate theoretical understanding of "computation." An adequate theoretical understanding of molecular manufacturing should likewise ease the problems of practical development by opening for consideration approaches and designs that might not otherwise be developed for decades or — as happened in the case of the computer — almost a century. In this paper we'll first address the question of what molecular manufacturing systems could look like, and then consider how to build them given our current technology. This second question involves not only a consideration of current experimental triumphs and how they might be extended, but also the definition of intermediate goals and systems.

The Advantages of Nanotechnology

What would it mean if we could inexpensively make things with every atom in the right place? For starters, we could continue the revolution in computer hardware right down to molecular gates and wires — something that today's lithographic methods (used to make computer chips) could never hope to do. We could inexpensively make very strong and very light materials: shatterproof diamond in precisely the shapes we want, by the ton, and over fifty times lighter than steel of the same strength. We could make a Cadillac that weighed fifty kilograms, or a full-sized sofa you could pick up with one hand. We could make surgical instruments of such precision and deftness that they could operate on the cells and even molecules from which we are made — something well beyond today's medical technology. The list goes on — almost any manufactured product could be improved, often by orders of magnitude.

The Advantages of Positional Control

One of the basic principles of nanotechnology is positional control. At the macroscopic scale, the idea that we can hold parts in our hands and assemble them by properly positioning them with respect to each other goes back to prehistory: we celebrate ourselves as the tool using species. Our wisdom and our knowledge would have done us scant good without an opposable thumb: we'd still be shivering in the bushes, unable to start a fire.

At the molecular scale, the idea of holding and positioning molecules is new and almost shocking. However, as long ago as 1959 *Richard Feynman, the Nobel prize winning physicist, said that nothing in the laws of physics prevented us from arranging atoms the way we want: "...it is something, in principle, that can be done; but in practice, it has not been done because we are too big."*

Before discussing the advantages of positional control at the molecular scale, it's helpful to look at some of the methods that have been developed by chemists — methods that don't use positional control, but still let chemists synthesize a remarkably wide range of molecules and molecular structures.

Self Assembly

The ability of chemists to synthesize what they want by stirring things together is truly remarkable. Imagine building a radio by putting all the parts in a bag, shaking, and pulling out the radio — fully assembled and ready to work! Self assembly — the art and science of arranging conditions so that the parts themselves spontaneously assemble into the desired structure — is a well established and powerful method of synthesising complex molecular structures. A basic principle in self assembly is selective stickiness: if two molecular parts have complementary shapes and charge patterns — one part has a hollow where the other part has a bump, and one part has a positive charge where the other part has a negative charge — then they will tend to stick together in one particular way. By shaking these parts around — something which thermal noise does for us quite naturally if the parts are floating in solution — the parts will eventually, purely by chance, be brought together in just the right way and combine into a bigger part. This bigger part can combine in the same way with other parts, letting us gradually build a complex whole from molecular pieces by stirring them together and shaking.

Many viruses use this approach to make more viruses — if you stir the parts of the T4 bacteriophage together in a test tube, they will self assemble into fully functional viruses.

Positional Devices and Positionally Controlled Reactions

While self assembly is a *path* to nanotechnology, by itself it would be hard pressed to make the very wide range of products promised by nanotechnology. We don't know how to self assemble shatterproof diamond, for example. (We'll later discuss a way of making diamond using positional control). During self assembly the parts bounce around and bump into each other in all kinds of ways, and if they stick together when we don't *want* them to stick together, we'll get unwanted globs of random parts. Many types of parts have this problem, so self assembly won't work for them. To make diamond, it seems as though we need to use indiscriminately sticky parts (such as radicals, carbenes and the like). These parts can't be allowed to randomly bump into each other (or much of anything else, for that matter) because they'd stick together when we didn't want them to stick together and form messy blobs instead of precise molecular machines.

We can avoid this problem if we can hold and position the parts. Even though the molecular parts that are used to make diamond are both indiscriminately and *very* sticky (more technically, the barriers to bond formation are low and the resulting covalent bonds are quite strong), if we can position them we can prevent them from bumping into each other in the wrong way. When two sticky parts do come into contact with each other, they'll do so in the right orientation because we're *holding* them in the right orientation. In short, positional control at the molecular scale should let us make things which would be difficult or impossible to make without it. Given our macroscopic intuition, this shouldn't be surprising. If we couldn't use our hands to hold and position parts, there are lots of things we'd have a very hard time making! If we are to position molecular parts we must develop the molecular equivalent of "arms" and "hands." We'll need to learn what it means to "pick up" such parts and "snap them together." We'll have to understand the precise *chemical reactions that such a device would use*.

One of the first questions we'll need to answer is: what does a molecular-scale positional device look like? Current proposals are similar to macroscopic robotic devices but on a much smaller scale. The illustrations (from *Nanosystems, the best technical introduction to nanotechnology*) show a design for a molecular-scale robotic arm proposed by *Eric Drexler*, a pioneering researcher in the field. Only 100 nanometers high and 30 nanometers in diameter, this rather squat design has a few million atoms and roughly a hundred moving parts. It uses no lubricants, for at this scale a lubricant molecule is more like a piece of grit. Instead, the bearings are "run dry" (following a suggestion by *Feynman*) as described in the following paragraph.

Running bearings dry should work both because the diamond surface is very slippery (see the coefficient of friction for diamond in the table) and because we can make the surface very smooth — so smooth that there wouldn't even be molecular-sized asperities or imperfections that might catch or grind against each other. Computer models support our intuition: *analysis of the bearings shown here using computational chemistry programmes* shows they should rotate easily.

Stewart Platforms

While Drexler's proposal for a small robotic arm is easy to understand and should be adequate to the task, more recent work has focused on the Stewart platform. This positional device has the great advantage that it is *stiffer than a robotic arm* of similar size. Conceptually, the Stewart platform is based on the observation that a polyhedron, all of whose faces are triangular, will be rigid. If some of the edges of the polyhedron can be adjusted in length, then the position of one face can be moved with respect to the position of another face. If we want a full six degrees of freedom (X, Y, Z, roll, pitch and yaw) then we must be able to independently adjust the lengths of six different edges of the polyhedron. If we further want one triangular face of the polyhedron to remain of fixed size and hold a "tool, " and a second face of the polyhedron to act as the "base" whose size and position is fixed, then we find that the simplest polyhedron that will suit our purpose is the octahedron.

In the Stewart platform, one triangular face of the octahedron is designated the "platform, " while the opposing triangular face is designated the "base." The six edges that connect the base to the platform can then be adjusted in length to control the position of the platform with respect to the base. Mechanically, this adjustment is often done using six hydraulic pistons. A picture of the Stewart platform (from *Nanosystems*) is shown above.

The advantage of the Stewart platform can now be seen: because the six adjustable-length edges are either in pure compression or pure tension and are never subjected to any bending force, this positional device is stiffer than a long robotic arm which can bend and flex. The Stewart platform is also conceptually simpler than a robotic arm, having fewer different types of parts; for this reason, we can reasonably expect that making one will be simpler than making a robotic arm.

The utility of diamond: why it's a dream material

From airplanes and space ships to cars and chairs, making products lighter and stronger is almost always an improvement. Sometimes, as *in the case of space flight, this is a quantum leap all by itself*. At other times it's just a convenience: moving heavy furniture is a chore most of us would avoid if we could.

The strength and lightness of materials depends on the number and strength of the bonds that hold their atoms together, and the lightness of those atoms. Light atoms that form many strong bonds are at the heart of strong, light stiff materials. Boron, carbon and nitrogen are lighter and form more and stronger bonds than other atoms. The carbon-carbon bond in particular is very strong, and carbon atoms can form four bonds to four neighboring atoms. In diamond, this makes a very dense network of very strong bonds, which creates a very strong, light and stiff material.

Besides light weight and great strength, diamond has a host of materials properties that make it an excellent choice for almost any application. (See the table for a list of diamond's material properties).

Another material with remarkable properties is graphite: carbon atoms arranged in a hexagonal lattice with each carbon atom having three bonds to its three neighbors. While graphite has fewer bonds per atom, the strength of these bonds is greater and so graphitic materials also have remarkable properties.

Diamond is also a wonderful material for making transistors and computer gates, though it takes a bit of explanation to understand why.

Computer gates should switch as quickly as possible: that's what makes computers so fast. To do this, the gates must be made of transistors in which the electrons move as fast as possible over the shortest possible distances. But as electrons move through a material, they generate heat: think of how hot the tungsten wire in a light bulb can get! So the faster we make our computer the faster the electrons move through the material and the hotter it gets. When it gets too hot the computer stops working.

Diamond excels in its electronic properties. Fundamentally, it lets us move charge around much faster before things stop working. There are several reasons for this. First, diamond transistors can operate at much higher temperatures because diamond has a larger "bandgap" than other materials (particularly silicon). Electrons in semiconductors (such as diamond or silicon) are either in the "conduction band" or the "valence band." An electron in the conduction band can move freely and transport charge. Moving an electron from the valence band to the conduction band requires adding a specific amount of energy to the electron. In silicon, this is 1.12 electron volts. In diamond, it is 5.47 electron volts. As the temperature increases, more and more electrons can jump into the conduction band from the valence band because of thermal noise. When too many electrons do this the semiconductor becomes conductive everywhere and the transistors short out and stop working. Because diamond has a wider bandgap, it shorts out at a proportionally higher temperature than silicon.

Diamond also has greater thermal conductivity, which lets us move heat out of a diamond transistor more quickly to prevent it from getting too hot.

To move an electron faster we want to "pull" on it harder. The stronger the electric field, the harder we're pulling the electrons. But as the electric field gets too strong, it tears electrons out of the valence band causing the transistors to short out. This "breakdown" field varies from material to material, and (you guessed it) diamond leads the pack.

Finally, electrons (and holes) move with different speeds through different materials, even when the electric field is the same. Again, electrons and holes in diamond move faster than in silicon.

Because diamond transistors can be hotter, are more easily cooled, can tolerate higher voltages before breaking down, and electrons move more easily in them; they make better transistors than other materials. Diamond would be ideal for electronic devices if only we could manufacture it inexpensively and with precisely the desired structure. In summary: from cars to airplanes to spaceships, from furniture to buildings, from pocket calculators to supercomputers, materials similar to diamond are just *better* than other materials, often a *lot* better. Just as the stone age, the bronze age and the steel age were named after the materials that we could make, this new age we are entering might be called *the diamond age*.

Positional Control and the Synthesis of Diamond

We can in principle make a great many very useful structures from diamond, including very powerful computers and molecular robotic devices able to position molecular components to within a fraction of an atomic diameter. But how can we synthesize diamond? One answer to this question comes from looking at how we grow diamond today using *Chemical Vapor Deposition (CVD)*. In a process somewhat reminiscent of spray painting, we build up layer after layer of diamond on a surface by holding that surface in a cloud of very reactive molecules and atoms — like H; CH_3, C_2H, etc. When these reactive molecules bump into the surface they change it, either by adding, removing, or re-arranging atoms. By carefully controlling the pressure, temperature, and the exact composition of the gas, we can create conditions that favour the growth of diamond on the surface. While these are the right chemical reactions to make diamond, randomly bombarding the growing surface with reactive molecules doesn't offer the finest control over the growth process. It would be like trying to build a wristwatch using a sandblaster. We want the chemical reactions to take place at precisely the places on the surface that we specify — not wherever random gusts of turbulent gas might dictate.

A second problem is that the hydrogenated diamond surface is chemically inert: it's difficult to add carbon (or anything else) to it. We could overcome this problem by removing a hydrogen atom from the surface, leaving behind a very reactive dangling bond. This is a critical first step during CVD diamond growth but it happens at random, anywhere on the growing diamond surface, when the right kind of reactive gas molecule happens to strike the surface in the right way. We want to do something similar but more controlled: remove a *specific* hydrogen atom from a *specific* spot on the diamond surface. To do this, we'll need a "hydrogen abstraction tool".

What might such a tool look like? If we base it on our understanding of diamond CVD, it should be a highly reactive radical with a high affinity for hydrogen. At the same time it must be possible to position it, so it must have a stable region which can serve as a "handle." The tool would be held by (for example) the molecular robotic arm discussed earlier, and would be positioned directly over the hydrogen we wish to abstract.

A simple strategy for finding such radicals is to look through the table of bond strengths in the *Handbook of Chemistry and Physics* and pick the molecule with the strongest bond to hydrogen. The radical created by removing the hydrogen should then have a very high affinity for hydrogen, as desired. Unfortunately, this strategy first produces fluorine. While atomic fluorine does indeed have a very high affinity for hydrogen, there is no obvious way to attach a "handle" to it. Interest therefore turns to the molecule with the second strongest bond to hydrogen: acetylene. Here, we are in luck. Not only does the acetylene radical have a high affinity for hydrogen, it also has a chemically stable region which can be modified into a "handle".

Drexler realised this sometime in the early 1980's. The author subsequently and independently followed the same line of logic and reached the same conclusions. An initial check of the validity of this idea was provided by semi-empirical calculations at Xerox PARC (easily done on a workstation using freely available programmes). Discussions with *Bill Goddard and members of his molecular modeling group at Caltech* resulted in *more accurate calculations using high level ab initio quantum chemistry methods* on a more powerful computer. (These methods have the interesting property that the calculations will converge on the correct answer if enough computing power is used). *Drexler published this observation in Nanosystems* in 1992. A *group at NRL interested in diamond growth* found the proposal intriguing and modeled the abstraction of a hydrogen at room temperature using a qualitatively different computational model of molecular behaviour (which permitted them to model the behaviour of several hundreds of atoms in the vicinty of the abstraction). As the results of three different computational approaches have all produced the same qualitative answer, and as the results are also in accord with our intuition about what *should* happen, it's safe to say that this hydrogen abstraction tool will, when finally implemented, be able to abstract hydrogen.

The hydrogen abstraction tool illustrates some important ideas common to other proposals for molecular tools for the synthesis of diamond. First, the tool has a reactive end which is brought into contact with the molecular workpiece and an inert end which is held by a molecular positional device. Second, the point of application of the tool is controlled by the molecular positional device. Third, the environment around the tool is itself inert (typical proposals involve the use of either vacuum or a noble gas). The inert environment prevents the tool (which is quite reactive) from reacting with anything undesired.

While the hydrogen abstraction tool creates a very reactive spot on the surface we also need a tool that will deposit one or more carbon atoms on the growing surface. One proposal is the dimer deposition tool. (a "dimer" is just two of something stuck together. In this case, two carbon atoms are stuck together by a triple bond, hence the two carbon atoms form a "dimer"). A dimer deposition tool is simply a dimer with two weak bonds to a supporting structure. A proposal by Drexler is shown to the left. A second proposed dimer deposition tool, which has particularly weak bonds to the dimer, is illustrated at the right. [Subsequent note: further investigation of the proposal at right suggests it might not be stable except at very low temperatures. A variety of alternative dimer deposition tools are possible, so even if this particular structure should prove unsatisfactory for room temperature operation, other tools able to carry out the same function should be feasible.]

To illustrate how we might combine the use of the hydrogen abstraction tool with the dimer deposition tool, we proceed in three steps. First, we start with the hydrogenated diamond (111) surface. (The notation (111) refers to a particular surface of the diamond crystal. When diamond shatters, it shatters along certain planes and not along others. As these different planes have surfaces with different properties, it's important to specify which plane you're talking about). Second, we use the hydrogen abstraction tool to remove two adjacent hydrogen atoms. Third, we use the dimer deposition tool to deposit two carbon atoms on the surface — the two ends of the carbon dimer are originally single-bonded to the rest of the tool. The two dangling bonds on the surface are very reactive, and react with the ends of the carbon dimer. When this happens, the bonds holding the dimer to the dimer deposition tool break, and the bonds in the deposition tool rearrange to eliminate what would otherwise be two radicals. Result: the carbon dimer is transferred from the tool to the surface. The following illustration shows the dimer deposition tool adding a dimer to a small cluster which represents the diamond (111) surface. (Because accurate ab initio calculations are computationally less expensive for a small cluster of atoms, the following illustrations show only the atoms directly involved in the reaction).

This sequence of steps starts with a flat diamond surface and adds two carbon atoms to it. It could be repeated at other sites. A computational investigation of the final step was reported by *Stephen Walch at the recent workshop on computational nanotechnology* sponsored by the NAS programme at NASA. Because the energy released during the reaction is much larger than thermal noise (e.g., the reaction is quite exothermic), the reverse reaction is very unlikely to occur. The barrier to the reaction is also small. Taken together, these mean the reaction can take place easily and irreversibly. The dimer will "snap" onto the surface and stay there. (Calculations at the 6-31G* MP2 level show all positive vibrational frequencies for this tool. Further investigation of its stability is in progress). A third proposed tool is the carbene insertion tool. "It is no exaggeration to claim a major role for carbenes in the modern chemists's attitude that he can very probably make anything he wants." (W. J. Baron). A positionally controlled carbene should be correspondingly more useful, as it could be made to react at any position on a growing molecular workpiece. Carbenes are very reactive and can insert readily into double and triple bonds. A plausible reaction would be to insert a carbene into the dimer described above.

A fourth proposal is for a hydrogen deposition tool. Where the hydrogen abstraction tool is intended to make an inert structure reactive by creating a dangling bond, the hydrogen deposition tool would do exactly the opposite: make a reactive structure inert by terminating dangling bonds. This tool could be used during the synthetic process to stabilize a structure that might otherwise undergo a spontaneous and undesired rearrangement or reconstruction, or it could be used at the end of the synthetic process to stabilize the finished product.

There are many possible candidates for such a tool: any molecular structure that has a weak bond to a hydrogen. One particularly attractive candidate is the use of tin. The hydrogen-tin bond is quite weak, and so a tin-based hydrogen deposition tool should be quite effective.

These four tools let us (a) make an inert surface reactive by removing a hydrogen from one or more specific sites, (b) add one or two carbon atoms to a surface at selected sites and (c) make a reactive surface inert by adding hydrogen to it — which lets us prevent undesired surface rearrangements.

Stiffness

Our molecular arms will be buffeted by something we don't worry about at the macroscopic scale: thermal noise. This makes molecular-scale objects wiggle and jiggle, just as Brownian motion makes small dust particles bounce around at random. *Can our molecular robotic arm maintain its position in the face of thermal noise?*

The critical property we need here is *stiffness*. Stiffness is a measure of how far something moves when you push on it. If it moves a lot when you push on it a little, it's not very stiff. If it doesn't budge when you push hard, it's very stiff.

Big positional devices, used in today's Scanning Probe Microscopes (SPMs), have been made stiff enough to image individual atoms despite thermal noise. In SPMs, a very sharp tip is brought down to the surface of the sample being scanned. Like a blind man tapping in front of him with his cane, we can tell that the tip is approaching the surface and so can "feel" the outlines of the surface in front of us. Many different types of physical interactions with the surface are used to detect its presence.

Some scanning probe microscopes literally push on the surface — and note how hard the surface pushes back. Others connect the surface and probe to a voltage source, and measure the current flow when the probe gets close to the surface. A host of other probe-surface interactions can be measured, and are used to make different types of SPMs. But in all of them, the basic idea is the same: when the sharp tip of the probe approaches the surface a signal is generated — a signal which lets us map out the surface being probed. The SPM can not only map a surface, in many cases the probe-surface interaction changes the surface as well. This has already been used experimentally to spell out molecular words, and the obvious opportunities to modify the surface in a controlled way are being investigated both experimentally and theoretically.

A few big SPMs making a few molecular structures won't let us make much — certainly not tons of precisely structured shatterproof diamond. We'll need vast numbers of very small positional devices operating in parallel. Unfortunately, as we make our positional devices smaller and smaller, they will be more and more subject to thermal noise. To make something that's *both* small *and* stiff is more challenging. It helps to get the stiffest material you can find. Diamond, as usual, is stiffer than almost anything else and is an excellent material from which to make a very small, very stiff positional device. *Theoretical analysis gives firm support to the idea that positional devices in the 100 nanometer size range able to position their tips to within a small fraction of an atomic diameter in the face of thermal noise at room temperature should be feasible*. Trillions of such devices would occupy little more than a few cubic millimeters (a speck slightly larger than a pinhead).

Stiff Hydrocarbons as a Model System for Study

These four molecular tools — by themselves — should be sufficient to make a remarkably wide range of stiff hydrocarbons. This suggests that we define a simplified model of nanotechnology that we might call "hydrocarbon-based nanotechnology." This simplified version would only be able to make things from hydrogen and carbon — a much less ambitious goal than making things from the approximately 100 elements found in the full periodic table. But in exchange for confining ourselves to this more limited class of structures, we make it much easier to analyze the structures that can be fabricated and the synthetic reactions needed to make them. This narrower proposal can be more readily and more thoroughly investigated than full nanotechnology.

At the same time, hydrocarbon-based nanotechnology retains several of the key features of general nanotechnology. Diamond and shatterproof variants of diamond still fall within its purview; and it can still make all the parts needed for the operation of basically mechanical devices, including: struts, bearings, gears, robotic arms, and the like.

It's often useful to define a somewhat broader class than the stiff hydrocarbons, but a class which is still much more restrictive than the class of "most structures consistent with physical law." We'd like to be able to build molecular structures that are similar to diamond but which incorporate some other elements. We might want to add impurities (as in diamond electronic devices) or terminate a bearing surface with fluorine. Perhaps we're bending the diamond structure, and want to add some silicon internally to relieve the strain, (the silicon-carbon bond is longer than the carbon-carbon bond, so adding silicon to a region will "puff it up"). Or we find that adding some nitrogen to the internal surface of a bearing will relieve some strain (the nitogen-carbon bond is slightly shorter than the carbon-carbon bond). The term "diamondoid" is used to cover these (and related) possiblities. Diamondoid materials are made primarily of the elements hydrogen; first row elements like carbon, nitrogen, oxygen or fluorine; and second row elements like silicon, phosphorous, sulfur or chlorine; all held together by many strong covalent bonds. While including only about a tenth of the elements in the periodic table, the diamondoid materials include many structures of striking utility.

Self Replication

Positional control combined with appropriate molecular tools should let us build a truly staggering range of molecular structures — but a few molecular devices built at great expense would hardly seem to qualify as a revolution in manufacturing. How can we keep the costs down?

Potatoes are a miracle of biology with tens of thousands of genes and proteins and intricate molecular machinery; yet we think nothing of eating this miracle, mashed with a little butter. Potatoes, along with many other agricultural products, cost less than a dollar a pound. The key reason: provide them with a little dirt, water and sunlight and a potato can make more potatoes.

If we could make a general purpose programmable manufacturing device which was able to make copies of itself (the *author does* work at *Xerox*, after all....), then the manufacturing costs for both the devices and anything they made could be kept quite low — likely no more than the costs for growing potatoes. Drexler called such devices "assemblers." The first serious analysis of self replicating systems was by *von Neumann* in the 1940's. *He carried out a detailed analysis of one such system in a theoretical cellular automata model* (The best known example of a cellular automata model is *Conway's Game of Life*, which is played out on a giant checkerboard and uses a simple set of rules to decide what happens at each square of the checkerboard at each time step). In von Neumann's cellular automata model he used a universal computer for control and a "universal constructor" to build more automata. The "universal constructor" was a robotic arm that, under computer control, could move in two dimensions and alter the state of the cell at the tip of its arm. By sweeping systematically back and forth, the arm could "build" any structure that the computer instructed it to. In his three-dimensional "kinematic" model, von Neumann retained the idea of a positional device (now able to position in three dimensions rather than two) and a computer to control it.

A NASA study in 1980 extended the general conclusions of von Neumann and concluded that a fully automated lunar mining and manufacturing operation, able to extend itself using its own mining and manufacturing capabilities, would be feasible and could be built given a multi-billion dollar budget and a few decades of work. (Note that this kind of budget and time frame was in keeping with other proposed aerospace projects of the time).

The architecture for Drexler's assembler is a specialisation of the more general architecture proposed by von Neumann. As before, there is a computer and constructor, but now the computer has shrunk to a "molecular computer" while the constructor combines two features: a robotic positional device (such as the robotic arm discussed earlier) and a well defined set of chemical operations that take place at the tip of the positional device (such as the hydrogen abstraction reaction and the other reactions involved in the synthesis of diamond).

The complexity of a *self replicating system* need not be excessive. In this context the complexity is just the size, in bytes, of a "recipe" that fully describes how to make the system. The complexity of an assembler needn't be beyond the complexity that can be dealt with by today's engineering capabilities. As shown in the following table, there are several self replicating systems whose complexity is well within current capabilities. Drexler estimated the complexity of his original proposal for an assembler at about 10, 000, 000 bytes. *Further work should reduce this.*

Complexity of self replicating systems (bytes)

Von Neumann's universal constructor	about 60, 000
Internet worm	60, 000
Mycoplasma genitalia	145, 018
E. Coli	1, 000, 000
Drexler's assembler	12, 000, 000
Human	800, 000, 000
NASA Lunar Manufacturing Facility	over 10, 000, 000, 000

What is the "complexity" of a living system? We'll take this to mean the number of bytes in the DNA "blueprints." As each base pair in DNA can be one of four possibilities, it encodes two bits. One byte (8 bits) can be encoded in four base pairs.

This means we count the number of base pairs in the DNA and divide by four to get the number of bytes in the "blueprints." For Mycoplasma genitalia, which has 580, 070 base pairs, this results in 145, 017.5 bytes. For humans, with roughly 3.2 billion base pairs, this results in 800 million bytes. (We're using the haploid base pair count: each cell in your body has DNA from your mother and DNA from your father — but much of this is similar. We're counting only the DNA from one parent). The complexity and sophistication of most living systems goes well beyond anything we might need just to achieve low cost manufacturing. Much of the complexity of the human genome is unrelated to issues of self-replication: people do more than just self-replicate!

The complexity of the internet worm is just an estimate of the number of bytes in its C programme. Because the environment in which it operates is highly structured and provides relatively easy access to complex and sophisticated software, it could be argued that its complexity might be much less than the complexity of a system that operated in a "simple" environment. We'll leave it in our table anyway, as it's an interesting data point. This argument is less applicable to von Neumann's universal constructor which operates in a simple environment: just a large two-dimensional checkerboard with a finite number of states at each square. Its complexity is an estimate of the number of bytes needed to describe the constructor. The complexity of the NASA Lunar Manufacturing Facility was estimated in the NASA study.

What will we be able to make?

Today, most airplanes are made from metal despite the fact that diamond has a strength-to-weight ratio over 50 times that of aerospace aluminum. Diamond is expensive, we can't make it in the shapes we want, and it shatters. Nanotechnology will let us inexpensively make shatterproof diamond (with a structure that might resemble diamond fibers) in exactly the shapes we want. This would let us make a Boeing 747 whose unloaded weight was 50 times lighter but just as strong.

Today, travel in space is *very* expensive and reserved for an elite few. *Nanotechnology will dramatically reduce the costs and increase the capabilities of space ships and space flight*. The strength-to-weight ratio and the cost of components are absolutely critical to the performance and economy of space ships: with nanotechnology, both of these parameters will be improved by one to two orders of magnitude. Improvements in these two parameters alone (ignoring other advantages provided by nanotechnology) should improve the overall cost/performance ratio by over three orders of magnitude. This has led the *National Space Society* (NSS) to adopt *a position paper supporting nanotechnology*; Dan Goldin, *NASA*'s chief administrator, to support nanotechnology; and *NAS at NASA Ames Research Center to start a project (now with perhaps half a dozen researchers) to examine molecular manufacturing systems and molecular machines using computational models*.

Beyond inexpensively providing remarkably light and strong materials for space ships, nanotechnology will also provide extremely powerful computers with which to guide both those ships and a wide range of other activities in space.

Today, computer chips are made using lithography — literally "stone writing." It's our finest manufacturing technology, but there seem to be fundamental limits in how much further we can improve it. In lithography, we draw fine lines on a silicon wafer by using methods borrowed from photography. A light-sensitive film — called a "resist" — is spread over the silicon wafer. The resist is exposed to a complex pattern of light and dark, like a negative in a camera, and developed. The exposed resist is then washed away, and a particular chemical is sprayed over the surface. Where the resist has been washed away, the chemical reaches the surface of the silicon and can diffuse a short distance into it. Where the resist has not been washed away, the spray is blocked. Finally, the remaining resist is washed away, along with any chemicals that it prevented from reaching the silicon surface. The result: a fine pattern of some desired chemical is laid out on the silicon surface. By repeating this process, an intricate set of interlocking patterns can be made that defines the complex logic elements of a modern computer chip.

If the computer hardware revolution is to continue at its current pace, in a decade or so we'll have to move beyond lithography to some new post lithographic manufacturing technology. Making patterns on a resist and spraying chemicals around simply can't arrange atoms with the ultimate precision that should be feasible. More precise methods will be needed. Ultimately, each logic element will be made from just a few atoms. Designs for computer gates with less than 1, 000 atoms have already been proposed — but each atom in such a small device has to be in exactly the right place. To economically build and interconnect trillions upon trillions of such small and precise devices in a complex three dimensional pattern we'll need a manufacturing technology well beyond today's lithography: we'll need nanotechnology. With it, we should be able to build mass storage devices that can store more than a hundred billion billion bytes in a volume the size of a sugar cube; RAM that can store a mere billion billion bytes in such a volume; and massively parallel computers of the same size that can deliver a billion billion instructions per second.

Today, "smart" weapons are fairly big — we have the "smart bomb" but not the "smart bullet." In the future, even weapons as small as a single bullet could pack more computer power than the largest supercomputer in existence today, allowing them to perform realtime image analysis of their surroundings and communicate with weapons tracking systems to acquire and navigate to targets with greater precision and control. We'll also be able to build weapons both inexpensively *and much more rapidly*, at the same time taking full advantage of the remarkable materials properties of diamond. Rapid and inexpensive manufacture of great quantities of stronger more precise weapons guided by massively increased computational power will alter the way we fight wars. Changes of this magnitude could destabilize existing power structures in unpredictable ways.

Military applications of nanotechnology raise a number of concerns that prudence suggests we begin to investigate before, rather than after, we develop this new technology. While molecular manufacturing will not arrive for many years, its obvious military potential will increasingly attract the interest of strategic planners. For example, in a talk titled *Nanotechnology and global security* at the *Fourth Foresight Conference on Molecular Nanotechnology*, *Admiral David E. Jeremiah, USN (Ret)*, former Vice Chairman of the Joint Chiefs of Staff, said "Military applications of molecular manufacturing have even greater potential than nuclear weapons to radically change the balance of power." As it seems implausible that military applications of this technology will never be developed and deployed, it would seem safer to encourage the relatively early interest of those organisations less prone to the abuse of power and more likely to curb its abuse by others.

To make power today we dig coal and oil from the ground, we dam rivers, and we burn nuclear fuel in nuclear power plants. Yet the sunshine all around us could provide orders of magnitude more power than we use — and do so more cleanly and less expensively — if only we could make low cost solar cells and batteries. We already know how to make solar cells that are efficient enough for this application: it just costs too much to make them. Nanotechnology will cut costs both of the solar cells and the equipment needed to deploy them, making solar power economical. In this application we need not make new or technically superior solar cells: making inexpensively what we already know how to make expensively would move solar power into the mainstream.

Today, our surgical tools are large and crude at the molecular scale — yet the cellular and molecular machinery in our tissue is small and precise. There is a fundamental mismatch between the capabilities of our tools and what's needed to cure the injuries in our tissue. Today's scalpels are, as seen by a cell, large crude scythes that are more suited to cut and tear than to heal and mend. Ripping through tissue, they leave dead and maimed cells in their wake. The only reason that modern surgery works is the remarkable ability of cells to regroup, bury their dead, and heal over the wound.

It is not modern medicine that does the healing, but the cells themselves: we are but onlookers. If we had surgical tools that were molecular both in their size and precision, we could develop a medical technology that for the first time would let us directly heal the injuries at the molecular and cellular level that are the root causes of disease and ill health. With the precision of drugs combined with the intelligent guidance of the surgeon's scalpel, *we can expect a quantum leap in our medical capabilities*.

Nanotechnology should let us make almost every manufactured product faster, lighter, stronger, smarter, safer and cleaner. We can already see many of the possibilities as these few examples illustrate. New products that solve new problems in new ways are more difficult to foresee, yet their impact is likely to be even greater. Could Edison have foreseen the computer, or Newton the communications satellite?

Summary of the Core Technology

We can now see the fundamental shape of a molecular manufacturing technology. Self replicating assemblers, operating under computer control, let us inexpensively build more assemblers. The assemblers can be reprogrammed to build other products. The assemblers use programmable positional control to position molecular tools and molecular components, permitting the inexpensive fabrication of most structures consistent with physical law. Diamondoid materials in particular become inexpensive and commonplace, and their remarkable properties usher in what has been called the Diamond Age.

Developmental Pathways

In the preceding discussion of nanotechnology we gave scant attention to possible developmental pathways that would let us migrate from our current technology to what we might call "mature" systems able to inexpensively manufacture most diamondoid structures. The proposals advanced so far have been driven largely by the desired goal: a system able to inexpensively synthesize most diamondoid structures. The proposal that emerged was a general purpose programmable manufacturing system which uses positionally controlled highly reactive tools in vacuum and is able to self replicate.

No constraint was imposed requiring that the proposed system be easy to fabricate given our current technology — and indeed, this problem appears non trivial. But we must eventually build such systems if they are to be of any use — how can we do this? To give the reader some feeling for the magnitude of the effort involved, we can compare the requirements for a mature system with present capabilities.

A mature system must be able to build an assembler with hundreds of millions or billions of atoms with no atom out of place. If we are to do this, and if each unit operation typically handles one or a few atoms (a hydrogen abstraction, a carbene insertion, etc), then either the error rate per unit operation must be low (less than one in a billion) or we must adopt error detection and correction methods.

A general rule of thumb is that making something right the first time is easier than making it wrong and fixing it. Given the choice between a manufacturing process that has a low enough error rate that most finished products will simply work correctly the first time versus a process which requires elaborate error detection and correction to make any working products at all, the former is greatly to be preferred. Because analysis of the fundamental causes of errors during the manufacture of diamondoid products in general and assemblers in particular supports the idea that error rates substantially lower than one in a billion should be achievable, and hence that an assembler can make another assembler with a high probability of success while using no error detection or correction, existing design proposals adopt the simpler and higher performance approach. Had the opposite conclusion been reached, i.e., that error rates could *not* be driven below one in one billion, then design proposals would have adopted a smaller module size (consistent with the error rate) and incorporated appropriate mechanisms for module testing, the rejection of bad modules, and the assembly of working modules into larger systems. This line of logic results in proposals for assemblers in which error rates are very low. These low error rates, however, are predicated on the prior existence of diamondoid assemblers. Unfortunately, we do not have assemblers today. Worse, current technology would be hard pressed to achieve such error rates. Chemists view a synthesis that provides 99% yield as very good. The synthesis of proteins from amino acids by ribosomes has an error rate of perhaps one in 10, 000. Replication of DNA, by using extensive error detection and correction, along with built in redundancy (DNA has two complementary strands), achieves an error rate of roughly one base in a billion (which varies depending on particular circumstances).

The size of structure that can be built by today's SPMs without error is smaller than has been achieved by chemists. The first such work *arranged 35 xenon atoms on a nickel surface at 4 kelvins in vacuum to spell out "IBM."* Later work in Japan spelled out "NANO SPACE" at room temperature in vacuum by removing individual sulfur atoms from the surface. More recently, *6 molecules were arranged in a hexagonal pattern at room temperature in vacuum*. As each molecule had 173 atoms, this modular approach was able to position over 1, 000 atoms. While these are great successes we have not yet seen a long sentence spelled out in this fashion, let alone a paragraph or a book. Further, SPMs can "see" the structure that's being built. Current successes use this ability to provide constant feedback to the human operator, who can detect and correct errors.

In short, today's SPMs can build structures that are only a small fraction of the size of proposed assemblers and have error rates high enough that they must use relatively sophisticated error detection and correction methods.

The situation is made more difficult by the additional requirement that unit operations be fast. If an assembler is to manufacture a copy of itself in about a day, and if this takes one hundred million to a billion operations, then each unit operation must take place in a fraction of a millisecond. Ribosomes take tens of milliseconds to add a single amino acid to a growing protein. Today's SPMs can take hours to arrange a few atoms or molecules.

At the same time, the use of vacuum prevents us from using self assembly or any of the other solution-based techniques developed by chemistry. Abandoning these powerful tools does not seem like a good first step.

While speed, reliability and operation in vacuum do not pose fundamental problems, the conclusion is obvious: directly building a diamondoid assembler using existing technology is a daunting task. Perhaps, rather than attempting to solve all the problems in a single giant leap, we should instead approach them in a more incremental fashion. To this end, we need to define one or more intermediate systems. Almost by definition, an intermediate system is easier to build from current technology, but is still able to make systems that are more powerful than itself. A series of intermediate systems, each able to build the next system in the chain, would let us adopt more incremental methods.

One approach is to eliminate the requirement that the assembler be made from diamondoid structures. We wanted to make diamondoid structures because of their remarkable strength, stiffness, electrical properties, etc. But an intermediate system need only be able to make a more advanced system, and perhaps products that are impressive in comparison with today's products. It doesn't have to be diamondoid itself.

This suggests what might be called "building block based nanotechnology." Rather than building diamond, we'll build some other material from molecular building blocks that are relatively large: tens, hundreds or even thousands of atoms in size. Larger building blocks reduce the number of assembly steps, so fewer unit operations are needed and they need not be as reliable. Soluble building blocks, with "linkage" groups that are selective (the building blocks only stick to other building blocks, not to the solvent or low concentrations of contaminants) eliminate the need for vacuum. We certainly have many choices: any of the wide range of molecules that chemists have synthesised or could reasonably synthesize which has the desired properties. *Krummenacker* concluded that each molecular building block should have at least three sites where it can link to other building blocks (two sites leads to the familiar polymers so ubiquitous in biological systems: proteins, DNA, RNA, etc. Addition of a third site makes the design of three dimensional stiff structures much easier). While such building blocks could be linked to each other using any one of a variety of reactions, a particularly attractive possibility is the Diels-Alder reaction. This reaction, well known to chemists, works in most solvents (or even in vacuum). It involves a reaction between a diene and a dieneophile.

The reaction is specific: the diene and the dieneophile react with each other but seldom with other groups (see illustration above, courtesy of Krummenacker). Because it works in vacuum and doesn't produce any small molecules (which would destroy the vacuum), it could be used both in intermediate systems in solution, and later in systems that use vacuum to make diamondoid materials. Solution based systems could use positional control to assemble the building blocks, but can also use the methods of self assembly. In particular, the self assembly of a positional device should be feasible. If you ask someone familiar with the field if it would be feasible to self assemble a robotic arm, the usual response is some variation of "no." If, instead, you ask whether it would be feasible to self assemble an octahedron, the usual response is to point out that *Nadrian Seeman* of New York University has already self assembled a *truncated octahedron from DNA* and plans to do the same for a regular octahedron (both for reasons unrelated to positional control). (The computer generated graphic of the truncated octahedron at right is courtesy of Nadrian Seeman). For this work he received *The 1995 Feynman Prize in nanotechnology*.

As we have seen, an octahedron is the basic structure required for the Stewart platform. The truncated octahedron, because it has sides that aren't triangular, is floppier than a regular octahedron. Even the regular octahedron is likely to be too floppy for this application because of the limited stiffness of DNA. Further, the length of the edges cannot be changed (a requirement if we wish to control the position of the platform with respect to the base). However, self assembling a Stewart platform able to assemble molecular building blocks seems much less difficult than directly building a diamondoid assembler able to synthesize diamond in vacuum.

Several ways to deal with the problem of inadequate stiffness and adjustable length edges are possible. One way would be to attach selectively sticky strands of DNA to a stiff molecular structure. This would let us use DNA (which has been intensively studied and whose selective stickiness is well known) to guide the process of self assembly, while the stiff molecular structure would would let us deal with concerns about inadequate stiffness. To change the length of an edge, we would need to use molecular structures that change shape in response to light, pressure, temperature, chemicals, or some other external signal. Many such molecular structures are known. Self assembling a Stewart platform whose edges are stiff enough to make a useful positional device, and the length of whose edges can be changed by a suitable signaling mechanism, is no longer a challenge that seems beyond our reach.

A second approach for positioning molecular building blocks in solution is to use an Atomic Force Microscope (AFM). This type of SPM relies on "touch" to create an image. By pushing on the structure being scanned, and feeling how hard it pushes back, the AFM can build up an image of stiff structures (it doesn't work very well when the applied force is strong enough to deform the structure). Because the AFM touches the surface of the structure being probed, it can also change that surface. To do this, it's very useful if the precise molecular structure of the tip can be controlled, so that the precise nature of the tip-surface interaction can be well defined.

This line of reasoning leads to the "molecular manipulator: " an AFM with a reactive fragment of an *antibody* bound to its tip. By changing the *antibody*, we can change what type of molecule will stick to the AFM tip. If the antibody both sticks selectively to a particular molecular building block, and also is bound to the AFM tip, we can now position the molecular building block by positioning the AFM tip. We can use any one of a wide range of possible molecular building blocks (*antibodies* can today be created that will bind to most small molecules). Again, this no longer seems beyond our reach.

Another developmental pathway is defined by the desire of the semiconductor industry to make ever smaller transistors despite the fact that optical lithography — the current workhorse manufacturing method for making computer chips — will reach a limit in a few more years. Transistors are made today by "drawing" very fine lines on silicon — but the optical methods in use today are limited by the wavelength of light: a few hundred nanometers. SPMs could let us make much smaller circuits by drawing much finer lines. Already demonstrated for making the finest and most critical lines in a transistor, if this approach can be made low cost and reliable it would let the entire semiconductor industry be retooled to use these finer lines to make more and smaller transistors. This, by itself, is not molecular nanotechnology; but once we learn to draw the smallest and finest lines of all, lines where every atom is in the right place (and there will be strong economic incentives to move in this direction), we could use the same techniques to make molecular machines. At first, these molecular machines would be very expensive and useful only for applications where their remarkable precision would justify the cost — but they should eventually lead to machines sophisticated enough to make simple assemblers.

Whether through self-assembly, by improvements in SPMs, some hybrid approach, or perhaps by some other path; we are moving from an era of expensive and imprecise products to an era of inexpensive products of molecular precision. We are going to replace most of the manufacturing base of the world with a fundamentally new and *much* better manufacturing technology.

Arguments that nanotechnology is infeasible are relatively rare and uniformly poor in technical quality. The best known are the arguments of David Jones, a *Nature* columnist and chemist. Quoted extensively in a *Scientific American* article on nanotechnology he advanced arguments like the following: "Single atoms ... are amazingly mobile and reactive. They will combine instantly with ambient air, water, each other, the fluid supporting the assemblers, or the assemblers themselves." However, as the proposals involving reactive molecular tools specify that the environment should be inert (i.e., vacuum) there is no "ambient air" to react with. As the molecular tools are positionally controlled, they will not react with each other or the assembler itself for the same reason that a hot soldering iron does not react with the skin of the person doing the soldering. Jones' criticisms raise greater questions about his understanding of the field than about the feasibility of nanotechnology. Other arguments against the feasibility of nanotechnology have likewise had obvious flaws, reminiscent of the 1920 New York Times editorial which said that rockets to the moon were impossible because there was no air to push against in space.

As should be clear from this article there are major technical challenges that must be overcome if we are to develop molecular manufacturing systems able to synthesize diamondoid structures. The serious questions, however, are about which development pathways should be pursued and how long it will take.

The single most frequently asked question about nanotechnology is: *how long?*. How long before it will let us make molecular computers? How long before inexpensive solar cells let us use clean solar power instead of oil, coal, and nuclear fuel? How long before we can explore space at a reasonable cost? Having said that, it is worth pointing out that the trends in the development of computer hardware have been remarkably steady for the last 50 years. Such parameters as

- the number of atoms required to store one bit
- the size of a transistor
- the energy dissipated by a single logic operation
- the resolution of the finest machining technology
- the cost of a computer gate

have all declined with great regularity, even as the underlying technology has changed dramatically. *From relays to vacuum tubes to transistors to integrated circuits to Very Large Scale Integrated circuits (VLSI) we have seen steady declines in the size and cost of logic elements and steady increases in their performance.*

Extrapolation of these trends suggests we will have to develop molecular manufacturing in the 2010 to 2020 time frame if we are to keep the computer hardware revolution on schedule.

Of course, extrapolating past trends is a philosophically debatable method of technology forecasting. While no fundamental law of nature prevents us from developing nanotechnology on this schedule (or even faster), there is equally no law that says this schedule will not slip.

Much worse, though, is that such trends imply that there is some ordained schedule — that nanotechnology will appear regardless of what we do or don't do. Nothing could be further from the truth. How long it takes to develop this technology depends very much on what we do. If we pursue it systematically, it will happen sooner. If we ignore it, or simply hope that someone will stumble over it, it will take much longer. And by using theoretical, computational and experimental approaches together, we can reach the goal more quickly and reliably than by using any single approach alone.

While some advances are made through serendipitous accidents or a flash of insight, others require more work. It seems unlikely that a scientist would forget to turn off the bunsen burner in his lab one afternoon and return to find he'd accidentally made a Space Shuttle.

Like the first human landing on the moon, the Manhatten project or the development of the modern computer the development of molecular manufacturing will require the coordinated efforts of many people for many years. How long will it take? A lot depends on when we start.

Era of Nanotechnology

We will use *molecular technology* to bring health because the human body is made of molecules. The ill, the old, and the injured all suffer from mis-arranged patterns of *atoms*, whether mis-arranged by invading viruses, passing time, or swerving cars. Devices able to rearrange atoms will be able to set them right. *Nanotechnology* will bring a fundamental breakthrough in medicine.

Physicians now rely chiefly on surgery and drugs to treat illness. Surgeons have advanced from stitching wounds and amputating limbs to repairing hearts and re-attaching limbs. Using microscopes and fine tools, they join delicate blood vessels and nerves. Yet even the best micro-surgeon cannot cut and stitch finer tissue structures. Modern scalpels and sutures are simply too coarse for repairing *capillaries*, *cells*, and molecules. Consider "delicate" surgery from a cell's perspective: a huge blade sweeps down, chopping blindly past and through the molecular machinery of a crowd of cells, slaughtering thousands. Later, a great obelisk plunges through the divided crowd, dragging a cable as wide as a freight train behind it to rope the crowd together again. From a cell's perspective, even the most delicate surgery, performed with exquisite knives and great skill, is still a butcher job. Only the ability of cells to abandon their dead, regroup, and multiply makes healing possible.

Yet as many paralysed accident victims know too well, not all tissues heal. Drug therapy, unlike surgery, deals with the finest structures in cells. Drug molecules are simple molecular devices. Many affect specific molecules in cells. Morphine molecules, for example, bind to certain receptor molecules in brain cells, affecting the neural impulses that signal pain. Insulin, beta blockers, and other drugs fit other receptors. But drug molecules work without direction. Once dumped into the body, they tumble and bump around in solution haphazardly until they bump a target *molecule*, fit, and stick, affecting its function.

Surgeons can see problems and plan actions, but they wield crude tools; drug molecules affect tissues at the molecular level, but they are too simple to sense, plan, and act. But molecular machines directed by nanocomputers will offer physicians another choice. They will combine sensors, programmes, and molecular tools to form systems able to examine and repair the ultimate components of individual cells. They will bring surgical control to the molecular domain.

These advanced molecular devices will be years in arriving, but researchers motivated by medical needs are already studying molecular machines and molecular engineering. The best drugs affect specific molecular machines in specific ways. Penicillin, for example, kills certain *bacteria* by jamming the nanomachinery they use to build their cell walls, yet it has little effect on human cells.

Biochemists study molecular machines both to learn how to build them and to learn how to wreck them. Around the world (and especially the Third World) a disgusting variety of viruses, bacteria, protozoa, fungi, and worms parasitize human flesh. Like penicillin, safe, effective drugs for these diseases would jam the parasite's molecular machinery while leaving human molecular machinery unharmed. Dr. Seymour Cohen, professor of pharmacological *science* at SUNY (Stony Brook, New York), *argues that biochemists should systematically study the molecular machinery of these parasites*. Once biochemists have determined the shape and function of a vital protein machine, they then could often design a molecule shaped to jam it and ruin it. Such drugs could free humanity from such ancient horrors as schistosomiasis and leprosy, and from new ones such as AIDS.

Drug companies are already redesigning molecules based on knowledge of how they work. Researchers at Upjohn Company have *designed and made modified molecules of vasopressin*, a hormone that consists of a short chain of *amino acids*. Vasopressin increases the work done by the heart and decreases the rate at which the kidneys produce urine; this increases blood pressure. The researchers designed modified vasopressin molecules that affected receptor molecules in the kidney more than those in the heart, giving them more specific and controllable medical effects. More recently, they designed a modified vasopressin molecule that binds to the kidney's receptor molecules without direct effect, thus blocking and *inhibiting* the action of natural vasopressin.

Medical needs will push this work forward, encouraging researchers to take further steps toward protein design and molecular engineering. Medical, military, and economic pressures all push us in the same direction. Even before the *assembler* breakthrough, molecular technology will bring impressive advances in medicine; trends in biotechnology guarantee it. Still, these advances will generally be piecemeal and hard to predict, each exploiting some detail of biochemistry. Later, when we apply assemblers and technical AI systems to medicine, we will gain broader abilities that are easier to foresee.

To understand these abilities, consider cells and their self-repair mechanisms. In the cells of your body, natural radiation and noxious chemicals split molecules, producing reactive molecular fragments. These can misbond to other molecules in a process called *cross-linking*. As bullets and blobs of glue would damage a machine, so radiation and reactive fragments damage cells, both breaking molecular machines and gumming them up.

If your cells could not repair themselves, damage would rapidly kill them or make them run amok by damaging their control systems. But *evolution* has favoured organisms with machinery able to do something about this problem. The self-replicating factory system sketched in this encyclopaedia repaired itself by replacing damaged parts; cells do the same. So long as a cell's *DNA* remains intact, it can make error-free tapes that direct ribosomes to assemble new protein machines.

Unfortunately for us, DNA itself becomes damaged, resulting in mutations. Repair *enzymes* compensate somewhat by detecting and repairing certain kinds of damage to DNA. These repairs help cells survive, but existing repair mechanisms are too simple to correct all problems, either in DNA or elsewhere. Errors mount, contributing to the aging and death of cells—and of people.

Life, Mind, and Machines

Does it make sense to describe cells as "machinery, " whether self-repairing or not? Since we are made of cells, this might seem to reduce human beings to "mere machines, " conflicting with a holistic understanding of life.

But *a dictionary definition of holism* is "the theory that reality is made up of organic or unified wholes that are greater than the simple sum of their parts." This certainly applies to people: one simpler sum of our parts would resemble hamburger, lacking both mind and life. The human body includes some ten thousand billion *billion* protein parts, and no machine so complex deserves the label—mere." Any brief description of so complex a system cannot avoid being grossly incomplete, yet at the cellular level a description in terms of machinery makes sense. Molecules have simple moving parts, and many act like familiar types of machinery. Cells considered as a whole may seem less mechanical, yet biologists find it useful to describe them in terms of molecular machinery.

Biochemists have unraveled what were once the central mysteries of life, and have begun to fill in the details. They have traced how molecular machines break food molecules into their building blocks and then reassemble these parts to build and renew tissue. Many details of the structure of human cells remain unknown (single cells have billions of large molecules of thousands of different kinds), but biochemists have mapped every part of some viruses. Biochemical laboratories often sport a large wall chart showing how the chief molecular building blocks flow through bacteria. Biochemists understand much of the process of life in detail, and what they don't understand seems to operate on the same principles. The mystery of heredity has become the industry of genetic engineering. Even embryonic development and memory are being explained in terms of changes in biochemistry and cell structure.

In recent decades, the very quality of our remaining ignorance has changed. Once, biologists looked at the process of life and asked, "How can this be?" But today they understand the general principles of life, and when they study a specific living process they commonly ask, "Of the many ways this could be, which has nature chosen?" In many instances their studies have narrowed the competing explanations to a field of one. Certain biological processes—the coordination of cells to form growing embryos, learning brains, and reacting immune systems—still present a real challenge to the imagination. Yet this is not because of some deep mystery about how their parts work, but because of the immense complexity of how their many parts interact to form a whole.

Cells obey the same natural laws that describe the rest of the world. Protein machines in the right molecular environment will work whether they remain in a functioning cell or whether the rest of the cell was ground up and washed away days before. Molecular machines know nothing of "life" and "death."

Biologists—when they bother—sometimes define life as the ability to grow, replicate, and respond to stimuli. But by this standard, a mindless system of replicating factories might qualify as life, while a conscious *artificial intelligence* modeled on the human brain might not. Are viruses alive, or are they "merely" fancy molecular machines? No experiment can tell, because nature draws no line between living and nonliving. Biologists who work with viruses instead ask about viability: "Will this *virus* function, if given a chance?" The labels of "life" and "death" in medicine depend on medical capabilities: physicians ask, "Will this patient function, if we do our best?" Physicians once declared patients dead when the heart stopped;

they now declare patients dead when they despair of restoring brain activity. Advances in cardiac medicine changed the definition once; advances in brain medicine will change it again.

Just as some people feel uncomfortable with the idea of machines thinking, so some feel uncomfortable with the idea that machines underlie our own thinking. The word "machine" again seems to conjure up the wrong image, a picture of gross, clanking metal, rather than signals flickering through a shifting weave of neural fibers, through a living tapestry more intricate than the mind it embodies can fully comprehend. The brain's really machinelike machines are of molecular size, smaller than the finest fibers.

A whole need not resemble its parts. A solid lump scarcely resembles a dancing fountain, yet a collection of solid, lumpy molecules forms fluid water. In a similar way, billions of molecular machines make up neural fibers and synapses, thousands of fibers and synapses make up a neural cell, billions of neural cells make up the brain, and the brain itself embodies the fluidity of thought.

To say that the mind is "just molecular machines" is like saying that the Mona Lisa is "just dabs of paint." Such statements confuse the parts with the whole, and confuse matter with the pattern it embodies. We are no less human for being made of molecules.

Protein Design as a Pathway to Molecular Manufacturing

Feynman's 1959 talk entitled "There's Plenty of Room at the Bottom" discussed microtechnology as a frontier to be pushed back, like the frontiers of high pressure, low temperature, or high vacuum. He suggested that ordinary machines could build smaller machines that could build still smaller machines, working step by step down toward the molecular level; he also suggested using particle beams to define two-dimensional patterns. Present microtechnology (exemplified by integrated circuits) has realised some of the potential outlined by Feynman by following the same basic approach: working down from the macroscopic level to the microscopic.

Present microtechnology handles statistical populations of atoms. As the devices shrink, the atomic graininess of matter creates irregularities and imperfections, so long as atoms are handled in bulk, rather than individually. Indeed, such miniaturisation of bulk processes seems unable to reach the ultimate level of microtechnology — the structuring of matter to complex atomic specifications. In this paper, I will outline a path to this goal, a general molecular engineering technology. The existence of this path will be shown to have implications for the present.

Although the capabilities described may not prove necessary to the achievement of any particular objective, they will prove sufficient for the achievement of an extraordinary range of objectives in which the structuring and analysis of matter are concerned. The claim that devices can be built to complex atomic specifications should not, however, be construed to deny the inevitability of a finite error rate arising from thermodynamic effects (and radiation damage). Such errors can be minimised through the use of free energy in error-correcting procedures (including rejection of faulty components before device assembly); the effects of errors can be minimised through fault-tolerant design, as in macroscopic engineering. The emphasis on devices that have general capabilities should be taken in the spirit of early work on the theoretical capabilities of computers, which did not attempt to predict such practical embodiments as specialised or distributed computation systems. The present argument, however, will proceed from step to step by close analogies between the proposed steps and past developments in nature and technology, rather than by mathematical proof. We commonly accept the feasibility of new devices without formal proof, where analogies to existing systems are close enough: consider the feasibility of making a clock from zirconium. The detailed design of many specific devices to render them describable by dynamical equations would be a task of another order (consider designing a clock from scratch) and appears unnecessary to the establishment of the feasibility of certain general capabilities.

Protein design

Biochemical systems exhibit a "microtechnology" quite different from ours: they are not built down from the macroscopic level but up from the atomic. Biochemical microtechnology provides a beachhead at the molecular level from which to develop new molecular systems by providing a variety of "tools" and "devices" to use and to copy. Building with these tools, themselves made to atomic specifications, we can begin on the far side of the barrier facing conventional microtechnology. What can be built with these tools? Gene synthesis and recombinant DNA technology can direct the ribosomal machinery of bacteria to produce novel proteins, which can serve as components of larger molecular structures. One might think assembly of such components into complex systems would require a preexisting technology able to handle molecules and assemble them; fortunately, biochemistry demonstrates that intermolecular attraction between complementary surfaces can assemble complex structures from solution. For example, the complex machinery of the ribosome self-assembles from more than 50 different protein molecules and can do so in vitro.

At present, the design of protein systems as complex as a ribosome seems an awesome task. Indeed, chemists cannot yet predict the three-dimensional conformation of a natural protein from its amino acid sequence, an ability that might seem requisite to the design of new proteins. Two considerations suggest that this obstacle is surmountable: first, the continuing improvement in protein science and, second, the difference between natural science and design engineering.

Regarding the first, computer simulation of protein molecules in solution (5) shows promise. As computer technology and chemical knowledge improve, simulations will increase in accuracy, speed, and size. Improvement promises new insight into protein behaviour and may permit the designer to modify (simulated) molecules quickly and to observe their behaviour directly.

Regarding the second consideration, natural scientists seek a more general understanding than design engineers require. Science seeks the ability to predict the conformations of all natural polypeptides. In attempting this, protein chemists can search for a minimum-energy chain conformation (in hope that the protein assumes not a local but a global minimum-energy conformation) (6) or can attempt to follow the chain-folding mechanism to find the final conformation (7). Prediction will be easier if the natural conformation has outstanding stability or if its folding mechanism proceeds in a sequence of strongly preferred steps. Unfortunately, natural selection accepts polypeptides that have natural conformations of low stability (in energetic terms) so long as they exhibit long lifetimes on the cellular time scale (or renature readily). Similarly, natural selection accepts any folding process so long as the chain reaches its natural conformation with essentially 100% yield. Moreover, random mutations are unlikely to enhance the stability of a particular conformation (or the predictability of its folding mechanism). Thus, natural proteins tend to accumulate disruptive changes until they reach the threshold of poor stability or reduced yield of the natural conformation; only then does natural selection come into play. Thus, it is little wonder that chemists cannot yet predict the conformations of natural proteins; they are not designed to fold predictably. Engineers (in contrast to scientists) need not seek to understand all proteins but only enough to produce useful systems in a reasonable number of attempts. An engineer designing a protein that has 1000 amino acids may choose among some 10^{1300} different amino acid sequences. It might be that only one in 10^{9} (or even 10^{700}) randomly selected sequences would yield a predictable conformation, yet this tiny fraction represents a vast number of proteins. Through use of strategically placed charged groups, polar groups, disulfide bonds, hydrogen bonds, and hydrophobic groups, the engineer should be able to design proteins that not only fold predictably to a stable structure (sometimes) but that serve a planned function as well. Even a low success rate will lead to an accumulation of successful designs. Thus, the difficulties encountered in predicting the conformations of natural proteins do not seem insurmountable obstacles to protein engineering. Computer modeling and chemical understanding of biological targets have already found use in pharmaceutical design (8), and an artificial 34-residue polypeptide designed to interact with RNA has been synthesised and found active (9). It has been proposed to give microcircuitry special sensitivities by adsorbing engineered proteins onto selected surfaces (10). The promise of enzyme design in chemical engineering is evident. As protein science has great promise and difficulties in understanding natural proteins need not block engineering, the substantial payoffs for improved capabilities should lead to development of protein design technology. It would be foolish to underestimate the time and effort that will be required to develop basic design capabilities and then a broad family of working molecular devices; still, the path seems clear to achieving the capabilities exhibited by existing biochemical systems, by copying their features if need be.

Molecular Machinery

A comparison f biochemical to macroscopic components will show the possibilities of the former by analogy to the latter. With structural members, moving parts, bearings, and motive power, versatile mechanical systems can be built. Molecular assemblages of atoms can act as solid objects, occupying space and holding a definite shape. Thus, they can act as structural members and moving parts. Sigma bonds that have low steric hindrance can serve as rotary bearings able to support ~ 10^{-9} N. A line of sigma bonds can serve as a hinge. Conformation-changing proteins (such as myosin) can serve as sources of motive power for linear motion; the reversible motor of the bacterial flagellum can serve as a source of motive power for rotary motion. The existence of this range of components in nature indicates that power-driven mechanical systems can be constructed on a molecular scale. By analogy with macroscopic devices, feasible molecular machines presumably include manipulators able to wield a variety of tools. Thermal vibrations in typical structures are a modest fraction of interatomic distances; thus, such tools can be positioned with atomic precision. As present microtechnology can lay down conductors on a molecular scale (10 nm) and molecular devices can respond to electric potentials (through conformation changes, etc.), such devices can be controlled by human operators or macroscopic machines. Further, by analogy with biological sensors, molecular scale instruments can evidently produce macroscopic signals, indicating the feasibility of feedback control in molecular manipulations.

Together, these arguments indicate the feasibility of devices able to move molecular objects, position them with atomic precision, apply forces to them to effect a change, and inspect them to verify that the change has indeed been accomplished. It would be foolish to minimize the time and effort that will be required to develop the needed components and assemble them into such complex and versatile systems. Still, given the components, the path seems clear.

Ordinary chemical synthesis relies on thermal agitation to bring reactant molecules in solution together in the correct orientation and with sufficient energy to cause the desired reaction. Enzyme-like molecular machines can hold reactants in the best relative positions as bonds are strained or polarised. Like some enzymes, they can do work on reactant molecules to drive reactions not otherwise thermodynamically favoured.

These are clearly techniques of great power, yet the synthetic capabilities of systems based on polypeptide chains might seem limited by amino acid properties. However, enzymes show that other molecular structures bound to the polypeptide (such as metal ions and complex ring structures) (11) can extend protein capabilities. The range of such tools is large and greater than found in nature. Thus, the synthetic capabilities of enzymes set only a lower bound on the capabilities of engineered protein systems. Indeed, as tool-wielding protein systems can control the chemical environment of a reaction site completely, they should be able, at a minimum, to duplicate the full range of moderate-temperature synthetic steps achieved by organic chemists. Further, where chemists must resort to complex strategies to make or break specific bonds in large molecules, molecular machines can select individual bonds on the basis of position alone. Conventional organic chemistry can synthesize not only one-, two-, and three-dimensional covalent structures but also exotic strained and fused rings. With the addition of controlled site-specific synthetic reactions, a broad range of large complex structures can doubtless be built.

Still, the synthetic abilities of protein machines will be limited by their need for a moderate temperature aqueous environment (although applied forces can sometimes replace or exceed thermal agitation as a source of activation energy and reaction sites and reactive groups can be protected from the surrounding water, as in some enzymatic active sites). These limits may be sidestepped by using the broad synthetic capabilities outlined above to build a second generation of molecular machinery whose components would not be coiled hydrated polypeptide chains but compact structures having three-dimensional covalent bonding. There is no reason why such machines cannot be designed to operate at reduced pressure or extreme temperatures; synthesis can then involve highly reactive or even free radical intermediates, as well as the use of mechanical arms wielding molecular tools to strain and polarize existing bonds while new molecular groups are positioned and forced into place. This may be done at high or low temperature as desired. The class of structures that can be synthesised by such methods is clearly very large, and one may speculate that it includes most structures that might be of technological interest.

Firmness of the argument

The development path described above should lead to advanced molecular machinery capable of general synthesis operations. As the results of this path can be shown to have consequences for the present, it is of interest to discuss the degree of confidence that should be placed in its feasibility.

It might be argued that complex protein or nonprotein machines are impossible or useless, on the grounds that, if they were possible and useful, organisms would be using them. A similar argument would, however, conclude that bone is a better structural material than graphite composite, that neurons can transmit signals faster than wires, and that technology based on the wheel is impossible or useless. Nature has been constrained less by what is physically possible than by what could be evolved in small steps. Thus, the absence of a proposed kind of molecular machinery in organisms in no way suggests its infeasibility.

To deny the feasibility of advanced molecular machinery, one must apparently maintain either (i) that design of proteins will remain infeasible indefinitely, or (ii) that complex machines cannot be made of proteins, or (iii) that protein machines cannot build second-generation machines.

In light of the expected improvements in computation, the simplified task of design engineers (compared with scientists), the possibilities offered by sheer trial-and-error modification of natural proteins, and the progress already made in protein design, the first seems difficult to maintain. Further, even if protein design were to prove intractible (because of difficulties in predicting conformations), this would in no way preclude developing an alternative polymer system with predictable coiling and using it as a basis for further development. In light of the presence of the needed components for mechanical devices in the cell, the second seems difficult to maintain. Indeed, the cytoskeleton provides a fair counterexample. In light of the results of synthetic organic chemistry and the ability of molecular machines to make reactions site specific, it seems difficult to maintain that nonprotein machine components cannot be built and assembled.

Each of the development steps outlined above seems closely analogous to past steps taken by nature or by technology. Each of these steps can be accomplished in many ways. To argue their infeasibility would seem to require some general principle precluding success, and it is difficult to see what such a principle might be like. Thus, the claim that advanced molecular technology can be developed seems well founded.

Although the existence of molecular machinery in cells indicates the feasibility of some sort of artificial molecular machinery, errors in assembly might limit the synthesis of structures of great complexity. In the cell, molecular machinery uses DNA to direct the assembly of DNA and other molecules. In some eukaryotic cells, DNA directs DNA synthesis with an error rate of ~ 10^{-11} per nucleotide added (12). As engineers commonly design systems to function reliably with many more failed components than 1 in 10^{11}, such an error rate seems no barrier to the construction of quite complex devices.

The possibility of low error rates is not surprising. For synthesis systems permitting error detection and correction (such as DNA synthesis), the net error rate in assembly can be reduced to roughly the product of the raw error rate in assembly and the rate at which errors are falsely identified as correct. As no uncertainty principle prohibits accurate discrimination between objects of different kinds (such as correctly and incorrectly assembled molecular structures), no limits to the detection and correction of errors are apparent.

Applications to computation

Molecular technology has obvious application to the storage and processing of information. A crude approach would involve literal "molecular machinery" patterned on the Babbage machine. In a more subtle approach, bits could be represented by protons, bound electrons, reactive groups, or conformation changes and transferred by movement of protons or of well-localised electrons (13), excitons, or phonons. The range of plausible device speeds is suggested by the 10^{-6} -sec turnover time for a fast enzyme, by the 10^{-13} -sec scale of collisional interactions (11), and by the 10^{-16} sec taken for an electron to cross an interatomic distance at a typical Fermi velocity.

It seems highly likely that a cubic cell 0.1 micrometers on a side (containing some 10^{8} optimally arranged atoms) can hold a bit or perform a logic operation and, at the same time, transmit bits through itself to provide communication from cell to cell in a lattice. If so, then computers can be built with at least 10^{15} active elements per cubic centimeter. In a well-designed computer (with elements closer to their true technological limit and not laid out in regular cubical cells), this volume estimate should prove quite conservative. Elements so small will be sensitive to radiation damage; to be reliable, systems will require a large measure of redundancy.

Concern might be raised about the cost of such intricately patterned matter, either because of labor or energy requirements. It seems clear, however, that molecular-scale production systems can be completely automated (what use is there for hands?). Thus, labor costs of production (including production of additional production equipment) can approach zero. The energy needed to produce molecularly engineered material will generally be greater than the energy needed to produce ordinary materials of similar bulk composition, but analogy suggests that the energy cost need not be vastly greater than for the production of biological materials. In many cases (e.g., advanced computers or any of a number of applications not discussed here), the unique value of the products would make such energy costs unimportant, even if energy costs remained high.

Some biological applications

Molecular devices can interact directly with the ultimate molecular components of the cell and thus serve as probes of unique value in studying processes within the cell. Further, molecular devices can characterize a frozen cell in essentially arbitrary detail by removal and characterisation of successive layers of material (atomically thin layers, if desired). Although the amount of data involved is large (a typical cell contains billions of protein molecules), the physical bulk of a device able to store and manipulate this amount of data will be quite small. The change of temperature and water distribution during freezing modifies cell structures in several ways, primarily by physical displacement of structures by ice crystals and denaturation of proteins by concentration of solutes in the residual liquid. With frozen tissue, knowledge of normal structures (membrane geometries, natural protein structures) and analysis of frozen structures (position of ice crystals, position of denatured proteins) should permit quite accurate reconstruction of the nature of the tissue before freezing.

Such procedures would have special utility in analysing the structure of tissue in the brain. Unlike, say, muscle or liver tissue, the function of brain tissue depends on the detailed three-dimensional structure of intertwined cells and their interfaces. The freezing process is far too slow to stop such dynamic processes as action potentials and synaptic transmission; short-term memory, however, is suspected to involve chemical modification of the neurons, and long-term memory is believed to involve the growth and modification of neuronal structures, particularly synapses. At the modest freezing rates possible in substantial pieces of tissue, ice crystals may be expected to nucleate and grow in the intercellular fluid, displacing the cell membranes as they do so. Electron micrographs, however, show that synapses (like many intercellular junctions) involve complementary structures on both sides of the intercellular gap, which should provide information enough to reconstruct the pre-freezing configurations of the cells almost regardless of ice crystal locations.

The ability to reconstruct the prefreezing structure of tissue, when combined with the general synthetic capabilities outlined above, will make feasible the physical restoration of tissue damaged by ordinary freezing through characterisation, reconstruction, and restoration of successive segments of frozen material. Although restored to a frozen condition, such tissue would lack the characteristic damage caused by the freezing process. As many tissues can survive the gross insult of ordinary freezing, it seems likely that most could survive freezing followed by repair. The remaining mode of damage would seem to be denaturation of proteins sensitive to cold alone during the thawing process. Should cell components of some species prove sensitive to short periods of cold, they could presumably be modified to resemble those of hardier species without changing either cell function or DNA.

Implications for the present

The existence of a path to an advanced molecular technology has implications for the present. As with all technologies, long-range promise should tend to increase interest in undertaking the early steps, even beyond the interest springing from more immediate benefits. The longer the expected wait, however, the less the interest.

On the other hand, molecular engineering of materials and devices can extend the capabilities of technology many fold in many areas. The implications of the feasibility of molecular technology are important to present day speculations concerning the probable behaviour (and likelihood of existence) of extraterrestrial technological civilisations. Similarly, those concerned with the long-range future of humanity must concern themselves with the opportunities and dangers arising from this technology. Finally, the eventual development of the ability to repair freezing damage (and to circumvent cold damage during thawing) has consequences for the preservation of biological materials today, provided a sufficiently long-range perspective is taken.

Development of the ability to design protein molecules will, by analogy between features of natural macromolecules and components of existing machines, make possible the construction of molecular machines. These machines can build second-generation machines able to perform extremely general synthesis of three-dimensional molecular structures, thus permitting construction of devices and materials to complex atomic specifications. This capability has implications for technology in general and in particular for computation and characterisation, manipulation, and repair of biological materials.

2

An Overview of Relationships Among Nano Chemistry, Nanotechnology and Nano-velcro: An In-depth Analysis

Co-relating Nanotechnology, Nanosystems, and Molecular Mechanics

Students interested in nanotechnology often ask what they should study. This web page provides a partial answer to that question. Foresight has a briefing on the subject by Eric Drexler.

The standard text in the field is *Nanosystems: molecular machinery, manufacturing and computation* by K. Eric Drexler. Buy a copy and study it.

Any manufacturing technology must move atoms from where they are to where we want them to be. How atoms move and the forces that act upon them during their motion are therefore critical areas of study in nanotechnology. This field is called molecular mechanics. A very brief discussion of molecular mechanics and its significance for nanotechnology is available on the web in *Computational Nanotechnology*. It contains references to further reading. A classic introduction to molecular mechanics is *Molecular Mechanics*, by Ulrich Burkert and Norman L. Allinger, ACS Monograph 177, American Chemical Society, 1982. While out of print, it can often be found in campus libraries.

Nanosystems introduces the basic concepts of molecular mechanics in this study. A great advantage of Drexler's treatment is the adoption of consistent SI units. A slow and careful reading of this chapter is worth the effort. Many other introductions to molecular mechanics are available. Software packages implementing specific approaches to molecular mechanics are available, and can be very useful in learning the concepts.

A central idea in nanotechnology is that of positional control. This can be provided by fairly standard robotic devices. A major difference between conventional robotic devices and the molecular robotic devices used to position molecular components is the need to consider thermal noise. At the molecular scale, components wiggle and jiggle because of Brownian motion. To control this, the component must be held stiffly, i.e., there must be a restoring force which acts to return the component to an equilibrium position if it is displaced. (The existence of a restoring force serves as a good abstract definition of "positional control.") The restoring force is usually assumed to be linear in the displacement: restoring force = k_s times displacement. The constant k_s is a measure of the stiffness of the system. The greater the stiffness, the greater the restoring force and the smaller the deviations of the system from its equilibrium position. The basic equation relating stiffness and positional uncertainty is:

$$\text{sigma}^2 = kT / k_s$$

This equation should be memorised and basic applications understood. To use it, it is essential to determine the stiffness k_s. The stiffness of a structure can be determined from its geometry and material properties. The basic concepts are introduced by Feynman in chapters 38 and 39 of *The Feynman Lectures on Physics*, by R.P. Feynman, R.B. Leighton, and M. Sands, M. Addison-Wesley (1964). Buying the Feynman Lectures is recommended.

Application of these equations to some robotic devices (including the Stewart platform which, because of its great stiffness, is very attractive for molecular robotic applications) is illustrated in *A New Family of Six Degree of Freedom Positional Devices*.

Self Replication

A second central idea in nanotechnology is that of self replication. The student should read the web page introducing self replication and select some of the references therein for further reading. The recursion theorem is basic to self replicating systems. This theorem should be understood. Exercise: write a programme which prints out an exact copy of itself. Buy a copy of *Kinematic Self-Replicating Machines*, it is an excellent survey of the literature.

There are, of course, a great many other subjects relevant to the development of nanotechnology. It seemed more useful to provide a short and focused list of critical subject areas that could be mastered with a reasonable effort than a longer and more unwieldy list which included everything of importance. The student can rest assured that there is no shortage of material to study relevant to this new field of research.

Understanding Nanotechnology and Worlds Stickiest Material

Scientists in Britain and USA are taking inspiration from the gecko to try and develop the ultimate in adhesive tape. The gecko has evolved with toe surfaces covered in the perfect adhesive that allows them to stick in the most difficult of circumstances—even upside down on wet glass. Geckos have micro and nano-scale hairs on their toes. Each main hair is a thousand times thinner than a human hair. On the end of each hair are more hairs—each a thousand times smaller again. Molecular attraction then occurs by way of Van der Waals forces that allow the gecko to cling on. Synthetic versions of these hairs have now been developed. Applications include incorporating the new adhesive material into lightweight rescue robots that can readily climb walls in search and rescue missions. NASA has plans to use the material on space flights as early as next year. The theory is that by being able to have astronauts stick to the exterior of a space station or the space shuttle, then spacewalks would become far less dangerous.

Nano Chemistry and Nano-velcro: Describing the Relationship

Researchers in the US are working on a Nano-velcro that could have a holding power tens of times greater than epoxy adhesives. The researchers at Michigan State University are working on the reusable material made from carbon nanotubes. Unlike conventional velcro that uses loops and hooks, nano-velcro only has hooks. It is believed that the material could become a strong, self-repairing, heat resistant micro-fastening system to hold together components in ultra-small robots and electrical devices.

Assuming that various elementary molecular parts fabrication and assembly operations can be performed for certain limited classes of structures, perhaps as described, then what structures would be interesting and useful to build? This Section briefly describes a small set of possibly useful molecular mechanical components and then concludes with a discussion of the concept of the molecular assembler.

In order to lay a foundation for molecular manufacturing, it is necessary to create and to analyze possible designs for nanoscale mechanical components that could, in principle, be manufactured. Because these components could not yet be built in 1998, such designs could not be subjected to rigorous experimental testing and validation. Designers were forced instead to rely upon ab initio structural analysis and computer studies including molecular dynamics simulations. Noted Drexler: "Our ability to model molecular machines (systems and devices) of specific kinds, designed in part for ease of modeling, has far outrun our ability to make them. Design calculations and computational experiments enable the theoretical studies of these devices, independent of the technologies needed to implement them."

In nanoscale design, building materials do not change continuously as they are cut and shaped, but rather must be treated as being formed from discrete atoms. A nanoscale component is a supermolecule, not a finely divided solid. Any stray atoms or molecules within such a structure may act as dirt that can clog and disable the device, and the scaling of vibrations, electrical forces, thermal expansion, magnetic interaction and surface tension with size lead to dramatically different phenomena as system size shrinks from the macroscale to the nanoscale.

Molecular bearings are perhaps the most convenient class of components to design because their structure and operation is fairly straightforward. One of the simplest examples is Drexler's overlap-repulsion bearing design, shown with end views and exploded views using both ball-and-stick and space-filling representations. This bearing has exactly 206 atoms including carbon, silicon, oxygen and hydrogen, and is comprised of a small shaft that rotates within a ring sleeve measuring 2.2 nm in diameter. The atoms of the shaft are arranged in a 6-fold symmetry, while the ring has 14-fold symmetry, a combination that provides low energy barriers to shaft rotation. An exploded view of a 2808-atom strained-shell sleeve bearing designed by Drexler and Merkle using molecular mechanics force fields to ensure that bond lengths, bond angles, van der Waals distances, and strain energies are reasonable. This 4.8-nm diameter bearing features an interlocking-groove interface which derives from a modified diamond surface. Ridges on the shaft interlock with ridges on the sleeve, making a very stiff structure. Attempts to bob the shaft up or down, or rock it from side to side, or displace it in any direction (except axial rotation, wherein displacement is-extremely smooth) encounter a very strong resistance. Whether these bearings would have to be assembled in unitary fashion, or instead could be assembled by inserting one part into the other without damaging either part, had not been extensively studied or modeled by 1998.

Molecular gears are another convenient component system for molecular manufacturing design-ahead. For example, Drexler and Merkle designed a 3557-atom planetary gear, shown in side, end, and exploded views. The entire assembly has twelve moving parts and is 4.3 nm in diameter and 4.4 nm in length, with a molecular weight of 51, 009.844 daltons and a molecular volume of 33.458 nm^3. An animation of the computer simulation shows the central shaft rotating rapidly and the peripheral output shaft rotating slowly.

The small planetary gears rotate around the central shaft, and they are surrounded by a ring gear that holds the planets in place and ensures that all of the components move in the proper fashion. The ring gear is a strained silicon shell with sulfur atom termination; the sun gear is a structure related to an oxygen-terminated diamond surface; the planet gears resemble multiple hexasterane structures with oxygen rather than CH_2 bridges between the parallel rings; and the planet carrier is adapted from a Lomer dislocation array created by R. Merkle and L. Balasubramaniam, and linked to the planet gears using C-C bonded bearings. View (c) retains the elastic deformations that are hidden in (a) — the gears are bowed. In the macroscale world, planetary gears are used in automobiles and other machines where it is necessary to transform the speeds of rotating shafts.

W. Goddard and colleagues at CalTech performed a rotational impulse dynamics study of this "first-generation" planetary gear. At the normal operational rotation rates for which this component was designed (e.g., <1 GHz for <10 m/sec interfacial velocities), the gear worked as intended and did not overheat. Started from room temperature, the gear took a few cycles to engage, then rotated thermally stably at ~400 K. However, when the gear was driven to ~100 GHz, significant instabilities appeared although the device still did not self-destruct. One run at ~80 GHz showed excess kinetic energy causing gear temperature to oscillate up to 450 K above baseline. One animation of the simulation shows that the ring gear wiggles violently because it is rather thin. In an actual nanorobot incorporating numerous mechanical components, the ring gear would be part of a larger wall that would hold it solidly in place and would eliminate these convulsive motions which, in any case, are seen in the simulation only at unrealistically high operating frequencies.

Drexler and Merkle later proposed a "second-generation" planetary gear design with 4235 atoms, a molecular weight of 72, 491.947 daltons and a molecular volume of 47.586 nm^3. This new version was indeed more stable but still had too much slip at the highest frequencies. Commenting on the ongoing design effort, Goddard suggested that an optimal configuration could have the functionality of a planetary gear but might have an appearance completely different from the macroscopic system, and offered an example: "Because a gear tooth in the xy plane cannot be atomically smooth in the z-direction, we may develop a Veè design so that the Vee shape of the gear tooth in the z-direction nestles within a Vee notch in the race to retain stability in the z-direction as the teeth contact in the xy plane. This design would make no sense for a macroscopic gear system since the gear could never be placed inside the race. However, for a molecular system one could imagine that the gear is constructed and that the race is constructed all except for a last joining unit. The parts could be assembled and then the final connections on the face made to complete the design" analogous to the ZARBI system of Rebek.

Another class of nanodevice that has been designed is a gas-powered molecular motor or pump. The pump and chamber wall segment contain 6165 atoms with a molecular weight of 88, 190.813 daltons and a molecular volume of 63.984 nm^3. The device can serve either as a pump for neon gas atoms or (if run backwards) as a motor that can convert neon gas pressure into rotary power. The helical rotor has a grooved cylindrical bearing surface at each end, supporting a screw-threaded cylindrical segment in the middle. In operation, rotation of the shaft moves a helical groove past longitudinal grooves inside the pump housing. There is room enough for small gas molecules only where facing grooves cross, and these crossing points move from one side to the other as the shaft turns, moving the neon atoms along. Goddard reported that preliminary molecular dynamics simulations of the device showed that it could indeed function as a pump, although "structural deformations of the rotor can cause instabilities at low and high rotational frequencies. The forced translations show that at very low perpendicular forces due to pump action, the total energy rises significantly and again the structure deforms." Merkle acknowledged that the pump moves neon atoms at an energy cost of 185 Kcal/mole-Angstrom (12, 900 zJ/atom-nm), which is not very energy-efficient. Further refinement of this crude design is clearly warranted.

Conveyor systems would also be useful. Employing a primitive molecular CAD software package called Crystal Sketchpad, G. Leach created a design for a 5-nanopart, ~2500-atom conveyor belt system comprised of two rollers, two axles, and a belt consisting of a strained thin-walled diamond sheet. The design has not been subjected to further computational analysis, either to minimize rotational energy barriers or to optimize rotational dynamics or operational stability.

Drexler and Merkle have also produced a preliminary design for a 2596-atom fine-motion controller. A general-purpose molecular assembler arm must be able to move its "hand" by many atomic diameters, position it with fractional-atomic-diameter accuracy, and then execute finely-controlled motions, perhaps to transfer one or a few atoms in a guided chemical reaction. Human arms use large muscles and joints for large motions and more finely-controlled finger motions for precision; the device presented here can execute precise finger-like motions over several atomic diameters with associated 90-degree rotations. The core of the device consists of a shaft linking two hexagonal endplates, sandwiching a stack of eight rings, making a modified Stewart platform. In a complete system, each ring would be rotated by a lever driven by a cam mechanism. Each ring supports a strut linked to a central platform (here shown raised, displaced, and twisted). Rotating a ring moves a strut; moving a strut moves the platform; positioning all eight rings (over)determines a platform position in x, y, z, roll, pitch, and yaw. (If the struts were rigid, six would do the job; here, two struts have been added to increase stiffness.) Notes Drexler: "The chief design problem is to enable an adequate range of motion without mechanical interference or unacceptable bond strains, and within the size constraints set by available modeling tools and patience."

Almost all current design research in molecular nanotechnology is restricted to computer simulation, which allows the design and testing of large structures or complete nanomachines, and the compilation of growing libraries of molecular designs. The work is relatively inexpensive and does not require the support of a large team. Of course, calculations of many-body systems are notoriously difficult, with many computer packages making a number of simplifying assumptions — e.g., nuclei as point masses, electrons treated as a continuous charge distribution, and 3-D potential energy functions derived semi-empirically from experimental data and treated as a classical field despite their true quantum mechanical character (for ease of computation). Notwithstanding these shortcuts, Tuzun and colleagues claim that classical simulations may overestimate the rate of energy transfer between vibrational modes even at low energies, in which case "current designs for various nanocomponents [would] actually perform better and be more stable than recent molecular dynamics simulations suggest."

Goddard notes that future nanosystem simulations may require 1-100 million atoms to be considered explicitly, demanding major improvements in molecular dynamics methodologies. By 1998, new algorithms for parallel processing of massive molecular dynamics simulations were being developed, producing methods and optimised parallelised computer programmes efficient for high capacity molecular dynamics simulations of 10, 000-1, 000, 000 atoms for finite molecular structures.

Ultimately, Computer-Aided (Molecular) Design (CAD) systems will be needed to efficiently design and analyze molecular components and their higher-order assemblies. CAD systems are commonplace in macroscale engineering and architecture, as are Computer-Aided Manufacturing (CAM) systems in macroscale manufacturing. In the molecular realm, Computer Aided Synthesis Design (CASD) has been avidly pursued by computational chemists since the 1980s, and by 1998 many university and commercial molecular modeling software packages were already well-known, including Alchemy (Tripos Inc.; http://www.tripos.com/), Cerius2 (Molecular Simulations Inc.; http://www.msi.com/), Chem3D (CambridgeSoft; http://www.camsoft.com/), Conformer (Princeton Simulations; http://www.conformer.com/), Gaussian94 (Gaussian Inc.; http://www.gaussian.com/), Hyperchem (Hypercube Inc.; http://www.hyper.com/), Molecular Operating Environment (Chemical Computing Group Inc.; http://www.chemcomp.com/), MOPAC97 (Fujitsu Ltd.; http://www.winmopac.com/), RealMol/CAVE/NAMD (Fraunhofer Institute for Computer Graphics; www.igd.fhg.de/www/igda4/research/chemie), Sculpt (Interactive Simulations Inc.; www.intsim.com/products/index.htm), and Spartan (Wavefunction Inc.; http://www.wavefun.com/).

In 1998, the capacity of such systems to design and manipulate large nanoscale mechanical components was extremely limited. A few very primitive molecular nanotechnology design packages had been attempted, including Crystal Sketchpad, DiamondCAD (www.zyvex.com/diamond.html), Molecular Assembly Sequence Software (www.carol.com/mass.shtml), Molecular Modelling Toolkit (starship.python.net/crew/hinsen/mmtk.html), and NanoCAD (world.std.com/~wware/ncad.html). The possible design of molecules for specific purposes using genetic software techniques had also been investigated.

Drexler has proposed the molecular assembler, a device resembling an industrial robot which would be capable of holding and positioning reactive moieties in order to control the precise location at which chemical reactions take place. This general approach would allow the construction of large atomically precise objects by a sequence of precisely controlled chemical reactions. Much like the ribosome in biology, an assembler would build various classes of useful molecular structures following a sequence of instructions. During this process, the assembler would provide three-dimensional positional and full orientational control over the molecular component (analogous to the individual amino acid in the ribosome model) that is being added to a growing complex molecular structure (analogous to the growing polypeptide in the ribosomal model). In one approach, a molecular assembler may be capable of forming any one of several different kinds of chemical bonds (e.g., by changing tool tips), not just a single kind such as the peptide bond that the ribosome makes. In bonding atoms or molecules to one another, the assembler would provide any needed energy (especially if the reaction happens not to be energetically favoured) through physical force, thus performing mechanosynthesis (as opposed to the traditional means of chemical synthesis in solution). In another approach, a molecular assembler might be capable only of noncovalent assembly operations, wherein nanoparts are fabricated by other means and then presented to the assembler, which assembles the nanoparts into working nanomachines.

The first simple molecular assembler will almost certainly be a macroscale device, perhaps a modified SPM system as was being pursued by Zyvex in 1998. Multiple SPM heads could be equipped with a small number of nanoscale tool tips. In one scenario, nanoparts fabricated using bulk chemistry techniques would be inspected and selected by the SPM, then assembled one by one into working nanomachines (e.g., the desired useful nanoscale products). Such assembly operations will be very slow, because the placement of each new component may require simultaneous rotations and translations of large macroscale SPM components. Assembly time scales roughly linearly with assembler size because smaller assembler components moving at a given velocity need to travel less distance to accomplish a given physical operation, hence consuming less time and energy per physical operation. An important early developmental goal thus will be to design and fabricate nanoscale molecular assemblers.

At its most basic level, the simplest possible nanoscale molecular assembler may be comprised of one or more nanoscale manipulators. For example, the diamondoid telescoping manipulator arm described in has about 4 million atoms excluding the base and control and power structures; doubling the size to account for support structures gives a molecular weight per arm of ~100 megadaltons and a total molecular volume of ~140, 000 nm^3. Designed for high strength and stiffness, this robot arm should be able to hold a molecular fragment stiffly enough to make it react with a chosen end of a carbon-carbon double bond with an error rate of only 10^{-15}. The robot arm could guide chemical reactions with high reliability at room temperature in vacuo, with little need for sensing the positions of the molecules with which it is working. Alternatively, with appropriate tool tips, such an arm could grasp and manipulate individual prefabricated nanoparts in solution phase. Assembly of nanoparts without fabrication may require only a very small tool set.

The manipulator arm must be driven by a detailed sequence of control signals, just as the ribosome needs mRNA to guide its actions. However, such detailed control signals can be provided by external acoustic, electrical, or chemical signals that are received by the robot arm via an onboard sensor or power transducer, using a simple "broadcast architecture", a technique which can also be used to import power. Such transducers may be extremely small, on the order of $(\sim 10\ nm)^3$ each. Interestingly, the biological cell may be regarded as an example of a broadcast architecture: the nucleus, located external to the cytoplasm, broadcasts mRNA chemical signals to millions of spatially diverse cytoplasmic ribosomes, thereby remotely controlling the construction of cellular proteins.

Thus a two-armed mechanical molecular assembler that receives power and instructions from some external agency may have a total molecular volume of ~300, 000 nm^3 (a cube ~67 nm on an edge), containing ~16 million atoms with a molecular weight of ~200 megadaltons — very roughly the mass and scale of a medium-size virus particle, such as an adenovirus. For comparison, the average enzyme (a biochemical crimping tool) weighs about ~0.1 megadalton, while a typical ribosome (a primitive "protein assembler") weighs ~4.2 megadaltons. A simple mechanical assembler in the 10-100 megadalton range cannot be ruled out. However, such a device might have a very small set of tool tips and an extremely limited manufacturing repertoire.

Multiple nanoscale assemblers each capable of independent simultaneous actuation may require control signals more conveniently provided by an onboard nanocomputer. This programmable nanocomputer must be able to accept stored instructions which are sequentially executed to direct the manipulator arm to place the correct moiety or nanopart in the desired position and orientation, thus giving precise control over the timing and locations of chemical reactions or assembly operations. The mechanical nanocomputer analysed by Drexler requires ~16 nm^3 per logic gate and ~40 nm^3 per data register. If such components could be used to construct the equivalent of the most primitive 4-bit Intel 4004 microprocessor, then the nanocomputer could process ~10^5 bits/sec (~25, 000 ops/sec) at a ~1 KHz clock speed within a mechanism volume of ~36, 000 nm^3, again neglecting power supply, I/O linkages, and the like. An additional ~160, 000 nm^3 of rod logic registers adds ~1 kilobyte of onboard RAM memory or ~40 kilobytes of internal tape memory (conservatively assuming a tape storage density comparable to linear DNA). (A memory tape containing all bits needed for complete self-description may be considerably longer, even allowing for substantial data compression.) Doubling the total volume to account for support structure and other overhead gives a minimum mechanical nanocomputer molecular volume of ~400, 000 nm^3, roughly 70 million atoms with a molecular weight of ~800 megadaltons. Thus the smallest nanocomputer-driven nanoscale molecular assembler with two manipulator arms may have a total molecular volume of ~700, 000 nm^3 (a cube ~88 nm on an edge), containing ~86 million atoms with a molecular weight of ~1 gigadalton.

By 1998, only a small amount of research targeted at actual assembler design had begun. Following Drexler's original discussions, during 1991-1998 R. Merkle authored or coauthored a continuing series of papers discussing various operational aspects and specific components of assembler design, including mechanosynthetic positional control, general design considerations for assemblers, the broadcast architecture, convergent assembly, binding sites, positioning devices, mechanosynthetic path sets, designs for a neon pump and a fine motion controller, and possible assembler casings. J.S. Hall has considered high-level designs for a nanoscale parts-fabrication and parts-assembly nanorobot. Zyvex, founded in 1996 by James von Ehr, has set itself the task of building the first programmable nanoassembler in a 5-10 year time frame (e.g., by ~2006). W. Goddard and colleagues have proposed a series of molecular dynamics simulations of simple assemblers, although by 1998 these studies evidently had not yet begun. According to Goddard's original proposal:

"Ultimately we need a programmable synthetic system to make a real device. Even though we may not have tools for all the chemical steps and may not have designs for all the pumps, engines, and transmissions needed, we propose to study the dynamics of simplified prototype assemblers. In these studies we anticipate having (1) a reservoir or supplies area for providing the various building units (atoms and fragments) required, (2) a work area in which we construct the nanomachine device (initially we will consider assembling the structure on top of a diamond surface), and (3) a molecular scale nanohand which will extract the atoms from (1) and carry them to (2). We will then use extensions of our massive molecular dynamics programme to operate the system: moving the tip from reservoir to work area, moving it to contact the appropriate surface site, moving it to regenerate the active tip, and then moving it back to add new atoms and molecules. This will include proper temperature effects, molecular vibrations, energy release upon the various chemical steps, etc. The ground rules here are that a realistic force field be used and that all pieces be treated at the atomic level (but some might be semi-rigid). This will use the force field developed for nanosynthesis. The purpose of these simulations is to examine issues of vibration caused by chemical forces as the tool picks up and delivers atoms to the growing surface. Also we want to consider the effect of energy release in the chemical steps on the thermal fluctuations in these systems (which may cause displacements and vibrations)."

Building mega-atom or giga-atom nanoproducts one at a time would be incredibly expensive and time consuming. For example, imagine that we wish to construct a simple medical nanorobot such as the 1-micron spherical respirocyte described in which consists of ~18 billion atoms (dry structure). A factory employing a coordinated team of 100 macroscale SPM assemblers, each able to place ~1 atom/sec-SPM on a convergently-assembled workpiece achieves a pitiful manufacturing throughput rate of ~2 respirocytes per decade. Nanoscale assemblers with appendages ~10^6 times smaller than macroscale SPM assemblers might plausibly achieve net assembly rates of ~10^6 atoms/sec-nanoassembler, but even a production line employing ~300 such nanoassemblers (which the abovementioned 100-SPM factory team could build in ~1 year, assuming ~10^7 atoms/nanoassembler) can only manufacture ~1 respirocyte nanorobot per minute. At that rate, it would take ~2 million years to build the first ~1 cm^3 therapeutic dosage containing ~10^{12} respirocyte nanorobots.

The necessary solution to this mass-production bottleneck is to employ any of several massively parallel approaches to manufacturing, including such techniques as self-assembly, convergent assembly, or, most usefully, self-replication. The basic advantage of self-replication is readily illustrated. Consider a single ~10^8 atom "seed" nanoassembler (having an onboard nanocomputer) that has been painstakingly built in ~10^6 sec using the abovementioned 100-SPM macroscale assembler team, working at ~1 atom/sec-SPM. The seed nanoassembler is first programmed to build a copy of itself, which it accomplishes (working at ~10^6 atoms/sec-nanoassembler) in ~100 sec. These two nanoassemblers then each build a copy of themselves in another ~100 sec; now there are four nanoassemblers. After ~48 generations, requiring a total of ~80 minutes to complete, there are ~3 x 10^{14} nanoassemblers. These nanoassemblers are then reprogrammed for the manufacture of 18-billion-atom respirocytes and are fed the appropriate, presumably different, feedstock. This vastly expanded nanoassembler manufacturing system can now produce ~10^{12} respirocytes (~1 cm^3 therapeutic dose) per minute.

The design of machines able to make copies of themselves was first described by von Neumann. Many variations on this theme have been reviewed by Freitas and Gilbreath and Sipper, and design for self-replication in the context of nanoscale assemblers has been considered by Drexler, Merkle, and Hall. As Drexler notes: "It may seem somehow paradoxical that a machine can contain all the instructions needed to make a copy of itself, including those selfsame complex instructions, but this is easily resolved. In the simplest approach, the machine reads the instructions twice: first as commands to be obeyed, and then as data to be copied. Adding more data does not increase the complexity of the data-copying process, hence the set of instructions can be made as complex as is necessary to specify the rest of the system. By the same token, the instructions transmitted in a replication cycle can specify the construction of an indefinitely large number of other artifacts."

The estimated information content of self-replicating systems — the length of the instruction tape — can be surprisingly small. Von Neumann's original analysis concluded that perhaps twelve different kinds of units of unknown complexity could be required as building materials, and Haldane inferred that as many as $\sim 10^5$ individual parts might be needed to make a replicator. This inference was refuted just three years later with the arrival of the first in a stream of very simple but ingenious designs for mechanical self-replicating machines that were assembled and operated in the late 1950s. In the first example, a pair of joined ratchet-like blocks, when placed in a "sea" of left and right blocks and then physically agitated, replicated itself from this well-ordered input substrate, making more block-pairs until all the single blocks were used up, a process similar to chemical autocatalysis. More complex congeries of blocks, including clever four-block, eight-block, and 12-block replicators that could replicate themselves in a sea of blocks were presented by Penrose. Jacobson demonstrated a 3-unit replicator consisting of toy train engines circulating on HO model railroad tracks, and Morowitz designed a simple two-unit device with one unit comprised of about a dozen components including switches, batteries and electromagnets, that could assemble copies of itself from parts floating on the water surface of a bathtub. In 1998, Lohn and colleagues gave two designs for simple self-replicating systems that could be constructed out of wood, batteries and electromagnets, with explicit analogies to nanoscale fabrication. Several important conclusions may be drawn from these examples:

First, replication is fundamentally so simple a task that machines capable of displaying this behaviour predate most of the modern electronic computer era. Second, assembly is an inherently simpler operation than fabrication. A complex part may embody hundreds or thousands of prior fabrication and assembly operations, yet may be installed within an assemblage in a single step. Hence nanopart assembly may be an easier candidate for early implementation in first-generation self-replicating molecular manufacturing systems than is molecular fabrication. Third, and most important, the simplest nanoreplicator may require the ability to assemble, but not to (atomically) fabricate, in order to replicate itself. A nanomachine capable of assembly alone can replicate itself only from a very limited set of well-ordered input materials of relatively high complexity, but such a machine can be extremely simple both in structure and function. Certainly a nanomachine capable of both assembly and fabrication can replicate itself from a more diverse and disordered set of more elementary input materials but only at the expense of far greater internal structural and functional complexity. At least two distinct models of replication have been identified in replicating systems design. The first model may be called the "unit replication" or organismic model, in which the replicator is an independent unit which employs the surrounding substrate to directly produce an identical copy of itself; both the original and the copy remain fertile and may replicate again, thus exponentiating their numbers. The second model may be called the "unit growth" or factory model, in which a population of specialist devices, each one individually incapable of self-replication, can collectively fabricate and assemble all necessary components comprising all specialist devices within the system, hence the factory is capable of expanding its size (or of manufacturing duplicate factory systems) indefinitely in an appropriate environment. The factory model is sometimes called "bootstrapping, " which is any production of more productive capacity, from less. T. McKendree [personal communication, 1999] likens the situation to an ant colony: "There necessarily is the line of reproducing queens, making a sufficient number of ants feasible; [however,] most of the queen's products are 'workers' that perform useful functions, but themselves are incapable of reproduction."

It cannot be emphasised too strongly that mechanical medical nanodevices will not self-replicate inside the human body, nor will they have any need for self-replication themselves. Machines that perform medical tasks are fundamentally different from machines that manufacture other machines. As R. Merkle explains: "While self-replicating systems are the key to low cost [manufacture], there is no need (and little desire) to have such systems function in the outside world. Instead, in an artificial and controlled environment they can manufacture simpler and more rugged systems that can then be transferred to their final destination. Medical devices designed to operate in the human body don't have to self-replicate: we can manufacture them in a controlled environment and then inject them into the patient as needed. The resulting medical device will be simpler, smaller, more efficient and more precisely designed for the task at hand than a device designed to perform the same function and self-replicate. This conclusion should hold generally: optimize [product] device design for the desired function, manufacture the [product] device in an environment optimised for manufacturing, then transport the [product] device from the manufacturing environment to the environment for which it was designed. A single device able to do everything would be harder to design and less efficient."

In an effort to stimulate scientific and engineering interest in constructing the first nanoassembler, in November 1995 the Foresight Institute created the $250, 000 Feynman Grand Prize, with funding contributed by Zyvex founder James von Ehr and St. Louis venture capitalist Marc Arnold. The prize will be awarded to the individual or group that first achieves both of two significant nanotechnology breakthroughs — first, the design and construction of a functional nanometer-scale robotic arm, and second, the design and construction of a functional 8-bit adder computing nanodevice. According to Grand Prize rules, the robotic nanomanipulator must fit entirely inside a 100-nm cube, carry out actions directed by input signals of specified types, be able to move to a directed sequence of positions anywhere within a 50-nm cube, complete all directed actions with a positioning accuracy of 0.1 nanometer or better, and perform at least 1, 000 accurate, nanometer-scale positioning motions per second for at least 60 consecutive seconds. The adder must fit entirely within a 50-nm cube, and must be capable of adding accurately any pair of 8-bit binary numbers, discarding overflow, accepting input signals of specified types, and producing its output as a pattern of raised nanometer-scale bumps on anatomically precise and level surface. (J.S. Hall notes that a conventional 8-bit adder may be constructed using a total of 94 AND, OR, and NOT gates; using XOR gates, the total may be reduced to 37 gates.) Both devices may accept inputs from acoustic, electrical, optical, diffusive chemical, or mechanical means, although any mechanical driving mechanism used for input must be limited to a single linkage that either slides or rotates on a single axis. To demonstrate the capacity for mass production, at least 32 copies of each device must be provided for analysis and destructive testing by judges.

Once nanoassemblers are available, the design of far more complex nanomachines each containing tens or hundreds of billions of precisely arranged atoms will require new molecular CAD tools and techniques, including automated hierarchical design decomposition of objects (that must be built up from an array of nanoparts) and a shape description language. Nanosystem design compilers, conceptually similar to silicon compilers that generate a complex pattern of transistors and conductors from an abstract specification of the properties of a digital circuit, will also be required for efficient nanorobot design. Given a fully specified design that meets all constraints necessary to physically permit assembly, all operations of the assembly process must be specified using an assembly process compiler. Explains Drexler: "The design of structures and assembly procedures by hierarchical decomposition directly generates a tree of assembly steps. If this tree has been chosen to generate parts of the appropriate sizes and numbers, then it can be mapped onto a manufacturing layout. The vast number of manipulator motions to be specified at the finest, earliest, and most-dispersed levels of the assembly process can be planned in parallel by identical software systems running almost independently on separate processors. The result of this parallel assembly process compilation is a set of instructions that, when executed by the transportation system and manipulator controllers of a manufacturing mechanism, will result in the assembly of an object corresponding to the initial design." Scale-up to larger systems, including the control, coordination, and programming of large aggregates of cooperating nanorobots.

Since molecular manufacturing systems can be used to make more molecular manufacturing systems, it is believed that the capital costs of production can be quite low. Drexler claims that an analysis of inputs, outputs, and productivity suggests the total cost of production can be in the range familiar in agriculture and in the production of industrial chemicals — on the order of tens of cents per pound.

Merkle has made similar assertions, for example: "Eventually (after amortisation of possibly quite high development costs), the price of assemblers (and of the objects they build) should be no higher than the price of other complex structures made by self-replicating systems. Potatoes — which have a staggering design complexity involving tens of thousands of different genes and different proteins directed by many megabits of genetic information — cost well under a dollar per pound." Such analogies from conventional agriculture or industrial chemistry are not strictly applicable to the medical products and pharmaceutical areas, where structural precision, product repeatability, and intensive quality control are of supreme importance, and where most products are subject to rigorous government regulation — unlike potatoes, which vary widely in size, shape, and molecular constitution, and generally do not require stringent product liability insurance, multiphase clinical trials, or FDA approval.

Still, it is likely that molecular manufacturing will eventually allow reasonably inexpensive production of large batches of designed medical nanorobots. As a very crude analogy, even in 1995 a typical mail-order bioscience vendor could manufacture 15 micromoles (~9 x 10^{18} product items) of customised oligonucleotides for $18/base up to at least 110 bases with a convenient 48-hour turnaround on orders; other online vendors offer similar or slightly higher prices up to 200-mer sequences with additional charges for necessary purifications. If a billion-base structure could be induced to self-assemble from 10 million different sets of 100-mer custom-built oligonucleotides (total ~20 billion atoms), forming a desired ~20-billion-atom nanomachine, then, optimistically assuming a ~100% yield, the cost would be ~$2 x 10^9 per nanomachine or ~$2000 per ~1 cm^3 of product volume which would include 10^{12} finished nanomachines. In 1998, many specialty biotechnology-related drug treatments were comparably expensive. For example, treatments for multiple sclerosis using two closely related ~22.5 kilodalton interferons were administered in dosages of ~1 milligram/week at a treatment cost of ~$10, 000/yr, which is equivalent to a treatment cost of ~$2 x 10^9 (2 "nanodollars") per giga-atom. The famous HeLa (Henrietta Lacks) cell line is an example of a once tiny population of unique cancerous cells that were purposely replicated in vitro (~24-hour replication time) over many decades and are now widely available worldwide at very low cost (e.g., $216 per 1-$cm^3$ ampule of the ATCC CCL-2 HeLa cell line). Swallowable therapeutic pills containing bacteria (i.e., natural biological nanomachines) are already widely available over the counter for gastrointestinal refloration, as for example Salivarex which "contains a minimum of ~2.9 billion beneficial bacteria per capsule", and Alkadophilus which "contains 1.5 billion organisms per capsule", both at a 1998 price of ~$0.2 x 10^9 per bacterium.

Another useful analogic approach to cost estimation is suggested by the plunging price of computer power during the 20th century. The cost, in constant 1998 U.S.

dollars, of a computer processing capability equivalent to ~1 TeraFLOP (10^{12} floating-point operations per second, ~10^{14} bits/sec assuming 64-bit words). The plotted data is from a compilation of 67 historical computers by Moravec through 1989, with an additional 27 data points for other computers spanning 1973-1999 added by the author. The chart shows that the cost of 1 TFLOP has fallen from \$2 x 10^{20} in 1908 (the mechanical Hollerith Tabulator) to just \$6 x 10^{6} in 1998 (the SGI/Cray T3E-1200E), a 90-year price decline of almost 14 orders of magnitude — on average, a halving every ~2 years, a century-long example of Moore's Law. This trend may give us some confidence that even if the first individual 1 micron^3 nanorobot costs ~\$100, 000 to assemble, eventually the unit price may fall at least 14 orders of magnitude to below ~\$$10^9$ per nanorobot or <\$1000 per ~1 cm^3 dosage (10^{12} nanorobots). Recycling or remanufacturing each nanomachine ~1000 times before final discard would then imply a net ~\$1/cm^3 treatment cost. Of course, in 1998 all such estimates were mere guesswork, but the ultimate expectation of relatively low-cost nanomedical treatments was certainly rational, if not provable.

This Chapter has demonstrated that there are many different pathways leading toward molecular nanotechnology, each one of which provides incremental benefits to motivate further travel down the paths. The likelihood of all the paths being blocked is low, even if we have little confidence that we can predict which approach will succeed first. Progress along the many pathways will provide precursor products, including some early medical applications, but the ultimate achievement of molecular manufacturing will finally make nanomedicine feasible.

As Drexler concluded Nanosystems: "Each step along [these pathways] will present great practical challenges, but each step will also bring valuable new capabilities. The long-term rewards, measured in terms of scientific and technological capabilities, appear large." This author agrees. For the remainder of the present work, we shall assume that a molecular manufacturing technology will someday exist that can economically manufacture macroscopic batches of microscopic machines comprised of atomically-precise nanoscale components.

REFERENCES

Alivisatos, A.P. 1996. Perspectives on the physical chemistry of semiconductor nanocrystals. *Journal of Physical Chemistry* 100: 13226-13239.

Alves, H., M. Ferreira, U. Koster, and B. Muller. 1996. *Materials Science Forum* 225–227: 769.

Andrea G. Bodnar, Michel Ouellette, *et al.*, "Extension of life-span by introduction of telomerase into normal human cells," *Science* 279(16 January 1998): 349-352.

Asada, M., Y. Mayamoto, and Y. Suematsu. 1986. *IEEE Journ. Quantum Electronics* QE- 22(9): 1915-1921.

B. W. Matthews, H. Nicholson, and W. J. Becktel, "Enhanced protein thermostability from site-directed mutations that decrease the entropy of unfolding." *Proc. Nat. Acad. Sci., 84:* 6663-6667 (1987), and included references.

B. Yurke, A.J. Turberfield, A.P. Mills, Jr., F.C. Simmel, J.L. Neumann, "A DNA-fuelled molecular machine made of DNA," *Nature* 406(10 August 2000): 605-608.

Ball, P.: 2002, 'Natural Strategies for the molecular engineer', *Nanotechnology*, 13, 15-28.

Ball, P.: 2003, 'Nanotechnology in the firing line', *Nanotechweb.org,* 23 December [http: // www.nanotechweb.org/articles/society/2/12/1/1].

Berger, R., C. Gerber, H.P. Lang, and J.K. Gimzewski. 1996. Micromechanics: a toolbox for femtoscale science: "Towards a laboratory on a tip." *Microelectronic Engineering*, 35: 373-9. (International Conference on Micro- and Nanofabrication, Glasgow, U.K., 22- 25 Sept. 1996).

Berkowitz, A.E., J.R. Mitchell, M.J. Carey, A.P. Young, S. Zhang, F.E. Spada, F.T. Parker, A. Hutten, and G. Thomas. 1992. *Phys. Rev. Lett.* 68: 3745.

Blackman, M. 1994. *An evaluation of the Link Nanotechnology Program and the National Initiative on Nanotechnology*. Cambridge.

Bohn, R., T. Haubold, R Birringer, and H. Gleiter. 1991. *Scripta Metall. Mater.* 25: 811.

Bowes, C.L., A. Malek, and G.A. Ozin. 1996. *Chem. Vap. Deposition* 2: 97.

Braun, P.V., P. Osenar, and S.I. Stupp. 1996. *Nature*. 368: 2. Brinker, C.J. 1996. *Curr. Opin. Solid State Mater. Sci.* 1: 798. Chen, C-Y., S.L. Burkett, H.-X. Li, and M.E. Davis. 1993. *Microporous Mater.* 2: 27.

Calcote, H.F., and D.G. Keil. 1997. Combustion synthesis of silicon carbide powder. In *Proc. of the Joint NSF-NIST Conf. on Nanoparticles*.

Canguilhem, G.: 1952, *Machine et organisme' in La connaissance de la vie*, Hachette, Paris [quoted from the fourth edition Vrin, Paris, 1971].

Carroll, J.S.: 2001, 'Social Science Research Methods for Assessing Societal Implications of Nanotechnology', in: M.C. Roco and W.S. Bainbridge (eds.), *Societal Implications of Nanoscience and Nanotechnology*, Dordrecht: Kluwer, pp. 188-192.

Carsley, J.E., R. Shaik, W.W. Milligan, and E.C. Aifantis. 1997. In *Chemistry and physics of nanostructures and related non-equilibrium materials*. ed. E. Ma, B. Fultz, R. Shull, J. Morral, and P. Nash. Warrendale, PA: TMS.

Chance, B., Mueller, P., DeVault, D. & Powers, L. (1980) *Phys. Today* 33 (10), 32-38.

Chang, K. "Smaller Computer Chips Built Using DNA as Template." New York Times, November 21, 2003.

Chantrell. 1994. *J. Appl. Phys.* 76: 6811. Erb, U., G. Palumbo, R. Zugic, and K.T. Aust. 1996. In *Processing and properties of nanocrystalline materials*, ed. C. Suryanarayana, J. Singh, and F.H. Froes. Warrendale, PA: TMS. Gertsman, V.Y., M. Hoffman, H. Gleiter and R. Birringer. 1994. *Acta Metall. Mater*. 42: 3539-3544.

Charles B. Musgrave, Jason K. Perry, Ralph C. Merkle, William A. Goddard III, "Theoretical studies of a hydrogen abstraction tool for nanotechnology," *Nanotechnology* 2(1991): 187-195; http: //www.zyvex.com/nanotech/Habs/Habs.html

Charles B. Musgrave, Jason K. Perry, Ralph C. Merkle, William A. Goddard III, "Theoretical studies of a hydrogen abstraction tool for nanotechnology," *Nanotechnology* 2(1991): 187-195; http: //www.zyvex.com/nanotech/Habs/Habs.html

Chen, J. and N.C. Seeman. 1991. The synthesis from DNA of a molecule with the connectivity of a cube. *Nature* 350: 631-633.

Chianelli, R.R. 1998. Synthesis, fundamental properties and applications of nanocrystals, sheets, and fullerenes based on layered transition metal chalcogenides. In *R&D status and trends,* ed. Siegel et al.

Chianelli, R.R. 1998. Synthesis, fundamental properties and applications of nanocrystals, sheets, and fullerenes based on layered transition metal chalcogenides. In *R&D status and trends,* ed. Siegel et al.

Chianelli, R.R., M. Daage, and M.J. Ledoux. 1994. *Advances in Catalysis* 40: 177.

Chidsey, E.E.D., and R.W. Murray. 1986. *Science*, 231: 25.

Chopra, N.G., R.J. Luyken, K. Cherrey, H. Crespi, M.L. Cohen, S.G. Louie, and A. Zettl. 1995. *Science* 269: 966.

Claeson, T., and Likharev, K.," Single Electronics," *Scientific American,* June 1992, pp. 80-85.

Clasen, R. 1990. *Int. Journal of Glass and Science Technology* 63: 291. Crandall, B.C. and J. Lewis, (eds.). 1992. *Nanotechnology: Research and perspectives*. Cambridge: MIT Press.

Coenen, C.: 2004, 'Nanofuturismus: Anmerkungen zu seiner Relevanz, Analyse und Bewertung', *Technikfolgenabschätzung ' Theorie und Praxis*, 13 (2), 78-85.

Crandall, B.C. and Lewis, J., *Nanotechnology: Research and Perspectives*, MIT Press, Cambridge, 1992.

Crow, M. & Sarewitz, D.: 2001, 'Nanotechnology and Societal Transformation', in: M.C. Roco and W.S. Bainbridge (eds.), *Societal Implications of Nanoscience and Nanotechnology*, Dordrecht: Kluwer, pp. 45-54.

Czekai, D., et al. 1994. *Use of smaller milling media to prepare nanoparticulate dispersions*. U.S. Patent application. Docket 69802(02) Filed 2/25/94.

D.B. Dusenbery, "Minimum size limit for useful locomotion by free-swimming microbes," *Proc. Natl. Acad. Sci. USA* 94(30 September 1997): 10949-10954.

D.W. Brenner, S.B. Sinnott, J.A. Harrison, O.A. Shenderova, "Simulated engineering of nanostructures," *Nanotechnology* 7(1996): 161-167; http: //www.zyvex.com/nanotech/nano4/brennerPaper.pdf

Dai, H., E.W. Wong, Y.Z. Lu, S. Fan, and C.M. Leiber. 1995. *Nature* 375: 769. deHeer, W.A., W.S. Bacsa, A. Chatelain, T. Gerfin, R. Humphrey-Baker, L. Forro, and D. Ugarte. 1995. *Science* 268: 845.

Decker, M.; Fiedeler, U.; Fleischer, Th.: 2004, 'Ich sehe was, was Du nicht siehst... zur Definition von Nanotechnologie', *Technikfolgenabschätzung ' Theorie und Praxis*, 13 (2), 10-16.

Dresselhaus, M.S., G. Dresselhaus, and P. Eklund. 1996. *Science of fullerenes and carbon nanotubes*. San Diego: Academic Press. Gleiter, H. 1989. *Prog. Mater. Sci.* 33: 223.

Drexler K.E.: 1995, 'Introduction to nanotechnology', in: Krummenacker, M. & Lewis, J. (eds.), *Prospects in Nanotechology. Proceedings of the 1st general conference on nanotechnology: developments, applications, and opportunities, November 11-14, 1992, Palo-Alto*, John Wiley & Sons, New York, pp. 1-20.

Drexler, E., Phoenix, C.: 2004, 'Self-Replication in Nanotechnology: Feasible, Potentially Safe, and Unnecessary' (unpublished paper, in preparation).

Drexler, K.E.: 2001, 'Machine-Phase nanotechnology', *Scientific American*, (Sept.), 66-67.

Dupuy, J.P.: 2004, 'Pour une évaluation normative du programme nanotechnologique', *Annales des Mines*, (February 2004), 27-32.

E. Francis J. Ring, Barbara Phillips, eds., *Recent Advances in Medical Thermology*, Plenum Press, New York, 1984.

E. Winfree, X. Yang and N.C. Seeman, Universal Computation via Self-assembly of DNA: Some Theory and Experiments, In: DNA Based Computers II, ed.by L.F. Landweber and E.B. Baum, Am. Math. Soc., Providence, pp. 191-213 (1998).

Einsiedel, E.F.; Goldenberg, L.: 2004, 'Dwarfing the Social? Nanotechnology Lessons from the Biotechnology Front', *Bulletin of Science, Technology & Society*, 24, 28-33.

Ellsberg, D. "Risk, Ambiguity and the Savage Axioms." Quarterly Journal of Economics 75 (1961): 643-669.

ETC: 2003b, 'No Small Matter II: The Case for a Global Moratorium ' Size Matters!' *Occasional Paper Series* 7(1) [www.etcgroup.org/documents/Occ.Paper_Nanosafety.pdf].

Etzkowitz, H., 2001, 'Nano-Science and Society: Finding a Social Basis for Science Policy', in: M.C. Roco and W.S. Bainbridge (eds.), *Societal Implications of Nanoscience and Nanotechnology*, Dordrecht: Kluwer, pp. 121-128.

European Comission: *European Workshop on Social and Economic Research on Nanotechnologies and Nanosciences, Brussels, 14-15 April 2004* [www.stage-research.net/STAGE/PAGES/Nano.html].

European Commission: 2004, *Converging Technologies: Shaping the Future of European Societies (Report of the High Level Expert Group "Foresighting the New Technology Wave")*, Brussels: European Commission Research [http: //europa.eu.int/comm/research/conferences/2004/ntw/index_en.html].

Evans, D.F.; Wennerstrom, H.: 1999, *The Colloidal Domain: Where Physics, Chemistry, Biology and Technology Meet*, John Wiley & Sons, New York.

F. Buot, "Mesoscopic Physics and Nanoelectronics: Nanoscience and Nanotechnology," *Physics Reports*, pp.73-174, 1993.

F. Capasso and S. Datta, "Quantum Electron Devices," *Physics Today*, pp.74-82, February 1990.

F. Liu, H. Wang and N.C. Seeman, Short Extensions to Sticky Ends for DNA Nanotechnology and DNA-Based Computation, Nanobiology 4, 257-262 (1999).

F. Mathieu, S. Liao, C. Mao, J. Kopatsch, T. Wang, and N.C. Seeman, Six-Helix Bundles Designed from DNA, NanoLetters 5, 661-665 (2005).

Faraday, M.: 1857, 'Experimental relations of gold (and other metals) to light', *Philosophical Transactions of the Royal Society London*, 147, 145-181.

Fedor N. Dzegilenko, Deepak Srivastava, Subhash Saini, "Simulations of carbon nanotube tip assisted mechano-chemical reactions on a diamond surface," *Nanotechnology* 9(December 1998): 325-330.

Feyerabend, P.: 1981, *Realism, Rationalism and Scientific Method* (Philosophical Papers, volume 1), Cambridge University Press, Cambridge.

Fielder, F.A.; Reynolds, G.H.: 1994, 'Legal Problems of Nanotechnology: An Overview', *Southern California Interdisciplinary Law Journal*, 3, 593-629.

Filler, M., C. Mui, G. Wang, C. Musgrave, and S. Bent, "Competition and Selectivity in the Reaction of Nitriles on Ge (100)-2x1," *Journal of the American Chemical Society* 125, 4928-4936 (2003).

Frazier, G., "An Ideology For Nanoelectronics," in *Concurrent Computations: Algorithms, Architecture, and Technology*, Plenum Press, New York, 1988.

Gallego-Juárez, J.A. (1989) Piezoelectric ceramics and ultrasonic transducers. J. Phys. E: Sci. Instrum. 22, 804-816

Glimell, H.: 2001, 'Challenging Limits' Excerpts from an Emerging Ethnography of Nano Physicists', in: H. Glimell & O. Johlin (eds.), *The Social Production of Technology: On the Everyday Life with Things*, Göteburg, SE: BAS Publisher, chapter 7, pp. 111-131 (reprinted in: H. Fogelberg & H. Glimell, *Bringing Visibility to the Invisible: Towards A Social Understanding of Nanotechnology*, Göteborg: Göteborg University, pp. 115-139.

Glimell, H.: 2001, 'Dynamics of the Emerging Field of Nanoscience', in: M.C. Roco and W.S. Bainbridge (eds.), *Societal Implications of Nanoscience and Nanotechnology*, Dordrecht: Kluwer, pp. 156-160.

Goldhaber-Gordon, D., Montemerlo, M.S., Love, J.C., Opiteck, G.J., and Ellenbogen, J.C., "Overview of Nanoelectronic Devices," submitted to the *Proceedings of the IEEE*, February 1997. For more information, please send e-mail to nanotech@mitre.org.

Gorman, M.E.; Groves, J.F. & Shrager, J.: 2004, 'Societal Dimensions of Nanotechnology as a Trading Zone: Results from a Pilot Project', in: D. Baird, A. Nordmann & J. Schummer (eds.), *Discovering the Nanoscale*, Amsterdam: IOS Press, pp. 63-73.

Grinbaum, A. & Dupuy, J.P.: 2004, 'Living with Uncertainty: Toward the Ongoing Normative Assessment of Nanotechnology', *Techne: Research in Philosophy and Technology*, 8(3) (forthcoming).

Grunwald, A.: 2004, 'Ethische Aspekte der Nanotechnologie. Eine Felderkundung', *Technikfolgenabschätzung ' Theorie und Praxis*, 13 (2), 71-78.

Guarnieri, F., M. Fliss, and C. Bancroft. 1996. Making DNA Add. *Science* 273: 220-223. Hanes, J., J.L. Cleland, and R. Langer. 1997.

Gund, P., Andose, J. D., Rhodes, J. B. & Smith, G. M. (1980) *Science* 208, 1425-1431.

Günther, B., A. Baalmann, and H. Weiss. 1990. *Mater. Res. Soc. Symp. Proc.* 195: 611-615.

Gupta, V.K. & Pangannaya, N.B.: 2000, 'Carbon nanotubes: bibliometric analysis of patents', *World Patent Information*, 22, 185-189.

Gutte, B., Dannigen, M. & Wittschieber, E. (1979) *Nature (London)* 281, 650-655.

Haruta, M. 1997. *Catalyst surveys of Japan* 1: 61 and references therein.

Hayles, N.K. (ed.): 2004, *Nanoculture: Implications of the New Technoscience*, Bristol, UK: Intellect Books.

Hessenbruch, A.: 2004, 'Nanotechnology and the Negotiation of Novelty', in: D. Baird, A. Nordmann & J. Schummer (eds.), *Discovering the Nanoscale*, Amsterdam: IOS Press, pp. 135-144.

Hietpas, P.B., S.D. Gilman, R.A. Lee, M.R. Wood, N. Winograd, and A.G. Ewing. 1996. Development of votammetric methods, capillary electrophoresis and tof sims imaging for constituent analysis of single cells. In *Nanofabrication and biosystems*, ed.

Hoh, J.H. and Hansma, P.K. (1992) Atomic force microscopy for high-resolution imaging in cell biology. Trends Cell Bio. 2, 208-213

Homer Jacobson, "On Models of Reproduction," *American Scientist* 46(1958): 255-284.

Imae, Y., and T. Atsumi. 1989. T. Na+-driven bacterial flagellar motors: A mini-review. *J. Bioenergetics and Biomembranes* 21: 705-716.

J. Chen and N.C. Seeman, The Synthesis from DNA of a Molecule with the Connectivity of a Cube, Nature 350, 631-633 (1991).

J.K. Gimzewski, C. Joachim, R.R. Schlittler, V. Langlais, H. Tang, I. Johannsen, "Rotation of a Single Molecule Within a Supramolecular Bearing," *Science* 281(24 July 1998): 531-533.

J.W. Lyding, K. Hess, G.C. Abeln, D.S. Thompson, J.S. Moore, M.C. Hersam, E.T. Foley, J. Lee, Z. Chen, S.T. Hwang, H. Choi, P.H. Avouris, I.C. Kizilyalli, "UHV-STM nanofabrication and hydrogen/deuterium desorption from silicon surfaces: implications for CMOS technology," *Appl. Surf. Sci.* 130(1998): 221-230.

Jaworek, T., D. Deher, G. Wegner, R.H. Wieringa, and A.J. Schouten. 1998. Electromechanical properties of an ultrathin layer of directionally aligned helical polypeptides. *Science* 279: 57-60.

Jena, P., S.N. Khanna, and B.K. Rao. 1996. In *Science and technology of atomically engineered materials*, ed. P. Jena. River Edge, NJ: World Scientific.

Johansson, M.: 2003, 'Plenty of room at the bottom: Towards an anthropology of nanoscience', *Anthropology Today*, 19 (No. 6), 3-6.

John Cumings, A. Zettl, "Low-friction nanoscale linear bearing realized from multiwall carbon nanotubes," *Science* 289(28 July 2000): 602-604.

Johnson, A.: 2004, 'The End of Pure Science: Science Policy from Bayh-Dole to the NNI', in: D. Baird, A. Nordmann & J. Schummer (eds.), *Discovering the Nanoscale*, Amsterdam: IOS Press, pp. 217-230.

Jonas, H. The Imperative of Responsibility. In Search of an Ethics for the Technological Age. Chicago: University of Chicago Press, 1985.

Justus, B.L., R.J. Tonucci, and A.D. Berry. 1992. *Appl. Phys. Lett.* 61: 3151.

K. E. Drexler, "Molecular engineering: An approach to the development of general capabilities for molecular manipulation." *Proc. Nat. Acad. Sci.,* 78: 5275-5258 (1981).

K. E. Drexler, *Engines of Creation*, Anchor/Doubleday (New York, 1986).

K. Eric Drexler, "Molecular engineering: an approach to the development of general capabilities for molecular manipulation," *Proc. Natl. Acad. Sci. (USA)* 78(1981): 5275-5278.

K. Eric Drexler, "Molecular nanomachines: Physical principles and implementation strategies," *Annu. Rev. Biophys. Biomol. Struct.* 23(1994): 377-405.

K. Eric Drexler, "Nanomachinery: Atomically precise gears and bearings," *IEEE Micro Robots and Teleoperators Workshop*, Hyannis, Cape Cod, November 1987.

K. Eric Drexler, David Forrest, Robert A. Freitas Jr., J. Storrs Hall, Neil Jacobstein, Ralph Merkle, Christine Peterson, "On Physics, Fundamentals, and Nanorobots: A Rebuttal to Smalley's Assertion that Self-Replicating Mechanical Nanorobots Are Simply Not Possible"; http: // www.imm.org/SciAmDebate2/smalley.html

Karak, N., and S. Maiti. 1997. Dendritic polymers: A class of novel material. *J. Polym. Mater.*14: 105.

Karger, J., and D.M. Ruthven. 1992. *Diffusion in zeolites*. New York: J. Wiley.

Karplus, M. & Weaver, D. L. (1976) *Nature (London)* 260, 404-406.

Kasuga, T., M. Hiramatsu, A. Hoson, T. Sekino, and K. Niihara. 1998. *Langmuir* 14: 3160.

Katari, J.E.B., V.L. Colvin, and A.P. Alivisatos. 1994. *J. Phys. Chem.* 98: 4109.

Ke, M., S.A. Hackney, W.W. Milligan, and E.C. Aifantis. 1995. *Nanostructured Mater.* 5: 689.

Kear, B., and G. Skandan. 1998. Nanostructural bulk materials: Synthesis, processing, properties and performance. *In R&D status and trends*, ed.

Kear, B.H., R.K. Sadangi, and S.C. Liao. 1997. Synthesis of WC/Co/diamond nanocomposites. In *Proc. of the Joint NSF-NIST Conf. on Nanoparticles.*

Keiper, A.: 2003, 'The Nanotechnology Revolution', *The New Atlantis*, 2, 17-34.

Keller, D.J. and Chih-Chung, C. (1992) Imaging steep, high structures by scanning force microscopy with electron beam deposited tips. Surf. Sci. 268, 333-339

Khushf, G.: 2004, 'A Hierarchical Architecture for Nano-scale Science and Technology: Taking Stock of the Claims About Science Made By Advocates of NBIC Convergence', in: D. Baird, A. Nordmann & J. Schummer (eds.), *Discovering the Nanoscale*, Amsterdam: IOS Press, pp. 21-33.

Khushf, G.: 2004, 'Systems Theory and the Ethics of Human Enhancement: A Framework for NBIC Convergence', *Annals of the New York Academy of Sciences*, 1013, 124-149.

Kirschvink, J.L., A. Koyayashi-Kirschvink, and B.J. Woodford. 1992. Magnetite biomineralization in the human brain. *Proc. Nat'l. Acad. Sci.* USA 89: 7683-7687.

Kishida, M., T. Fujita, K. Umakoshi, J. Ishiyama, H. Nagata, and K. Wakabayashi. 1995. *Chem. Commun.* 763.

Klein, J.D., et al. 1993. *Chem. Mater.* 5: 902. Koch, C.C. 1989. Materials synthesis by mechanical alloying. *Annual Review of Mater. Sci.* 19: 121-143.

Knight, F.H. Risk, Uncertainty and Profit. London School of Economics and Political Science, London, 1933.

Koch, C.C. 1998. Bulk behavior. In *R&D status and trends*, ed. Siegel et al. 32 *Evelyn L. Hu and David T. Shaw.*

Kortan, A.R., R. Hull, R.L. Opila, M.G. Bawendi, M.L. Steigerwald, P.J. Carroll, and L.E. Brus. 1990. *J. Am. Chem. Soc.* 112: 1327.

Kourilsky, Ph.&Viney, G. Le Principe de précaution. Report to the Prime Minister, Paris, Éditions Odile Jacob, 2000.

Krätschmer, W., L.D. Lamb, K. Fostiropoulos, and D.R. Huffman. 1990. *Nature* 347: 354.

Krejchi, M.T., E.D.T. Atkins, A.J. Waddon, M.J. Fournier, T.L. Mason, and D.A. Tirrell. 1994. Chemical sequence control of beta-sheet assembly in macromolecular crystals of periodic proteins. *Science* 265: 1427-1432.

Krejchi, M.T., S.J. Cooper, Y. Deguchi, E.D.T. Atkins, M.J. Fournier, T.L. Mason, and D.A. Tirrell. 1997. Crystal structures of chain-folded antiparallel beta-sheet assemblies from sequence-designed periodic polypeptides. *Macromolecules* 30: 5012-5024.

Kresge, C.T., M.E. Leonowicz, W.J. Roth, J.C. Vartuli, and J.S. Beck. 1992. *Nature* 359: 710.

Kresge, C.T., M.E. Leonowicz, W.J. Roth, J.C. Vartuli, and J.S. Beck. 1992. *Nature* 359: 710. Kupperman, A., S. Nadimi, S. Oliver, G. Ozin, J. Garcés, and M. Olken. 1993. *Nature* 365: 239.

Kroes, P. & Meijers, A. "The Dual Nature of Technical Artifacts—Presentation of a New Research Programme." Techné 6, 2 (2002): 4 - 8.

Krstic, V., U. Erb, and G. Palumbo. 1993. *Scripta Metall. et Mater*. 29: 1501.

Krumhansl, J. A. & Pao, Y. H. (1979) *Phys. Today* 32 (11), 25-32.

Kuhn, T.: 1970, *The Structure of Scientific Revolutions* (second edition), University of Chicago Press, Chicago.

Kumar, A., and G.M. Whitesides. 1993. Features of gold having micrometer to centimeter dimensions can be formed through a combination of staming with an elastomeric stamp and an alkanethiol ink followed by chemical etching. *App. Phys. Lett*. 63: 2002-2004.

Kupperman, A., S. Nadimi, S. Oliver, G. Ozin, J. Garcés, and M. Olken. 1993. *Nature* 365: 239.

L. J. Perry and R. Wetzel, "Disulfide bond engineered into T4 lysozyme: stabilization of the protein toward thermal inactivation." *Science*, 226: 555-557 (1984).

Landon, B.: 2004, 'Less is More: Much Less is Much More: The Insistent Allure of Nanotechnology Narratives in Science Fiction', in: N.K. Hayles (ed.), *Nanoculture: Implications of the New Technoscience*, Bristol, UK: Intellect Books, pp. 131-146.

Larrère, R.: 2002, 'Agriculture: artificialisation ou manipulation de la nature?', *Cosmpolitiques*, 1 (June), 158-173.

Lehn, J.M., 2004: 'Une chimie supramoléculaire foisonnante', *La lettre de l'Académie des sciences*, 10, 12-13.

Lenhard, J.: 2004, 'Nanoscience and the Janus-Faced Character of Simulations', in: D. Baird, A. Nordmann & J. Schummer (eds.), *Discovering the Nanoscale*, Amsterdam: IOS Press, pp. 93-100.

Lent, C.S., Tougaw, P.D., Porod, W., Bernstein, G.H., "Quantum Cellular Automata," *Nanotechnology*, Vol. 4, p. 49, 1993.

Leonard, D., M. Krishnamurthy, C.M. Reaves, S.P. Denbaars, and P.M. Petroff. 1993. Direct formation of quantum-sized dots from uniform coherent islands of InGaAs on GaAs surfaces. *Appl. Phys. Lett*. 63: 3203-5.

Lewak, S.E.: 2004, 'What's the Buzz? Tell Me What's A-Happening: Wonder, Nanotechnology, and Alice's Adventures in Wonderland', in: N.K. Hayles (ed.), *Nanoculture: Implications of the New Technoscience*, Bristol, UK: Intellect Books, pp. 201-.

Liang, G., Z. Li, and E. Wang. 1996. *J. Mater. Sci*. Makino, A., A. Inoue, T. Hatanai, and T. Bitoh. 1997. *Materials Science Forum* 235-238: 723.

Lukens, R.: 2004, 'Historic Nanotechnology', *Chemical Heritage,* 22, 29.

M.C. Hersam, G.C. Abeln, J.W. Lyding, "An approach for efficiently locating and electrically contacting nanostructures fabricated via UHV-STM lithography on Si(100)," *Microelectronic Engineering* 47(1999): 235-.

M.L. Petrillo, C.J. Newton, R.P. Cunningham, R.-I. Ma, N.R. Kallenbach and N.C. Seeman. The Ligation and Flexibility of 4-Arm DNA Junctions. Biopolymers, 27, 1337-1352 (1988).

Martin, T.P., U. Naher, H. Schaber, U. Zimmerman. 1993. *Phys. Rev. Lett*. 70: 3079.

Mayer, S.: 2002, 'From genetic modification to nanotechnology: the dangers of 'sound science'', in: T. Gilland (ed.), *Science: Can We Trust the Experts?*, London: Hodder and Stoughton, pp. 1-15.

Meyer, G. and Amer, N.M. (1990) Simultaneous measurement of lateral and normal forces with an optical-beam-deflection atomic force microscope. Appl. Phys. Lett. 57(20), 2089-2091

Meyer, M. & Kuusi, O.: 2004, 'Nanotechnology: Generalizations in an Interdisciplinary Field of Science and Technology', *Hyle: International Journal for Philosophy of Chemistry*, 10(2), 153-168.

Michael E.J. Holwill, Peter Satir, "II.4 Generation of Propulsive Forces by Cilia and Flagella," in J. Bereiter-Hahn, O.R. Anderson, W.-E. Reif, eds., *Cytomechanics: The Mechanical Basis of Cell Form and Structure*, Springer-Verlag, Berlin, 1987, pp. 120-130.

Michael Page, Donald W. Brenner, "Hydrogen abstraction from a diamond surface: *Ab initio* quantum chemical study using constrained isobutane as a model," *J. Am. Chem. Soc.* 113(1991): 3270-3274.

Mnyusiwalla, A.; Abdallah, S.D.; Singer, P.A.: 2003, 'Mind the gap: science and ethics in nanotechnology', *Nanotechnology*, 14, R9-R13.

Mody, C.C.M.: 2004, 'How Probe Microscopists Became Nanotechnologists', in: D. Baird, A. Nordmann & J. Schummer (eds.), *Discovering the Nanoscale*, Amsterdam: IOS Press, pp. 119-133.

Mui, C. and C. Musgrave, "The Hydroxylation and Oxidation of the Ge(100)-2x1 Surface by Hydrogen Peroxide and Water," *Langmuir*, 20, 7604-7609 (2004).

Mui, C., S. Bent and C. Musgrave, "A Quantum Chemistry Based Statistical Mechanical Model of Hydrogen Deposition from Si(100)-2x1, Ge(100)-2x1, and SiGe Alloy Surfaces," *Journal of Physical Chemistry B*, 108, 18243-18253, (2004).

N. Koumura, R.W. Zijlstra, R.A. van Delden, N. Harada, B.L. Feringa, "Light-driven monodirectional molecular rotor," *Nature* 401(9 September 1999): 152-155.

N.C. Seeman & A.M. Belcher, Emulating Biology: Nanotechnology from the Bottom Up, Proceedings of the National Academy of Sciences (USA) 99 (supp. 2), 6451-6455 (2002).

N.C.Seeman, J. Chen, Y. Zhang, T.-J. Fu, X. Li, X. Yang, Y. Wang, J. Qi, B. Liu and F. Liu, Control of Structure and Topology in DNA Nanotechnology. In: Molecular Nanotechnology: Biological Approaches and Novel Applications, ed. by S.A. Minden, IBC Libraries, Southborough, MA, Chapter 2.2 (31 pages) (1997).

Nieman, G.W., J.R. Weertman, and R.W. Siegel. 1991a. *Mater. Res. Soc. Symp. Proc.* 206: 581-586.

Nomura, M. & Held, W. (1974) in *Ribosomes,* eds. Nomura, M., Tissiers, A. & Lengyel, P. (Cold Spring Harbor Laboratory, Cold Spring Harbor, NY), pp. 193-203.

Nordmann, A.: 2003, 'Shaping the World Atom by Atom: Eine nanowissenschaftliche WeltBildanalyse', in: A. Grunwald (ed.), *Technikgestaltung zwischen Wunsch und Wirklichkeit*, Berlin: Springer, pp. 191-199.

Nordmann, A.: 2004, 'Molecular Disjunctions: Staking Claims at the Nanoscale', in: D. Baird, A. Nordmann & J. Schummer (eds.), *Discovering the Nanoscale*, Amsterdam: IOS Press, pp. 51-62.

Nordmann, A.: 2004, 'Nanotechnology's WorldView: New Space for Old Cosmologies', *IEEE Technology and Society Magazine*, 23 (forthcoming).

Nordmann, A.: 2004, 'Social Imagination for Nanotechnology', in: European Commission (Community Health and Consumer Protection): *Nanotechnologies: A Preliminary Risk Analysis on the Basis of a Workshop, Brussels, 1-2 March 2004*, pp. 111-113 [www.europa.eu.int/comm/health/ph_risk/documents/ev_20040301_en.pdf].

Nordmann, A.: 2004, 'Was ist TechnoWissenschaft? ' Zum Wandel der Wissenschaftskultur am Beispiel von Nanoforschung und Bionik', in: T. Rossmann & C. Tropea (eds.), *Bionik 'Neue Forschungsergebnisse aus Natur-, Ingenieur- und Geisteswissenschaften*, Berlin: Springer, 2004 (forthcoming)

P. Livingston, "Ganglioside vaccines with emphasis on GM2," *Semin. Oncol.* 25(December 1998): 636-645.

P.S. Lukeman, A. Mittal, & N.C. Seeman, Two Dimensional PNA/DNA Arrays: Estimating the Helicity of Unusual Nucleic Acid Polymers, Chemical Communications 2004, 1694-1695 (2004).

Parker, J.C., et al. 1995. U.S. Patent 5,460,701. POST (Parliamentary Office of Science and Technology). 1995.

Paschen, H.; Coenen, C.; Fleischer, T.; Grünwald, R.; Oertel, D. & Revermann, C.: 2004, *Nanotechnologie: Forschung, Entwicklung, Anwendung*, Berlin: Springer (*Nanotechnologie, TAB-Arbeitsbericht 92*, Berlin: Büro für Technikfolgen-Abschätzung beim Deutschen Bundestag).

Paul A. Schulte, Frederica P. Perera, eds., *Molecular Epidemiology: Principles and Practices*, Academic Press, San Diego, CA, 1993.

Philip G. Collins, Michael S. Arnold, Phaedon Avouris, "Engineering carbon nanotubes and nanotube circuits using electrical breakdown," *Science* 292(27 April 2001): 706-709.

Philip G. Collins, Phaedon Avouris, "Nanotubes for electronics," *Scientific American* 283(December 2000): 62-69.

Phoenix, C.; Drexler, E.: 2004, 'Safe Exponential Manufacturing', *Nanotechnology*, 15, 869-72.

Pitt, J.C.: 2004, 'The Epistemology of the Very Small', in: D. Baird, A. Nordmann & J. Schummer (eds.), *Discovering the Nanoscale*, Amsterdam: IOS Press, pp. 157-163.

Pressman, J.: 2004, 'Nano Narrative: A Parable from Electronic Literature', in: N.K. Hayles (ed.), *Nanoculture: Implications of the New Technoscience*, Bristol, UK: Intellect Books, pp. 191-200.

Prigogine, I., and S. Rice. 1988. *Advances in chemical physics*, Vol. 70, Parts 1 & 2. New York: J. Wiley.

Putman, C.A.J., De Grooth, B.G., Van Hulst, N.F., and Greve, J. (1992) A detailed analysis of the optical beam deflection technique for use in atomic force microscopy. J. App. Phys. 72(1), 6-12

R. Sha, F. Liu and N.C. Seeman, Atomic Force Measurement of the Interdomain Angle in Symmetric Holliday Junctions, Biochemistry 41, 5950-5955 (2002).

R. Sha, F. Liu, D.P. Millar N.C. Seeman, Atomic Force Microscopy of Parallel DNA Branched Junction Arrays, Chemistry & Biology 7, 743-751 (2000).

R. Sha, F. Liu, M. Bruist and N.C. Seeman, Parallel Helical Domains in DNA Branched Junctions Containing 5', 5' and 3', 3' Linkages, Biochemistry 38, 2832-2841 (1999).

R. T. Bate, "Nanoelectronics," *Nanotechnology*, Vol. 1, pp. 1-7, 1990.

R.G. Postier, "Antibiotic-resistant organism infection," *Am. Surg.* 66(February 2000): 112-116.

R.I. Ma, N.R. Kallenbach, R.D. Sheardy, M.L. Petrillo and N.C. Seeman, 3-Arm Nucleic Acid Junctions Are Flexible. Nucleic Acids Research 14, 9745-9753 (1986).

R.J. Jackson, A.J. Ramsay, C.D. Christensen, S. Beaton, D.F. Hall, I.A. Ramshaw, "Expression of mouse interleukin-4 by a recombinant ectromelia virus suppresses cytolytic lymphocyte responses and overcomes genetic resistance to mousepox," *J. Virol.* 75(February 2001): 1205-1210.

R.N. Jones, "Resistance patterns among nosocomial pathogens: trends over the past few years," *Chest* 119(February 2001): 397S-404S.

Rademann, K., B. Kaiser, U. Even, F. Hensel. 1987. *Phys. Rev. Lett.* 59: 2319.

Rao, C.N.R., B.C. Satishkumar, and A. Govindaraj. 1997. *Chem. Commun.* 1581.

Rao, M.B., and S. Sircar. 1993. *Gas Separation and Purification* 7: 279.

Rechard, R.P. "Historical Relationship Between Performance Assessment for Radioactive Waste Disposal and Other Types of Risk Assessment." Risk Analysis 19, 5 (1999): 763-807.

Reed, M.A., "Quantum Dots," *Scientific American,* January 1993, pp. 118-123.

Richard P. Feynman, "There's Plenty of Room at the Bottom," *Engineering and Science* (California Institute of Technology), February 1960, pp. 22-36. Reprinted in B.C. Crandall, James Lewis, eds., *Nanotechnology: Research and Perspectives*, MIT Press, 1992. pp. 347-363, and in D.H. Gilbert, ed., *Miniaturization*, Reinhold, New York, 1961, pp. 282-296; http: // www.zyvex.com/nanotech/feynman.html

Rietman, E.A.: 2001, 'Drexler hypothesis of a universal assembler is supported not by theoretical arguments alone but by existence proof in the form of biological life', in: *Molecular Engineering of Nanosystems*, Springer, New York & Berlin.

Rip, A., Misa, Th. J.,&Schot, J. W. (eds.). Managing Technology in Society. The Approach of Constructive Technology Assessment. London: Pinter Publishers, 1995.

Robert A. Freitas Jr., "Clottocytes: Artificial Mechanical Platelets," Foresight Update No. 41, 30 June 2000, pp. 9-11; http: //www.imm.org/Reports/Rep018.html

Robert A. Freitas Jr., "Exploratory design in medical nanotechnology: A mechanical artificial red cell," *Artificial Cells, Blood Substitutes, and Immobil. Biotech.* 26(1998): 411-430; http: // www.foresight.org/Nanomedicine/Respirocytes.html

Robert A. Freitas Jr., "Microbivores: Artificial Mechanical Phagocytes using Digest and Discharge Protocol," Zyvex preprint, March 2001; http: //www.zyvex.com/Publications/articles/ Microbivores.html. See also: Robert A. Freitas Jr., "Microbivores: Artificial Mechanical Phagocytes," Foresight Update No. 44, 31 March 2001, pp. 11-13; http: //www.imm.org/ Reports/Rep025.html.

Robert A. Freitas Jr., "Nanodentistry," *J. Amer. Dent. Assoc.* 131(November 2000): 1559-1566.

Robert A. Freitas Jr., "Some Limits to Global Ecophagy by Biovorous Nanoreplicators, with Public Policy Recommendations," Zyvex preprint, April 2000; http: //www.foresight.org/NanoRev/ Ecophagy.html

Robert F. Service, "Is nanotechnology dangerous?" *Science* 290(24 November 2000): 1526-1527.

Roberts, J.A.: 2004, 'Deciding the Future of Nanotechnologies: Legal Perspectives on Issues of Democracy and Technology', in: D. Baird, A. Nordmann & J. Schummer (eds.), *Discovering the Nanoscale*, Amsterdam: IOS Press, pp. 247-255.

Robinson, C.: 2004, 'Images in NanoScience/Technology', in: D. Baird, A. Nordmann & J. Schummer (eds.), *Discovering the Nanoscale*, Amsterdam: IOS Press, pp. 165-169.

Robison, W.L.: 2004, 'Nano-Ethics', in: D. Baird, A. Nordmann & J. Schummer (eds.), *Discovering the Nanoscale*, Amsterdam: IOS Press, pp. 285-300.

Robot Factory, http: //www.fanuc.co.jp/eannai/servo.htm; citation courtesy of Tihamer Toth-Fejel, 2000.

Roco, M.C. & Bainbridge, W.S. (eds.): 2001, *Societal implications of nanoscience and nanotechnology*, (Proceedings of a workshop organized by the National Science Foundation, September 28-29, 2000), Kluwer: Dordrecht [available online at http: //itri.loyola.edu/nano/societalimpact/nanosi.pdf]

Roco, M.C. & Tomellini, R. (eds.): 2002, *Nanotechnology: Revolutionary Opportunities and Societal Implications, Workshop, Lecce (Italy), 31 January - 1 February 2002*, Luxemburg: Office for Official Publications of the European Communities, [200 pp.]

Roco, M.C.; Bainbridge, W.S. (eds.): 2002, *Converging Technologies for Improving Human Performance: Nanotechnology, Biotechnology, Information Technology and the Cognitive Science*, Arlington, VA: National Science Foundation.

Rod Logic and Thermal Noise in the Mechanical Nanocomputer, by K. Eric Drexler, in *Proceedings of the 3rd International Symposium on Molecular Electronic Devices*, F.L. Carter, R.E. Siatkowski, H. Wohltjen, eds., North-Holland 1988.

Rofagha, R., R. Langer, A.M. El-Sherik, U. Erb, G. Palumbo, and K.T. Aust. 1991. *Scr. Metall. Mater*. 25: 2867.

Rohlfing, E.A., D.M. Cox, and A. Kaldor. 1984. *J. Chem. Phys*. 81: 3846.

Romanov, A.E., V.I. Vladimirov. 1992. In *Dislocations in solids*, ed. F.R.N. Nabarro, Vol. 9. Amsterdam: North-Holland. Salishchev, G.A., O.R. Valiakhmetov, V.A. Valitov, and S.K. Mukhtarov. 1994. *Materials Science Forum*. 170-172: 121.

Rosen, A. 1998. A periodic table in three dimensions: A sightseeing tour in the nanometer world. In *Advances in quantum chemistry*. In press.

Rosenberg, N. "Why Technology Forecasts Often Fail." Futurist July/August 1995, 16- 21.

Ruthven, D.M., S. Farooq, K.S. Knaebel. 1994. *Pressure swing adsorption*. New York: VCH Publishers. Sayari, A. 1996. *Chem. Mater*. 8: 1840.

S. Xiao, F. Liu, A. Rosen, J.F. Hainfeld, N.C. Seeman, K.M. Musier-Forsyth & R.A. Kiehl, Self-Assembly of Nanoparticle Arrays by DNA Scaffolding, J. Nanoparticle Research 4, 313-317 (2002).

Schuler, E.: 2004, 'Perception of Risks and Nanotechnology', in: D. Baird, A. Nordmann & J. Schummer (eds.), *Discovering the Nanoscale*, Amsterdam: IOS Press, pp. 279-284.

Schummer, J.: 2004, 'Interdisciplinary Issues in Nanoscale Research', in: D. Baird, A. Nordmann & J. Schummer (eds.), *Discovering the Nanoscale*, Amsterdam: IOS Press, pp. 9-20.

Schummer, J.: 2004, 'Multidisciplinarity, Interdisciplinarity, and Patterns of Research Collaboration in Nanoscience and Nanotechnology', *Scientometrics*, 59, 425-465.

Schummer, J.: 2004, 'Why do Chemists Perform Experiments?', in: D. Sobczynska, P. Zeidler, E. Zielonacka-Lis (eds.), *Chemistry in the Philosophical Melting Pot*, Peter Lang (Frankfurt/M.), 2004, pp. 395-410.

Sha, R., Zhang, X., Liao, S., Constantinou, P.E., Ding, B., Wang, T., Garibotti, A.V., Zhong, H., Israel, L.B., Wang, X., Wu, G., Chakraborty, B., Chen, J., Zhang, Y., Mao, C., Yan, H.,

Kopatsch, J., Zheng, J., Lukeman, P.S., Sherman, W.B. and Seeman, N.C., Motifs and Methods in Structural DNA Nanotechnology, Proc. Intl. Conf. Nanomaterials, NANO 2005, July 13-15, 2005, Mepco Schlenk Engineering College, Srivakasi, India, V. Rajendran, ed., pp. 3-10 (2005).

Siegel, H.: 1980, 'Objectivity, Rationality, Incommensurability, and More', *British Journal for the Philosophy of Science*, 31, 359-384.

Susan Smith, Izabela Giriat, Anja Schmitt, Titia de Lange, "Tankyrase, a poly(ADP-ribose) polymerase at human telomeres," *Science* 282(20 November 1998): 1484-1487.

Taylor, K.J., C.L. Pettiette-Hall, O. Cheshnovsky, and R.J. Smalley. 1992. *J. Chem. Phys.* 96: 3319.

Trudeau, M.L., and J.Y. Ying. 1996. *Nanostructured Materials* 7: 245.

Trudeau, M.L., and J.Y. Ying. 1996. *Nanostructured Materials* 7: 245. Tschope, A.S., W. Liu, M. Flyzani-Stephanopoulos, and J.Y. Ying 1995. *J. Catal.* 157: 42.

Vermaas, P.E.: 2004, 'Nanoscale Technology: A Two-Sided Challenge for Interpretations of Quantum Mechanics', in: D. Baird, A. Nordmann & J. Schummer (eds.), *Discovering the Nanoscale*, Amsterdam: IOS Press, pp. 77-91.

W.B. Sherman and N.C. Seeman, A Precisely Controlled DNA Bipedal Walking Device, NanoLetters 4, 1203-1207 (2004); Erratum, 4, 1801-1801.

W.C. Albrich, M. Angstwurm, L. Bader, R. Gartner, "Drug resistance in intensive care units," *Infection* 27(1999): S19-S23.

Wald, C.A.: 2004, 'Working Boundaries on the NANO Exhibition', in: N.K. Hayles (ed.), *Nanoculture: Implications of the New Technoscience*, Bristol, UK: Intellect Books, pp. 83-108.

Whitesides G.: 1995, 'Self-Assembling Materials,' *Scientific American*, (Sept.), 146-9.

Widjaja, Y. and C. Musgrave, "Atomic Layer Deposition of Hafnium Oxide: A Detailed Reaction Mechanism from First Principles," *Journal of Chemical Physics* 117, 1931-1934 (2002).

Wilder, K., H.T. Soh, T. Soh, A. Atalar, and C.F. Quate. 1997. Hybrid atomic force/scanning tunneling lithography. *J. of Vacuum Science & Technol. B.* 15: 1811-17.

Williams, R.S. 1998. Functional nanostructures. In *R&D status and trends*, ed. Siegel et al. Wiltzius, P. 1998.

Winningham, M.J. and D.Y. Sogah. 1997. A modular approach to polymer architecture control via catenation of prefabricated biomolecular segments: Polymers containing parallel beta-sheets templated by a phenoxathiin-based reverse turn mimic. *Macromolecules* 30: 862 - 876.

Woods, S.; Jones, R. & Geldart, A.: 2003, *The Social and Economic Challenges of Nanotechnology*, Report to the Economic and Social Research Council (ESRC), Swindon, UK.

3

Nanoscale Chemistry, Nano Synthesis and Assembly: Towards Knowing the Various Implications

Nanoscale Chemistry

Chemistry is one of the best developed, all encompassing and well defined fields of science today. Since ancient times, mankind has been using chemical techniques to manipulate matter on the atomic and molecular levels. As chemists improve our ability to organise atoms into atomically precise structures known as 'molecules,' fields including medicine, materials science, computer science and engineering reap the benefits. It may seem that there is little difference between the achievements of chemistry and the goals of nanotechnology, that is, to understand and manipulate molecular scale phenomena. However, nanotechnology takes a novel and revolutionary approch to this goal.

While chemistry deals with molecules in a statistical sense, nanotechnology deals with them as discrete entities, each requiring special attention. As Richard Feynman accurately *predicted in 1959*, 'The problems of chemistry and biology can be greatly helped if our ability to see what we are doing, and to do things on an atomic level, is ultimately developed—a development which I think cannot be avoided.' Indeed, *Scanning Probe Microscopy* has allowed sub-angstrom resolution of surfaces, leading to radical advancements in chemistry, biology and other fields.

For example, consider chemical synthesa that, and he shakes it, and he fiddles around. And, at the end of a difficult process, he usually does succeed in synthesising what he wants.' Despite enormous progress in chemical methods since that time, the mixing of reagents described by Feynman is still the primary method used by synthetic chemists today. Due to the random nature of the molecular collisions in a solution, synthesis of complex (atomically precise) molecules suffers from the statistical law of entropy. Thus no reaction gives 100% yield. Furthermore, since most syntheses require multiple reactions, valuable starting materials are lost in unwanted side reactions and purification processes.

A biomimetic solution to this problem involves the idea of a molecular assembly line. Biological systems often carry out multistep reactions by passing a substrate from one enzyme to the next, the method used by Henry Ford to mass produce Model-T automobiles with nearly 100% yield. Molecular Manufacturing is a sub-field of nanotechnology that seeks to realize synthetic methods such as this one.

Nanoscale Reaction Vessels

Vesicles are abundant in biological organisms, where they often serve the function of transporting various molecules. They are bubble-like structures composed of a lipid bilayer membrane, which can be created by means of biomolecular *self-assembly*. Vesicles of diameters ranging from 25 nm to 20 microns can be fabricated in a laboratory with some self-assembly technique. For instance, 25 nm vesicles can be made by sonicating a bilayer membrane at high frequencies, while 20 micron vesicles are produced when a low frequency alternating current is applied to the solution.

Zare et. al. has demonstrated the possiblility of carrying out reactions in vesicles containing attoliter to zeptoliter volumes. Each vesicle can contain as little as one molecule of a reagent. In this case, the reagents were deposited by electroporation or by electrofusion, methods involving application of an electrical pulse to propel charged particles through the solution.

It may seem that that shrinking reaction vessels and using molecular assembly lines could be considered a simple extension of current chemical techniques. However, the discrete nature of nanoscale chemistry requires an adaption of the fundamental principles defined by traditional chemistry. For instance, in a previous article entitled '*quantum biology*' it was noted that our statistical method for measuring temperature breaks down at the nanoscale. The current definition of temperature is especially meaningless in systems that are not at equilibrium, such as a biological cell. Indeed, the term 'equilibrium' itself loses meaning (breaks down) at the nanoscale due to its statistical definition. In order to study and design nanoscale systems, one cannot rely only on macroscale statistics that have proven invaluable to nearly all fields of science. Since temperature is one of the 7 fundamental units upon which other units rely, it is becoming necessary for nano-scientists to tweak this and other units to better reflect nanoscale phenomena.

Nanoscale Temperature

Temperature is defined as 'the average kinetic energy of all particles in the system.' According to the theory, there should be a standard deviation to go along with this average, but I have yet to see a thermometer that gives standard deviation. When one considers temperature in mixtures of different kinds of molecules or systems that are not at equilibrium, at the nanoscale there should be localised regions of various temperatures contributing to this average.

Temperature distribution is frequently calculated in non-equilibrium molecular dynamics simulations. In these theoretical models, the kinetic energy of each particle at various points in time is taken into account in order to provide a better understanding of molecular scale events. One might then define local temperature as—the average kinetic energy of a specified set of atoms within the system. While this is still a statistical method, it allows discrete statements to be made about localised parts of a system. Take, for intance, the *myosin* molecule in the previous article. If one were to specify as the set of atoms those included in the beta sheet adjacent to the ATP binding site, then, considering the 6 trillion myosin molecules [*myology*] in a muscle cell may even provide a sufficient sample size for spectroscopic measurements.

While local temperature is relatively easy to describe theoretically, measurements are more difficult to make. However, novel methods are being developed that have the potential to verify theoretical models experimentally. For example, local temperatures and temperature gradients can now be directly measured by *Scanning Thermal Microscopy (SThM)*. While the lateral resolution is currently limited to 150-200nm (sufficient for numerous microscale applications), it is reasonable to expect this resolution to improve in the future, perhaps eventually reaching the level of single molecules.

Molecular and Supramolecular Chemistry

A second broad category of enabling technology for molecular manufacturing includes molecular chemistry — the conscious design of completely artificial, nonbiological chemical structures that could potentially serve as molecular "parts" or which have specific devicelike functionality. For example, catalysts may very loosely "be thought of as rudimentary assemblers that are slightly programmable through changes in the reaction milieu (i.e., changes in pH, temperature, etc.)". The number of natural molecular "parts" is immense — perhaps a few hundred thousand bioorganic compounds have been separated, purified and identified — but by 15 February 1999 chemists had already registered 19, 245, 458 well-characterised artificial molecules or other "substances" in the CAS Registry. Synthesis of natural molecules containing ~100 atoms (e.g., a ~1 nm^3 molecular "part"), using methods of classical organic synthesis often requiring ~1 synthesis step per atom with ~90% yield per step, was state-of-the-art in 1998.

"What is exciting about modern nanotechnology, " says Nobel chemist Roald

Hoffmann, "is (a) the marriage of chemical synthetic talent with a direction provided by 'device-driven' ingenuity coming from engineering, and (b) a certain kind of courage provided by those incentives, to make arrays of atoms and molecules that ordinary, no, extraordinary chemists just wouldn't have thought of trying. Now they're pushed to do so. And of course they will. They can do anything."

Molecular chemists study how small numbers of atoms combine covalently with each other to produce molecules with a wide variety of different properties. Supramolecular chemists are primarily concerned with the much weaker noncovalent forces that can arise between molecules — such as hydrogen bonds and van der Waals interactions — that can be sufficient to bind collections of molecules tightly enough to form functional nanometer-sized structures. Supramolecular chemistry studies "chemistry beyond the molecule" or "the chemistry of the noncovalent bond". Its subject matter includes atomically precise supermolecules (well-defined discrete oligomolecular species resulting from the intermolecular association of a few components, such as a receptor and its substrate) and polymerlike supramolecular assemblies (polymolecular entities that result from the spontaneous association of a large undefined number of components).

Self-assembly is a key concept in supramolecular chemistry as it is in nature, allowing the manufacture of large numbers of compound objects simultaneously and in parallel, rather than sequentially. Molecular self-assembly is a strategy for nanofabrication that involves designing molecules and supra-molecular entities so that complementarity causes them to aggregate into desired structures. Self-assembly of atomically precise supermolecules thus demands well-defined adhesion between selected molecules, which may require steric (shape and size) complementarity, interactional complementarity, large contact areas, multiple interaction sites, and strong overall binding.

G.M. Whitesides explains that self-assembly has a number of advantages as a strategy. First, it carries out many of the most difficult steps in nanofabrication — those involving the smallest atomic-level modifications of structure — using the very highly developed techniques of synthetic chemistry. Second, it draws from the enormous wealth of examples in biology for inspiration; self-assembly is one of the most important strategies used in biology for the development of complex, functional structures, such as the well-studied examples of complex biomachine self-assembly of whole bacteriophage virions and flagellar rotor motors from their smaller molecular "parts". Third, self-assembly can incorporate biological structures directly as components in the final systems. Fourth, self-assembly requires target structures to be among the most thermodynamically stable available to the system, thus tends to produce structures that are relatively defect-free and self-healing. As Lehn points out: "By increasing the size of its entities, nanochemistry works its way upward towards microlithography and microphysical engineering, which, by further and further miniaturisation, strive to produce ever smaller elements". Hierarchical self-assembly of many billions of micron-scale spherical components into a periodic honeycomb-structured photonic crystal optical device ~30 microns thick and ~1 cm^2 in area was demonstrated in 1998.

There is a wide range of different molecular systems that can self-assemble, including those which form ordered monomolecular structures by the coordination of molecules to surfaces, called self-assembled monolayers (SAMs), self-assembling thin films and Langmuir-Blodgett films, and self-organising nanostructures. In these systems, a single layer of molecules affixed to a surface allows both thickness and composition in the vertical axis to be adjusted to 0.1 nm by controlling the structure of the molecules comprising the monolayer, although control of in-plane dimensions to <100 nm is very difficult. Fluidic self-assembly of microscale parts and the dynamics of Brownian self-assembly have also been described, and the theory of designable self-assembling molecular machine structures is beginning to be addressed.

Another approach using solid-phase peptide synthesis and a massively convergent self-assembly process was employed by M.R. Ghadiri and colleagues, who designed and synthesised a number of self-assembling peptide-based nanotubes using cyclic peptides with an even number of alternating D- and L-amino acids for the building blocks of the nanotubes. The alternating stereo-chemistry of the cyclic peptides allowed all the side chains of the amino acids to be pointing outwards which would not be possible in an ordinary all L-cyclic peptide. In this conformation, the amide backbone can H-bond in a direction perpendicular to the plane of the cyclic peptide. The stacking of two cyclic peptides forms an H-bonding network resembling an antiparallel beta-sheet as is commonly found in natural proteins. The H-bonding lattice quickly propagates perpendicular to the plane of the cyclic peptide, forming a tubular microcrystalline structure with 0.75-nm pores, or, in another experiment, 1.3-nm pores. Another group of nanotubes was designed with a highly hydrophobic outer surface and a hydrophilic inner pore. These nanotubes were easily inserted into a lipid bilayer and have been shown to be highly efficient ion channels; nanotubes with slightly larger pores transport small molecules such as glucose as well. Stable flat nanodisks with diameters continuously (chemically) adjustable from 30-3000 nm have been self-assembled in surfactant solutions.

Self-assembled crystalline solids such as zeolitesincompletely fill space, leaving substantial voids that may be occupied by solvent molecules or other guest molecules, producing solid-state host-guest complexes known as clathrates (from the Latin *clathratus*, meaning "enclosed by the bars of a grating"). MacNicol and coworkers reported the first rationally designed clathrate host in 1978, but the first true de novo design of an organic clathrate was in 1991 and involved the formation of an extensive three-dimensional diamondlike porous lattice built from a single Tinkertoy-type subunit containing four tetrahedrally arrayed pyridone groups that acted as connectors to assemble the units together in a well-defined geometry. However, the crystal was held together only by weak hydrogen bonds and the links had a rotational degree of freedom, rendering the exact crystal molecular arrangement unpredictable. Engineered nanoporous molecular crystals can provide molecular-scale voids with controlled sizes, shapes, and embedded chemical environments at resolutions of 0.3-4.0 nm. Other self-assembling crystal structures have been described including the crystal engineering of diamondoid networks, template-directed colloidal crystallisation or colloidal epitaxy, organic templating to form crystalline zeolite-type structures with ordering lengths <3 nm, 3-D polymer channels with chemically functionalizable channel linings, and micromolding combined with templating and cooperative assembly of block copolymers to produce hierarchical ordering over discrete and tunable characteristic length scales of ~10 nm, ~100 nm, and ~1000 nm in a single body.

Dendrimers, also known as arborols or fractal polymers, are a well-known set of self-assembling structures that possibly could be used as tools to assist in the assembly of early nanomachines. Dendrimers are large regularly-branching macro-molecules resembling fractal patterns, made by an iterative process in which small linear molecules are allowed to bind to each other at a certain number of sites along their length, building up branches upon branches with each iteration working outward from a core molecule having at least two chemically reactive arms, somewhat mimicking tree growth. Different core molecules or building-block chains produce macromolecules with different shapes, and the outer surface can be terminated with chemical groups of specific functionality, producing hundreds of differently-surfaced branched molecules that have already found use in medical research. Growth is regulated, so size can be accurately controlled — dendrimers are typically a few nanometers wide but have been constructed with masses exceeding a million protons (~10^5 atoms) and with diameters larger than 30 nm. In 1999, convergent assembly of specified 6-mer dendrimers, via trimerisation of precursor dimer "parts, " was demonstrated.

Autocatalysis or self-replication is yet another approach to the chemical self-assembly of subunits. In 1989, J. Rebek reported the synthesis of super-molecules that could generate copies of themselves when placed in a sea of simpler molecular parts, with each component part consisting of up to a few dozen atoms. Molecular self-replication has continued to be studied by others. Ghadiri reported a self-replicating peptide in 1996, and by 1998 his laboratory had moved on to studies of more complex autocatalytic cycle networks. It is unknown how complex such assembly cycles may be made, but Ghadiri was investigating a self-replicating 256-component molecular ecosystem incorporating ~32, 000 possible binary interactions, in 1998.

By 1998, chemists had already synthesised a vast variety of simple molecular parts of which only a tiny sampling can be mentioned here. For instance, carbon rings have been linked edge-on or by corners to create propeller-shaped molecules called propellanes or rotanes, or stacked with short hydrocarbon links to make gear-looking molecules such as cyclophane and superphane. Hydrocarbon molecular polyhedra in the shape of triangular (e.g., triangulane), cubic (e.g., cubane, first made by P. Eaton in 1964), pentagonal and hexagonal prisms, collectively known as prismanes, have been created, along with more unusual geometrical forms such as paddlane, housane, basketane, churchane, pagodane, and bivalvane. Cryptands, spherands, cryptaspherands and carcerands, corannulene baskets, and calixarenes are bowl-shaped molecules whose benzene ring walls hold their cavities rigid. The bowl rim of a calixarene can be lined with chemical groups to determine which guests it will accept, or two bowls can be bonded together, making a hollow nanoscale reaction vessel, and various molecular cages and capsules with precise sizes and characteristics have been assembled. A variety of nonchiral and chiral square molecules have been synthesised, as well as molecular "bottlebrushes"; DNA-polymer constant-force springs; the ferric wheel ; molecular barrels and molecular tennis balls;various helical, boxlike, and other hollow organic structures; molecular "turnstiles" and "molecular tweezers"; simple and complex trefoil knot molecules; and stable carbyne molecular rods with chain lengths up to ~300 carbon atoms having alternating single and triple covalent bonds. Chemists are justifiably awed by the creation of structures like the ferric wheel; "for me, " observed Nobel chemist Roald Hoffmann, "this molecule provides a spiritual high....The molecule is beautiful because its symmetry reaches directly into the soul, [playing] a note on a Platonic ideal."

Molecular gear systems first became a popular object of chemical study during the 1980s. For example, G. Yamamoto described "compounds that exist in conformations which are regarded as static meshed gears with two-toothed and three-toothed wheels and some of them behave as dynamic gears." H. Iwamura prepared one system that formed a chain of beveled molecular gears with ~GHz rotation rates, and another "doubly connected bevel gear system...[Transfer] of information from one end of the molecular system to the other end could take place in large molecules via cooperativity of the torsional motions of the chain." Kurt Mislow gave numerous examples of molecular gear systems that he had synthesised using the methods of traditional solution chemistry, which "resemble to an astonishing degree the coupled rotations of macroscopic mechanical gears." He added: "It is possible to imagine a role for these and similar mechanical devices, molecules with tiny gears, motors, levers, etc., in the nanotechnology of the future."

In 1998, Gimzewski and coworkers synthesised and observed a 1.5-nm diameter gearlike single molecule (hexatertbutyl decacyclene) rotating within a supramolecular bearing (a void in a 2-D lattice) at room temperature on the surface of a Cu{100} lattice. The authors suggest that added nonthermal noise (e.g., temperature differences produced by tunnel current heating of the rotor) could be rectified by asymmetries in the rotor/neighbor potential energy curve, allowing the rotor to turn unidirectionally. Says Gimzewski: "Our wheel is frictionless in the conventional sense of the word, and it doesn't wear out." Gimzewski's experimental work confirms Drexler's calculations that van der Waals bearings can operate in molecular systems, that multiatomic bearing surfaces with higher load capacities than sigma bond bearings can be built, that these bearings can run with no lubricants, and that such bearings have sufficiently low energy barriers to allow turning by thermal vibrations alone.

By 1998, limited control of the mechanical action of molecular parts had been demonstrated. For example, in 1994 T. Ross Kelly's group created a "paddlewheel" molecule (a spinning propeller-shaped wheel) with a built-in chemically-controlled brake. In solution, the wheel spins freely, but when Hg^{++} ions are added, a blocking ligand physically rotates into a new position (tripping the brake) and stopping the rotation; removal of the mercury causes the wheel to spinning. In 1997, Kelly's group used similar techniques to synthesize the first molecular ratchet, a propeller with three benzene ring "blades" that serve as gear teeth. A row of four rings (the pawl) sits between two of the blades, such that the propeller cannot turn without pushing it aside; because of a twist in the pawl, it is easier to turn clockwise than counterclockwise under thermal agitation, providing mechanical rectification without net motion or energy extraction. Shinkai and colleagues fabricated a pair of molecular tongs for grasping metal ions that uses two crown ether rings as the jaws, linked by two benzene rings joined through two double-bonded nitrogen atoms. Irradiation with ultraviolet light induces photoisomerisation of the double bond between the nitrogen atoms, closing the jaws; heat reverses the process, opening the jaws. In related systems, collectively called "butterfly molecules, " irradiation with visible light or pH changes can also induce opening. Seeman's chemically-operated mechanical DNA system provides similar control of molecular mechanical movement. Numerous controllable molecular electronics devices and possible nanocomputer components are reviewed.

Noncovalent self-assembly of molecular parts has also been demonstrated. For example, Fraser Stoddart and colleagues have extensively investigated the rotaxanes, molecules in which one part is threaded through a hole or loop in another, and kept from unthreading by end-groups, like a ring trapped on a barbell. In one system, a ring-shaped molecule slides freely along a shaft-like chain molecule, moving back and forth between stations at either end with a frequency of ~500 Hz, making an oscillating "molecular shuttle". This shuttling behaviour can be controlled by a series of different chemical, electrochemical, or photochemical external stimuli. Using a different approach, G. Wenz threaded ~120 molecular beads onto a single, long, poly-iminooligomethylene polymer chain, making a "molecular necklace." Other groups are pursuing related research regarding [n]rotaxanes and pseudorotaxanes (mechanically-threaded molecules), while still others are using the necklace technique as a means, for example, of self-assembling ~2 nm wide cyclodextrin nanotubes.

Another interesting molecular system that can be made entirely by self-assembly is the catenanes, which have two or more closed rings joined like the links of a chain. The rings are mechanically linked — there are no covalent bonds between separate links. The first catenane (2 linked rings) was constructed by Edel Wasserman of AT&T Bell Laboratories in 1960, using simple hydrocarbon rings; the first catenane was synthesised in 1977. In 1994, Fraser Stoddart and David Amabilino constructed the first catenane, having ~372 atoms, dubbed "olympiadane" because of its resemblance to the Olympic rings, and in 1997 the same group announced the first deterministically-ordered 7-ring heptacatenane; another group later reported the spontaneous self-assembly of a 10-component catenane. By 1998, much research on larger polymeric chains of linked rings (e.g., oligocatenanes), supramolecular "daisy chains", fullerene-containing catenanes, and supramolecular weaving was in progress. The smallest catenane constructed to date has dimensions 0.4 nm x 0.6 nm. Self-assembled mechanically-interlocked 2-dimensional and 3-dimensional "infinite" arrays of single molecular ring-like species are known. Interestingly, in 1998 it was discovered that many viral coats (which also self-assemble) appear to be a 2-dimensional "chain-mail" weave of mechanically-interlocked protein rings — for instance, the spherical bacteriophage HK97 capsid shell consists of exactly 72 interlinked protein rings, specifically 60 hexamers and 12 pentamers.

Complex molecular parts also can be built up from simpler molecular parts by covalent bonding as low-symmetry shape-invariant molecular object polymers, or via processes variously known as modular chemistry, heterosupramolecular chemistry, chemical assembly, or structure-directed synthesis — including the well-known Diels-Alder reactions which may be employed repetitively to fabricate a series of atomically-precise molecular rods, rings, and nanocages (e.g., "beltenes" and "collarenes") composed of polyacenequinone units with nanometer-scale dimensions. Such efforts have been described as steps toward "molecular LEGO", "molecular Meccano", "molecular Tinkertoys", or "molecular building blocks". Some structure-directed assembly has been achieved with simple DNA forms. Other simple techniques for synthesising nanowires, nanorods, nanotubes, and nanocages, or for selecting polymers for length, are well-known.

For instance, Jean-Marie Lehn has created rectangular grid complexes by mixing rodlike ligands that make use of different numbers of binding sites — one example is a 4 x 5 array consisting of nine ligands and 20 metal ions, nanoscale grids which he claims might one day find use "as components within a futuristic information storage and processing nanotechnology". C.M. Drain has described a 21-component

$(5 \text{ nm})^2$ square array of nine porphyrins tethered together by 12 palladium ions shaped like a four-pane window. The structure is built from three different kinds of porphyrins — (1) an X-shaped unit that coordinates to four metal ions, forming the center of the array, (2) a T-shaped unit that coordinates to three metals and forms each side of the array, and (3) an L-shaped unit that coordinates to only two metals and forms each corner of the array. When these components are placed in solution in the correct ratio, they form the square array within a half hour at room temperature with about 90% yield. The same porphyrin units can be combined in different ratios and induced to form wires or tapes. The properties of such arrays can be fine-tuned by choosing the appropriate metal ion linker and functionalised porphyrin unit.

Jeffrey Moore has investigated a three-dimensional nanoscaffolding comprised of a molecular lattice that could serve as a framework for catalysts, photosynthetic molecules, or more complex molecular devices. Moore's original objective was to synthesize modular building blocks with physical and chemical properties that would dictate a "programmed assembly" protocol for "nanoarchitectures". The characteristics of each modular unit shaped the weak intermolecular attractions — electrostatic, van der Waals, and hydrogen-bonding forces — between it and its neighbors so that units would array themselves in a larger structure with the desired spacing and geometry. Moore's group first linked phenylacetylene subunits into oligomers of various lengths which bend and twist like wire sculptures until their ends meet to form closed polyhedral molecules — the basic scaffold components (of which there were at least half a dozen shapes). These building blocks then self-assembled or folded into geometries at least partly controlled by designed hydrogen-bonding interactions. The molecular polyhedra each possessed several phenylene groups, "chemical handles" upon which other molecular groups could be hung. Moore then focused on ~10-nm dendrimeric forms and later began attaching photosensitive antenna molecules to these structures. By 1997, Moore had begun investigating norbornadiene on Si{100} in an effort to develop ways to covalently attach molecules onto surfaces with subnanometer spatial precision, and had suggested a packing model for interpenetrated diamondoid structures. Robson's group has also investigated possible scaffolding molecules.

Josef Michl is attempting to build a molecular construction kit using molecular rods and connectors, pursuing an explicit vision of "molecular Tinkertoys" by working with simple molecular structures that form stiff, flexible rods. Michl has assembled rods from a mixture of carbon-boron molecules (e.g., 10-vertex or 12-vertex carboranes) and carbon-hydrogen molecules, providing fine control over the total rod length. The rods are built up from more primitive molecular parts, such as propellane (a strained form of C_5H_6) and cubane (a strained form of C_8H_8), to make "staffanes" which are a series of cage units in a linear series. (Strained molecules are constructed with bonds that are forced out of their normal angles, as for example 90° in the case of cubane, compared with carbon's normal orientation of 109.5°.) Michl has fabricated rods whose lengths vary from 0.5 nm to 2.5 nm, in precise 0.1-nm steps. There are other ways of making rodlike molecules, but Michl's are highly inert, do not absorb visible or UV light, are stable up to at least 200°C, and do not react with the oxygen in air even at high temperatures.

Related work is underway in Michl's laboratory on connectors to join the rods together. Metal atoms would be the simplest solution, offering the useful quality of strong joints that can be easily disassembled. Different metals give different binding geometries — square, octahedral, and so on. Many of the connectors are metal-containing species; some are symmetrically trisubstituted or hexasubstituted benzenes or tetrasubstituted cyclobutadiene complexes. Michl must build the right chemical groups onto the rod termini to make them adhere to the connectors as desired. In one example, tiny crosses were built by attaching carboxylate groups to one end of the staffanes, allowing them to bind to a connector made of two rhodium atoms. The groups at the other end of the rods were converted to an ester group, which doesn't bind to rhodium, and so the crosses self-assembled correctly. Michl has also fabricated "star connectors" in which the rods are built into the molecules as covalently bonded arms of a star. For instance, one 3-arm star connector uses three large carboranes coupled to a central benzene ring, allowing a pedestal to be attached vertically to the benzene ring via a ruthenium "sandwich" bond. Additional devices could be installed on this molecular scaffolding, such as optically active molecules or even molecular mechanical "windmills". Nanoscale square and hexagonal planar grids have been fabricated. Extensive ab initio quantum mechanical calculations accompany the experimental work and aid in the interpretation of the results. Michl's ultimate goal is "the production of thin layers of solids of completely controlled aperiodic structure consisting of an inert covalent scaffolding carrying selected active groups."

G. Leach and colleagues used computers to simulate various diamond and graphite nanostruts using molecular dynamics calculations, analysing solid rectangular struts with varying aspect ratios, cross-sections, terminating atoms (for diamond), potential energy functions, and temperatures. The most dramatic differences were seen between struts with 100:1 aspect ratios and struts with lower (10:1 or 1:1) aspect ratios — the diamond strut with a 100:1 aspect ratio and a unit cell $(\sim 0.4\ nm)^2$ cross-section begins to curl (with end-to-end distance decreasing 1.2 nm after 20 picosec) due to thermal vibration (both at 150 K and 300 K), while a unit strut with an aspect ratio of 10:1 or a 100:1 strut with a $(1\ nm)^2$ cross-section showed end-to-end length fluctuations of only ~0.1 nm. The authors concluded that support struts with a cross-section of at least $(1\ nm)^2$ would be sufficient to keep both end-to-end distance and transverse fluctuations below ~0.1 nm in struts with 10:1 aspect ratios. Other work by the same group claims that many nanomachine designs may actually perform better than molecular dynamics results might suggest, or may require fewer atoms for the same positional stability. The elastic and wear characteristics of diamond are being studied computationally by others.

No discussion of molecular "parts" could be complete without mention of the fullerenes, first discovered in 1985 and vigorously investigated ever since. Fullerenes are one of the three known allotropic forms of carbon. In graphite, the most common allotrope, carbon atoms are arranged in hexagonal rings and strongly bonded into parallel planar sheets, with much weaker bonding between the sheets, giving graphite its excellent lubricating properties. In diamond, the second allotropic form, carbon atoms are arranged in a symmetric, tetrahedral structure, giving immense strength. Exposed to air, both diamond surfaces and graphite edges are quickly coated or "passivated" with hydrogen or other atoms that tie up the dangling bonds. In fullerenes, the third allotropic form, carbon atoms form large and hollow cage-like structures, often roughly spherical or tubular, made up of closed curved one-atom-thick sheets of carbon atoms arranged in a number of five-, six-, and higher-membered rings. Interestingly, the fullerenes need no passivating hydrogens or other atoms to satisfy their chemical bonding requirements at the surface — in this sense, "fullerenes are the first and only stable forms of pure, finite carbon".

A new journal, *Fullerene Science and Technology*, appeared in 1993, and by the mid-1990s ~3000 papers on fullerenes and their properties had been published and 149 fullerene-related patents had been issued in the U.S. alone (through 1996). By 1998, an incredible variety of fullerenes of many shapes and sizes had been synthesised.

The original fullerene, C_{60}, consists of exactly 60 carbon atoms arranged in a soccerball structure with 20 hexagons and 12 pentagons (with each pentagon entirely surrounded by hexagons). In theory, small molecules such as H_2 or CO, and probably CH_4, would fit inside C_{60}. A slightly larger variant, C_{70}, is the same except that an extra belt of carbon atoms has been inserted around the equator. Smaller and larger variants have been observed; as cages get bigger, the corners (where the 12 pentagons needed for closure reside) get sharper. The topologically smallest possible fullerene is a dodecahedron consisting of 12 pentagons and 20 carbon atoms, but the ring isomer of C_{20} (a hoop of single carbons) is apparently energetically favoured over the bowl or spherical (fullerene) isomers; in 1998, C_{28} was the smallest fullerene that had been observed experimentally, and C_{36} could be made in significant quantities by arc-discharge. The approximate diameter (measured to atomic centers) of a spheroidal fullerene of formula C_n is $D_{ball} \sim 1.1\ (n/60)^{1/2}$ nanometers. In 1998, purified C_{60} cost \$27.50/gm (99.5% pure) or \$60/gm (99.9%), C_{70} \$250/gm (98%), and C_{84} \$3, 750/gm (90%) from Dynamic Enterprises Ltd. of London.

Insertion of many additional equatorial belts of carbon atoms (incorporated as hexagons) into a spheroidal fullerene results in a long cylinder with graphite-like ("graphene") walls and spherical endcaps, making a class of fullerenes known as the capsular fullerenes or single-walled carbon nanotubes (SWNT) that come in many sizes and chiral forms. The molecular form of carbon nanotubes has been likened to rolled chicken wire. SWNTs may be synthesised by vaporising graphite spiked with 1% catalyst from the nickel, cobalt and iron group above 3000°C, then allowing the vapor to slowly condense. Tubes form because the metal atoms interact with dangling bonds at the end of a tube, favouring tube extension over hemispherical capping by newly arriving carbon vapor atoms. (For any given mass number, ball-shaped fullerenes are energetically favoured over tube-shaped fullerenes during high-temperature synthesis in the absence of catalyst.) SWNTs typically self-assemble in 1.1 nm wide tubes, although smaller single-walled nanotubes and larger single-walled and multi-walled nested nanotubes exist. SWNT carbon atoms are bonded in virtually flawless hexagonal arrays. Simulations show that isolated flaws migrate to the ends of the tube and are eliminated, a phenomenon known as "self-healing". Single-molecule nanotubes of $C_{1,000,000}$ or larger, with lengths >100, 000 times longer than their widths (e.g., ~1 mm long), had been synthesised by 1998. While graphite is a very brittle material, the graphene walls of a carbon nanotube are quite resilient. Computer simulations and experiments confirm that nanotubes kink when bent, then snap back when released. Nanotube diameter is $D_{tube} = 0.078\ (n^2 + nm + m^2)^{1/2}$ nanometers, where (n, m) is the "rollup vector" defined by the number of steps required for a repeat pattern along two crystallographic cylinder wall axes. The cohesive energy per atom required to curve a flat graphene sheet into a cylinder is $E_{rollup} = (13\ zJ\text{-}nm^2 / D_{tube}{}^2)$. (1 zeptojoule (zJ) = 10^{21} joules.) As an example, a (10, 10) nanotube has $D_{tube} = 1.36$ nm and $E_{rollup} = 7$ zJ/atom. Young's modulus for individual SWNTs, assuming a hollow (open-ring) cross-section, has been estimated as high as ~5.5×10^{12} N/m^2; elastic bending modulus (measured as 1×10^{12} to 0.1×10^{12} N/m^2) decreases sharply with increasing diameter (from 8-40 nm). In 1998, purified carbon singlewalled nanotubes were available via the Internet for $1400/gm from Tubes@Rice, or for $200/gm at lesser purity from CarboLex.

From spatial geometry, it is known that hexagonal tiling produces flat sections (like cylinder walls), and inserting pentagons into an hexagonal array produces positive curvature (like spheres), making endcaps. But 7-sided heptagons can also be inserted into hexagonal arrays to induce negative curvature, thus permitting concave surface deformations such as saddle-shaped fullerenes or possibly helical tubular structures. Indeed, note Colbert and Smalley, "the use of hexagonal, pentagonal, and heptagonal substructures are sufficient to produce caged carbon structures of any topology". Nature makes use of this same geometrical principle in the radiolarians, tiny protozoans with fullerene-like siliceous skeletons, and in the conical nucleoprotein core particle of the HIV-1 virus. R. Smalley observes that a graphene sheet has the highest tensile strength of any known 2-dimensional network, and the packing density of atoms in the sheet (e.g., atoms/m^2) is higher than any other network made of any atoms in the periodic table and higher than the packing density in any 2-dimensional slice through any 3-dimensional object — even diamond which has the highest known 3-D packing density. Hence the graphene sheet is effectively impermeable under normal chemical conditions.

Carbon nanotubes have also been observed with a variety of endcap shapes and with tubes that reduced or increased their diameter for some distance before terminating. Toroidal fullerenes can be produced experimentally; conelike, spindlelike, and helical fullerene objects have also been examined. By 1998, the mechanical properties of fullerenes and carbon nanotubes were being extensively investigated and there were experimental demonstrations of:

1. a limited ability to cut fullerene nanotubes to specific desired lengths;
2. dissolving derivatised SWNTs in common organic solvents;
3. the synthesis of self-oriented nanotube arrays; and
4. the trapping of individual C_{60} molecules in a perforated Langmuir-Blodgett film "workpiece holder" (though an even simpler means for producing 2-nm hole arrays with 78 nm hole spacings was later reported).

Nanotubes were also being investigated as nanoscale sensor components.

Given an ability to synthesize a wide variety of fullerene shapes and sizes, the next task is to find ways to join them together in specific desired molecular architectures. Two hydrogens were readily added to fullerene carbon atoms, making specific isomers of $C_{60}H_2$ and $C_{70}H_2$, and in the 1990s fullerene chemistry began to be explored. A little more than a decade after the discovery of C_{60}, gram quantities of buckyballs became widely available for chemical experimentation. Progress in covalent fullerene chemistry exploded. By the late 1990s, fullerene surfaces could be regioselectively functionalised in many interesting ways.

Theory and experiment suggest that SWNTs may be joined at 30° angles to create complex structures, including helices and three-way nanotube junctions as suggested.

In 1998, the idea of using fullerenes as "Tinkertoys" for molecular construction was just starting to be considered. For example, given the synthesis of C_{60} exohedral monoadducts and multiple adducts described above, it may be possible to fabricate simple gears by bonding rigid ligands onto the external surfaces of carbon nanotubes in the manner of gear teeth, although the chemical synthesis of nanotubes with precisely positioned teeth will not be easy. Given the success in 1996 of IBM scientists in positioning individual organic molecules (each having a total of 173 atoms and a 1.5-nm diameter) at room temperature by purely mechanical means, it might also be possible to align and maintain these molecular gear teeth in atomically precise meshed positions. If this could be done, would such devices actually work in the manner of macroscale gears?

J. Han and colleagues at NASA/Ames performed a detailed 2000-atom molecular dynamics simulation to investigate the properties of molecular gears fashioned from carbon nanotubes with teeth added via a benzyne reaction known to occur with C_{60}. Computationally, one gear is powered by forcing the atoms near the end of the nanotube to rotate, and a second gear is allowed to rotate by keeping the atoms near the end of its nanotube constrained to a cylinder (i.e., the ends of the shaft were constrained to not elongate but were allowed to move within a plane transverse to the tube symmetry axis). The meshing aromatic gear teeth transfer angular momentum from the powered gear to the driven gear. Each gear is made of a 1.1-nm diameter (14, 0) nanotube with seven benzyne teeth. The spacing between two nanotubes is 1.8 nm and the smallest distance between a tooth atom and a tube atom is ~0.4 nm. The results show that the gears can operate up to 70 GHz in vacuo at room temperature without overheating or slipping. As rotational speed rises above 150 GHz, the gears overheat and begin to slip with tooth tilting up to 20°, but no bond or tooth breaking occurs up to at least 3000 K and slipping gears can always be returned to proper operation by lowering the temperature or the rotation rate. A related nanotube gear system (58 sprockets, 290-464 atoms) simulated by Robertson and colleagues at the Naval Research Laboratory (NRL) showed similar overheating at 500 GHz but was stable when accelerated to only 20 GHz.

The NASA group also simulated several other types of nanotube-based gear systems. For instance, a rack and pinion system was designed using a gear made from a (14, 0) nanotube with teeth separated by two hexagon rings and a shaft made from a (9, 9) tube with teeth separated by three rings. Gear and shaft are 1.94 nm apart, with tooth face normal to the radial direction of nanotube (14, 0) for the gear, but in the axis direction of nanotube (9, 9) for the shaft. The gear could receive power, driving the shaft, or vice versa, and worked well for shaft translational velocities up to ~100 m/sec. (Most devices described in Drexler typically move at 1 m/sec.) Since shaft mass is almost twice gear mass, it takes more power for the gear to drive the shaft. In another simulation involving a large 1.4-nm gear coupled to a smaller 0.8-nm gear, the large gear drives the smaller gear smoothly, but if power is instead applied to the smaller gear at a sufficiently large acceleration, then the smaller gear does not drive the larger one but instead "bounces back and forth several times, like elastic collisions of a small ball between two boards."

Another research group at Oak Ridge National Laboratory (ORNL) used classical molecular dynamics to investigate the properties of molecular bearings consisting of an inner and an outer carbon nanotube. The graphite bearings ranged in size from inner shafts between 0.4-1.6 nm in diameter up to 12 nm long, and outer cylinders between 1.0-2.3 nm in diameter up to 4 nm long. The original simulations found excessive vibrational motion, but subsequent work using a more complete quantum approach found that under certain conditions the nanobearing is "frictionless" and undergoes superrotation, a classical dynamical behaviour reminiscent of superfluidity. The regime of superrotary motion is somewhat restricted when the nanobearing is under load, which suggests that a very careful design is required to ensure optimum performance.

The ORNL group simulated a fullerene motor consisting of two concentric graphite cylinders (shaft and sleeve) with one positive and one negative electric charge attached to the shaft. Rotational motion of the shaft was induced by applying one, or sometimes two, oscillating laser fields. The shaft cycled between periods of undesirable rotational pendulum-like behaviour and good unidirectional motor-like behaviour. The NASA/Ames group simulated a pulsed-laser-powered carbon nanotube gear-motor system which rotated consistently in one direction, although that direction could be either clockwise or counterclockwise.

Nanotechnology and Nanostructured Materials

Nanomaterial technology is receiving a great deal of attention in the coatings industry today. Several approaches within this technology can be used to achieve organic-inorganic nanocomposite or nanostructured coatings. These approaches include incorporation of preformed nanoparticles in organic resin systems, in-situ generation of nanoparticles or nanophases, and other nanostructuring mechanisms. Following an overview of benefits and critical issues of nanomaterial technology, this article reviews advancements in the application of nanomaterial technology to coatings. Several commercial applications of the technology are described.

Conventional materials have grains sizes ranging from microns to several millimeters and contain several billion atoms each. Nanometer sized grains contain only about 900 atoms each. As the grain size decreases, there is a significant increase in the volume fraction of grain boundaries or interfaces. This characteristic strongly influences the chemical and physical properties of the material. For example, nanostructured ceramics are tougher and stronger than the coarser grained ceramics. Nanophase metals exhibit significant increases in yield strength and elastic modulus. It has also been shown that other properties(electrical, optical, magnetic...) are influenced by the fine grained structure of these materials. Using a variety of synthesis methods, it is possible to produce nanostructured materials in the following forms: thin films, coatings, powders and as a bulk material. There is also considerable interest in the generation of carbon nanostructures, which are related to the famous Buckyball. In addition, the use of nano-sized materials as fillers for composite materials is generating interest. specifically in the case of polymer nanocomposites.

There are several different types of nanostructured materials. These range from zero dimensional atom clusters to three dimensional equiaxed grain structure. Each class has at least one dimension in the nanometer range. Atom clusters and filaments are defined as zero modulation dimensionality and can have any aspect ratio from 1 to ¥2. Any multilayered material with layer thickness in the nanometer range is classified as one-dimensionally modulated. Layers in the nanometer thickness range consisting of ultrafine grains (nanometer range diameter) are two-dimensionally modulated. This class includes coatings, buried layers and thin films. The last class is that consisting of three dimensionally modulated microstructures or nanophase materials.

The methods employed to produce nanostructured materials are numerous, with each method having advantages and disadvantages depending on the desired properties or application. Atom clusters are typically synthesised via a vapor condensation route which is essentially evaporation of a solid metal followed by rapid condensation to form nano-sized clusters. These resulting powders are essentially agglomerates of the nano-sized atom clusters. The powders can be then be used as fillers for composite materials or consolidated into bulk material. The cluster assembly method can be used to produce ceramic or metal nanostructured powders. The main advantage of this method is the extremely low contamination levels and the ability to control the final cluster size by manipulation of the process parameters such as temperature, gas environment and evaporation rate. Several different methods have been developed using the vapor condensation concept. These include inert gas condensation, chemical vapor condensation, laser ablation, electron beam deposition, arc evaporation, etc.

Both metals and ceramics can be produced using a variety of chemical approaches such as sol-gel or thermal decomposition. These methods provide large quantities of nano-sized agglomerates at a low cost. Chemical processes also allow effective control of the stoichiometry of the final product. However, the precursor chemicals required may leave residual contamination on the particle surfaces, which leads to difficulties in compaction and sintering. Additionally, powders produced through wet chemical techniques often have difficulties with agglomeration. Other methods which implement chemical reactions include Examples: A common method of producing nanostructured powders is through mechanical deformation, i.e. milling or shock deformation. These processes produce nanostructured materials through severe mechanical deformation of a coarse grained precursor material. The nanometer sized grains nucleate within the dislocation cell structures located in the shear bands. The final grain size is a function of the amount of energy input during the milling as well as time, temperature during milling and milling atmosphere. The precursor materials can be crystalline or amorphous materials in the form of powders or tapes. Through mechanical milling, it is possible to produce nanostructured powders of systems which are otherwise immicsible. A major disadvantage of this method is the possibility of contamination from the milling media due to the large forces and energies involved in the milling process. Other synthesis methods based on *mechanical deformation*: friciton/sliding wear. Examples: Three dimensional nanostructured materials are also synthesised through thermal crystallisation of an amorphous material. By controlling the nucleation and growth during annealing of an amorphous material, one can produce a bulk material with average grain sizes of less than 20 nm without the need for consolidation and sintering steps. However, this process is limited to material composition which are readily available in the form of a metallic glass, which has an amorphous microstructure.

The atomic structure of nanostructured materials is unlike that seen in glasses or crystals because of the large volume fraction (greater than 50 per cent) of grain boundaries or interfaces. Coarse-grained materials have less than 3 per cent of all atoms associated with grain boundaries or interfaces. However, in materials with nanosized grains between 5 per cent and 50 per cent of all atoms are associated with grain boundaries or interfaces. The grain boundary structure in these nanostructured materials is similar to that of the conventional coarse grained materials.

The material properties of the nanostructured materials show remarkable improvement or deviation from the properties exhibited by the coarser grained material. These unique properties are attributed to the significant increase in grain boundary area due to the small grain size. In terms of the mechanical properties, nanostructured metals have shown increases in hardness values (ultimate and yield strengths). Specifically, pure nanophase metals have shown a clear increase in hardness with decreasing grain size, following the well known Hall-Petch equation. Nanostructured ceramics have exhibited superplastic properties at low temperatures. This is significant as ceramics are conventionally brittle materials. Further deviations from trends observed in conventional materials are observed in creep of nanostructured metals.

Nanostructure Science and Technology

Nanostructure science and technology is a broad and interdisciplinary area of research and development activity that has been growing explosively worldwide in the past few years. It has the potential for revolutionising the ways in which materials and products are created and the range and nature of functionalities that can be accessed. It is already having a significant commercial impact, which will assuredly increase in the future. A worldwide study of research and development status and trends in nanoparticles, nanostructured materials, and nanodevices (or more concisely, nanostructure science and technology) was carried out during the period 1996-98 by an eight-person panel under the auspices of the World Technology (WTEC) Division of Loyola College. Led by the National Science Foundation, a wide range of U.S. government agencies commissioned and funded this study: the Air

Force Office of Scientific Research, the Office of Naval Research, the Department of Commerce (including the National Institute of Standards and Technology and the Technology Administration), the Department of Energy, the National Institutes of Health, and the National Aeronautics and Space Administration. Their support indicates the breadth of interest in and the far-reaching potential of this burgeoning new field. The purpose of the study was to assess the current status and future trends internationally in research and development in the broad and rapidly growing area of nanostructure science and technology. The goals were fourfold:

1. provide the worldwide science and engineering community with a broadly inclusive and critical view of this field
2. identify promising areas for future research and commercial development
3. help stimulate development of an interdisciplinary international community of nanostructure researchers
4. encourage and identify opportunities for international collaboration

This report is the principal volume in a three-part publication of the activities and findings of the WTEC panel; it is an overview of the panel's observations and conclusions regarding nanostructure science and technology worldwide. It includes reviews of panel workshops held in Germany and Sweden, as well as site reports of panel visits to university, government, and industry laboratories in Europe, Japan, and Taiwan. An earlier volume, published in January 1998, reported the proceedings of a WTEC workshop on R&D Status and Trends in Nanoparticles, Nanostructured Materials, and Nanodevices in the United States (Baltimore: Loyola College, International Technology Research Institute, NTIS #PB98- 117914). A third volume to be published by WTEC reports the proceedings of a workshop held in St. Petersburg, Russia on related work.

Findings

There are two overarching findings from this WTEC study:

First, it is abundantly clear that we are now able to nanostructure materials for novel performance. That is the essential theme of this field: novel performance through nanostructuring. It represents the beginning of a revolutionary new age in our ability to manipulate materials for the good of humanity. The synthesis and control of materials in nanometer dimensions can access new material properties and device characteristics in unprecedented ways, and work is rapidly expanding worldwide in exploiting the opportunities offered through nanostructuring. Each year sees an ever increasing number of researchers from a wide variety of disciplines enter the field, and each year sees an ever increasing breadth of novel ideas and exciting new opportunities explode on the international nanostructure scene.

Second, there is a very wide range of disciplines contributing to the developments in nanostructure science and technology worldwide. The rapidly increasing level of interdisciplinary activity in nanostructuring is exciting and growing in importance, and the intersections between the various disciplines are where much of the novel activity resides. The field of nanostructure science and technology has been growing very rapidly in the past few years, since the realisation that creating new materials and devices from nanoscale building blocks could access new and improved properties and functionalities. While many aspects of the field existed well before nanostructure science and technology became a definable entity in the past decade, it has only become a coherent field of endeavor through the confluence of three important technological streams:

1. new and improved control of the size and manipulation of nanoscale building blocks
2. new and improved characterisation (spatial resolution, chemical sensitivity, etc.) of materials at the nanoscale
3. new and improved understanding of the relationships between nanostructure and properties and how these can be engineered

As a result of these developments, a wide range of new opportunities for research and applications in the field of nanotechnology now present themselves. Some examples of present and potential applications with significant technological impact that were identified in the course of this study. Considerable resources are being expended around the world for research and development aimed at realising these and a variety of other promising applications. Government funding alone approached half a billion dollars per year in FY 1997: $128 million in Western Europe; $120 million in Japan; $116 million in the United States; and $70 million altogether in other countries such as China, Canada, Australia, Korea, Taiwan, Singapore, and the countries of the former Soviet Union.

An overall comparison of the current levels of activity among the major regions assessed (Europe, Japan, and the United States) in the various areas of the WTEC study. These broad areas— synthesis and assembly, biological approaches and applications, dispersions and coatings, high surface area materials, nanodevices, and consolidated materials—constitute the field of nanostructure science and technology. These are the areas around which the study was crafted. In the synthesis and assembly area, the United States appears to be ahead, with Europe following and then Japan. In the area of biological approaches and applications, the United States and Europe appear to be rather on a par, with Japan following. In nanoscale dispersions and coatings, the United States and Europe are again at a similar level, with Japan following. For high surface area materials, the United States is clearly ahead of Europe, followed by Japan. On the other hand, in the nanodevices area,

Japan seems to be leading quite strongly, with Europe and the United States following. And finally, in the area of consolidated materials, Japan is a clear leader, with the United States and Europe following. These comparisons are, of course, integrals over rather large areas of a huge field and therefore possess all of the inevitable faults of such an integration. At best, they represent only a snapshot of the present, and the picture is admittedly incomplete.

More detailed findings in each of these major areas are included in the individual chapters of this report, along with additional general findings and observations compares the scope and funding levels for the relevant nanostructure science and technology R&D programmes around the world. The appendices give details on the site visits and workshops of the panel: B contains the Europe site reports, C contains notes on workshops held in Germany and Sweden, D contains the Japan site reports, and E contains the Taiwan site reports.

Challenges

We are now at the threshold of a revolution in the ways in which materials and products are created. How this revolution will develop, how great will be the opportunities that nanostructuring can yield, and how rapidly we progress, will depend upon the ways in which a number of challenges are met.

Among the challenges facing us are those concerned with making the necessary advances in enabling technologies in order for rapid progress to continue in this field. We must increase characterisation capabilities in visualisation and chemical analysis at ever finer size scales. We must be able to manipulate matter at ever finer size scales, and we must eventually use computational approaches in directing this, if we are really going to take full advantage of the available opportunities. Experiment simply cannot do it alone. Theory and modeling are essential. Fortunately, this is an area in which the sizes of the building blocks and their assemblies are small enough that one can, with the ever increasing capabilities in computational sciences, now start doing very serious controlled modeling experiments to guide us in the nanostructuring of matter. Hence, multiscale modeling of nanostructuring and the resulting materials properties across the hierarchy of length scales from atomic, to mesoscopic, to macroscopic is an absolute necessity as we go down the road in the next decades to realising the tremendous potential of nanostructure science and technology.

Furthermore, we need to understand the critical roles that surfaces and interfaces play in nanostructured materials. Nanoparticles have very high specific surface areas, and thus in their assembled forms there are large areas of interfaces. We need to know in detail not only the structures of these interfaces, but also their local chemistries and the effects of segregation and interaction between the nanoscale building blocks and their surroundings. We need to learn more about the control of nanostructure size and size distribution, composition, and assembly. For some applications, there are very stringent conditions on these parameters; in other applications less so. We must therefore understand the relationships between this stringency and the desired material or device properties.

We also need to be concerned with the thermal, chemical, and structural stability of nanostructured materials and the devices made therefrom, in the face of both the temperature and changing chemistries of the environments in which these nanostructures are asked to function. A nanostructure that is only a nanostructure at the beginning of a process is not of much use to anybody, unless the process is over in a very short time or the process itself is the actual nanostructure advantage. So for many applications, stability is an important consideration, and we must investigate whether natural stability is sufficient or whether we must additionally stabilize against changes that we cannot afford.

To effectively commercialize and utilize the nanostructuring of matter we also need enhancements in statistically driven process control. Achieving reproducibility and scalability of nanoparticle synthesis and consolidation processes in nanostructuring are paramount if successful scale-up is to be effected and if what we do in the laboratory is to contribute to the society that pays for this research. Given a commercial need, the viability of nanostructure production and utilisation is wrapped up in the costs of precursors or raw materials, processing costs, and also the costs of dealing with effluent. It is the total integrated cost, in terms of raw materials, synthesis of the building blocks, manufacture of parts from those building blocks, and effluent clean-up costs, that is important and that will ultimately determine commercial viability.

Finally, in order for the field of nanostructure science and technology to truly reach fruition, it is an absolute necessity to create a new breed of researchers who can work across traditional disciplines and think "outside the box." Educating this new breed of researchers, who will either work across disciplines or know how to work with others in the interfaces between disciplines, is vital to the future of nanostructure science and technology. People must start thinking in unconventional ways if we are to take full advantage of the opportunities in this new and revolutionary field.

Nano Synthesis and Assembly

The common theme of this WTEC study is the engineering of materials with novel (i.e., improved) properties through the controlled synthesis and assembly of the material at the nanoscale level. The range of applications is extremely broad, and these will be described in further detail in subsequent chapters in this report. The corresponding means of synthesis and assembly are similarly wide-ranging. But however multifaceted the synthesis approaches and the ultimate applications, there are common issues and unique defining features of these nanostructured materials. First, there is the recognition of critical scale lengths that define the material structure and organisation, generally in the nanometer range, and that ultimately determine the fundamental macroscopic properties of the material. Research in nanostructured materials is motivated by the belief that ability to control the building blocks or nanostructure of the materials can result in enhanced properties at the macroscale: increased hardness, ductility, magnetic coupling, catalytic enhancement, selective absorption, or higher efficiency electronic or optical behaviour. Synthesis and assembly strategies accommodate precursors from liquid, solid, or gas phase; employ chemical or physical deposition approaches; and similarly rely on either chemical reactivity or physical compaction to integrate nanostructure building blocks within the final material structure. The variety of techniques is shown schematically

Schematic of variety of nanostructure synthesis and assembly approaches. The "bottom-up" approach first forms the nanostructured building blocks and then assembles them into the final material. An example of this approach is the formation of powder components through aerosol techniques (Wu et al. 1993) and then the compaction of the components into the final material. These techniques have been used extensively in the formation of structural composite materials. One "top-down" approach begins with a suitable starting material and then "sculpts" the functionality from the material. This technique is similar to the approach used by the semiconductor industry in forming devices out of an electronic substrate (silicon), utilising pattern formation (such as electron beam lithography) and pattern transfer processes (such as reactive ion etching) that have the requisite spatial resolution to achieve creation of structures at the nanoscale. This particular area of nanostructure formation has tremendous scope, warranting its own separate study, is a driving issue for the electronics industry, and will not be a principal theme of this study. Another top-down approach is "ball-milling," the formation of nanostructure building blocks through controlled, mechanical attrition of the bulk starting material (Koch 1989). Those nano building blocks are then subsequently assembled into a new bulk material. In fact, many current strategies for material synthesis integrate both synthesis and assembly into a single process, such as characterizes chemical synthesis of nanostructured materials (Murray et al. 1993; Katari et al. 1994). The degree of control required over the sizes of the nanostructure components, and the nature of their distribution and bonding within the fully formed material varies greatly, depending on the ultimate materials application. Achieving selective optical absorption in a material (e.g., UVblocking dispersions) may allow a wide range of sizes of the component nanostructure building blocks, while quantum dot lasers or single electron transistors require a far tighter distribution of size of the nanostructure components. Compaction methods may provide excellent adhesion for nanocomposite materials of improved structural performance (e.g., ductility), but such interfaces may be unsatisfactory for electronic materials. The intention of this chapter of the report is not to recapitulate in detail the various synthesis and assembly techniques that have been and are being employed in the fabrication of nanostructured materials; that detail can be found in succeeding chapters as well as in excellent summary descriptions provided in the May 8-9, 1997 WTEC workshop proceedings (Siegel, Hu, and Roco 1998). Rather, in attempting to capture the salient features of a new impetus for and interest in a field of nanostructure science and technology, it is more useful to identify the emerging commonalities than the differences among synthesis and assembly approaches.

Critical Issues for Nanostructure Syntehsis and Assembly

However broad the range of synthesis approaches, the critical control points fall into two categories:

1. control of the size and composition of the nanocluster components, whether they are aerosol particles, powders, semiconductor quantum dots, or other nanocomponents

2. control of the interfaces and distributions of the nanocomponents within the fully formed materials These two aspects of nanostructure formation are inextricably linked; nevertheless, it is important to understand how to exercise separate control over the nucleation of the nanostructure building blocks and the growth (for example, minimising coagulation or agglomeration) of those components throughout the synthesis and assembly process. This latter issue is related to the importance of the following:

- the chemical, thermal, and temporal stability of such formed nanostructures
- the ability to scale-up synthesis and assembly strategies for low-cost, large-scale production of nanostructured materials, while at the same time maintaining control of critical feature size and quality of interfaces (economic viability is a compelling issue for any nanostructure technology) All researchers in this area are addressing these issues.

Common Enabling Technologies

There has been steady technological progress in all fields of nanostructure synthesis and assembly, in no small part because of the more general availability of characterisation tools having higher spatial, energy, and time resolution to clearly distinguish and trace the process of nanostructure formation. As transmission electron microscopy and X-ray diffraction techniques helped in an earlier period to relate the improved properties of "age-hardened" aluminum alloys to their nanostructure (Koch 1998), today's technological advances in materials characterisation are providing new insights into the role of the nanostructure in determining macroscopic properties. The tightly-coupled iteration between characterising the nanostructure, understanding the relationship between nanostructure and macroscopic material properties, and improved sophistication and control in determining nanostructure size and placement have accelerated the rate of progress and helped to define the critical components of this "new" field of nanostructure science and technology. Tightly focused (1-2 μm), high brightness synchrotron X-ray sources provide detailed structural information on colloids, polymers, alloys, and other material structures, highlighting the inhomogeneities of the material with suitable spatial resolution (Hellemans 1998).

Improved characterisation - higher spatial resolution - higher sensitivity leads t o motivat es Better control of size and placement Understanding of - structure-property link - influence of interfaces stimulates.

Interactive cycle of characterisation, understanding and enhanced control in the synthesis and assembly of nanostructures. Another important enabling technology has been the now widely available scanning probe technology, including scanning tunneling microscopy and atomic force microscopy. The power of these techniques has provided impetus for developing even higher performance scanning probe tips, fabricated through microfabrication techniques. Development of different tip structures in various materials has given rise to an entire family of powerful scanning probe techniques that encompass such a wide range of characterisation capabilities that one can envision "a laboratory on a tip" (Berger et al. 1996). The development of a tip technology also impacts the synthesis and assembly processes themselves: scanning probe technologies have been used as the basis of materials patterning and processing at nanometer scales (Held et al. 1997; Snow et al. 1997; and Wilder et al. 1997) and have provided information on the mechanical and thermal properties of materials at the nanoscale (Nakabeppu et al. 1995; Tighe et al. 1997; and Zhang et al. 1996). More sophisticated in situ monitoring strategies have provided greater understanding and control in the synthesis of nanostructured building blocks, particularly those formed in vacuum environments. Molecular beam epitaxy (MBE) represents a physical vapor (gas phase) deposition technique where sub-monolayer control can be imposed on the formation of two-dimensional and, more recently, three-dimensional nanostructured materials (Leonard et al. 1993). A great deal of the understanding and control derives from the ability to carry out sensitive monitoring of the growth process in situ: reflection high energy electron diffraction (RHEED) details the nature of the surface and surface bonding, and oscillations of the RHEED intensity provide information on the growth rate (Neave et al. 1983). The improvements brought about by these advances in technology have been substantial, but perhaps of greater importance for this nascent field of nanostructure science and technology has been the development of strategies and technologies that have been formed across the former disciplines. More reliable means of controlling nanostructure size and placement, with an end view of being able to scale up the production of such materials while maintaining the control over the nanostructure, have given impetus to a common search for novel synthesis and assembly strategies. In that search, it is apparent that the naturally occurring synthesis and assembly of biological materials can provide us with some critical insights.

Major efforts in nanoparticle synthesis can be grouped into two broad areas: gas phase synthesis and sol-gel processing. Nanoparticles with diameters ranging from 1 to 10 nm with consistent crystal structure, surface derivatisation, and a high degree of monodispersity have been processed by both gas-phase and sol-gel techniques. Typical size variances are about 20%; however, for measurable enhancement of the quantum effect, this must be reduced to less than 5% (Murray et al. 1993). Initial development of new crystalline materials was based on nanoparticles generated by evaporation and condensation (nucleation and growth) in a subatmospheric inert-gas environment (Gleiter 1989; Siegel 1991, 1994). Various aerosol processing techniques have been reported to improve the production yield of nanoparticles (Uyeda 1991, Friedlander 1998). These include synthesis by combustion flame (Zachariah 1994, Calcote and Keil 1997, Axelbaum 1997, Pratsinis 1997); plasma (Rao et al. 1997); laser ablation (Becker et al. 1997); chemical vapor condensation (Kear et al. 1997); spray pyrolysis (Messing et al. 1994); electrospray (de la Mora et al. 1994); and plasma spray (Berndt et al. 1997). Sol-gel processing is a wet chemical synthesis approach that can be used to generate nanoparticles by gelation, precipitation, and hydrothermal treatment (Kung and Ko 1996). Size distribution of semiconductor, metal, and metal oxide nanoparticles can be manipulated by either dopant introduction (Kyprianidou-Leodidou et al. 1994) or heat treatment (Wang et al. 1997). Better size and stability control of quantum-confined semiconductor nanoparticles can be achieved through the use of inverted micelles (Gacoin 1997), polymer matrix architecture based on block copolymers (Sankaran et al. 1993) or polymer blends (Yuan et al. 1992), porous glasses (Justus et al. 1992), and ex-situ particle-capping techniques (Majetich and Canter 1993; Olshavsky and Allcock 1997).

Additional nanoparticle synthesis techniques include sonochemical processing, cavitation processing, microemulsion processing, and highenergy ball milling. In sonochemistry, an acoustic cavitation process can generate a transient localised hot zone with extremely high temperature gradient and pressure (Suslick et al. 1996). Such sudden changes in temperature and pressure assist the destruction of the sonochemical precursor (e.g., organometallic solution) and the formation of nanoparticles. The technique can be used to produce a large volume of material for industrial applications. In hydrodynamic cavitation, nanoparticles are generated through creation and release of gas bubbles inside the sol-gel solution (Sunstrom et al. 1996). By rapidly pressurising in a supercritical drying chamber and exposing to cavitational disturbance and high temperature heating, the sol-gel solution is mixed. The erupted hydrodynamic bubbles are responsible for nucleation, growth, and quenching of the nanoparticles. Particle size can be controlled by adjusting the pressure and the solution retention time in the cavitation chamber. Microemulsions have been used for synthesis of metallic (Kishida et al. 1995), semiconductor (Kortan et al. 1990; Pileni et al. 1992), silica (Arriagada and Osseo-Assave 1995), barium sulfate (Hopwood and Mann 1997), magnetic, and superconductor (Pillai et al. 1995) nanoparticles. By controlling the very low interfacial tension (~10-3 mN/m) through the addition of a cosurfactant (e.g., an alcohol of intermediate chain length), these microemulsions are produced spontaneously without the need for significant mechanical agitation. The technique is useful for large-scale production of nanoparticles using relatively simple and inexpensive hardware (Higgins 1997). Finally, high energy ball milling, the only top-down approach for nanoparticle synthesis, has been used for the generation of magnetic (Leslie- Pelecky and Reike 1996), catalytic (Ying and Sun 1997), and structural (Koch 1989) nanoparticles. The technique, which is already a commercial technology, has been considered dirty because of contamination problems from ball-milling processes. However, the availability of tungsten carbide components and the use of inert atmosphere and/or high vacuum processes have reduced impurities to acceptable levels for many industrial applications. Common drawbacks include the low surface area, the highly polydisperse size distributions, and the partially amorphous state of the asprepared powders.

Other Synthesis Issues Means to Achieve Monodispersity

One of the most challenging problems in synthesis is the controlled generation of monodispersed nanoparticles with size variance so small that size selection by centrifugal precipitation or mobility classification is not necessary. Among all the synthesis techniques discussed above, gas-phase synthesis is one of the best techniques with respect to size monodispersity, typically achieved by using a combination of rigorous control of nucleationcondensation growth and avoidance of coagulation by diffusion and turbulence as well as by the effective collection of nanoparticles and their handling afterwards. The stability of the collected nanoparticle powders against agglomeration, sintering, and compositional changes can be ensured by collecting the nanoparticles in liquid suspension. For semiconducting particles, stabilisation of the liquid suspension has been demonstrated by the addition of polar solvent (Murray et al. 1993); surfactant molecules have been used to stabilize the liquid suspension of metallic nanoparticles.

Alternatively, inert silica encapsulation of nanoparticles by gas-phase reaction and by oxidation in colloidal solution has been shown to be effective for metallic nanoparticles (Andres et al. 1998). New approaches need to be developed for the generation of monodisperse nanoparticles that do not require the use of a size classification procedure. An example of this is a process developed in Japan where very monodispersed gold colloidal nanoparticles with diameters of about 1 nm have been prepared by reduction of metallic salt with UV irradiation in the presence of dendrimers (Esumi et al. 1998). Poly(amidoamine) dendrimers with surface amino groups of higher generations have spherical 3-D structures, which may have an effective protective action for the formation of gold nanoparticles. Although the specific role of dendrimers for the formation of monodispersed nanoparticles has yet to be defined, good monodispersity is thought to come from the complex reaction accompanying the decomposition of dendrimers, which eventually leads to the conversion of solution ions to gold nanoparticles.

Scaleup production is of great interest for nanoparticle synthesis. High energy ball milling, already a commercial high-volume process, as mentioned above, has been instrumental in generating nanoparticles for the preparation of magnetic, structural, and catalytic materials. However, the process produces polydispersed amorphous powder, which requires subsequent partial recrystallisation before the powder is consolidated into nanostructured materials. Although gas-phase synthesis is generally a low production rate process (typically in the 100 milligrams per hour range) in research laboratories, higher rates of production (about 20 grams per hour) are being demonstrated at Ångström Laboratory at Uppsala University in Sweden. Even higher production rates (about 1 kg per hour) are now being achieved commercially. For sol-gel processing, the development of continuous processing techniques based on present knowledge of batch processing has yet to be addressed for economical scaleup production of nanoparticles. Other related sol-gel issues concern the cost of precursors and the recycling of solvent. Overall, sol-gel processing is attractive for commercial scale-up production.

A recent paradigm shift envisioned for optoelectronics and computational devices involves the assembly of molecular or quantum wires (Chidsey and Murray 1986). Large polymeric molecules have been used as nano building blocks for nanoporous molecular sieves, biocompatible materials, optical switching, data processing, and other nonlinear optical components. Chain aggregates of nanoparticles can be considered as polymer-like units with their primary particles composed of a few hundred to a few thousand molecules. Thus, these chain aggregates can be considered "heavy" quantum wires. In fact, nanoparticle chain aggregates have been studied extensively as magnetic materials (Zhang and Manthiram 1997), as reinforced elastomers (Pu et al. 1997), and as additives in concrete (Sabir 1997). These aggregates have been shown to have chemical and mechanical properties different from those of individual primary particles (Friedlander et al. 1998). Depending on the particle size and its compositional material, the bonding force responsible for holding the aggregates together varies from weak van der Waals force for micrometer particles to strong chemical bonds for nanometer particles to very strong magnetic dipolar bonds for nanosized magnetic particles. The mechanical, optical, and electronic transport properties can be varied by controlling the diameter and the monodispersity of the primary particles, the crystalline structure and morphology, aggregate length, interfacial properties, and material purity. These chain aggregates can be formed by allowing agglomeration of nanoparticles generated by any of the synthesis techniques discussed above, with the exception of high energy ball milling, which generates particles with low surface area and high anisotropic morphologies, both of which are detrimental for the formation of chain aggregates. Depending on the magnetic and electric charging properties of the nanoparticles, an external applied magnetic or electric field can be used to control the fractal dimension of aggregates. For optical applications of chain aggregates, lower fractal dimensions (i.e., relatively straight chain aggregates with few branches) are desirable.

Recent advances in the fabrication of nanometer fibers or tubes offer another form of building blocks for nanostructured materials. An effective way to generate nanometer fibers (or tubes) is based on the use of membrane-template techniques (Martin 1994). Membranes, with nanochannels generated by fission-fragment tracks or by electrochemical etching of aluminum metal, are used as templates for either chemical or electrochemical deposition of conductive polymers (Pathasarathy and Martin 1994), metal (van de Zande et al. 1997), semiconductor (Klein et al. 1993), and other materials for the generation of nanofibers or tubes. Since the nanochannels on membranes are very uniform in size, the diameter and the aspect ratio of the nanofibers (or tubes) synthesised by the membranetemplate technique can be precisely controlled. This greatly facilitates the interpretation of optical data and the processing of these fibers (or tubes) into 2-D nanostructured materials (de Heer et al. 1995). Single-crystal semiconductor nanofibers can also be grown catalytically by metalorganic vapor phase epitaxy and laser ablation vapor-liquid-solid techniques (Hiruma et al. 1995; Morales and Lieber 1998).

The synthesis of these onedimensional structures with diameters in the range of 3 to 15 nm holds considerable technological promise for optoelectronic device applications, such as the p-n junctions for light emission at Hitachi Central Research Laboratory in Japan. The advent of carbon-based nanotubes has created yet another way to fabricate nanometer fibers and tubes. These nanotubes have been used as templates for the fabrication of carbide and oxide nanotubes (Dai et al. 1995; Kasuga et al. 1998). Synthesis of nanotubes based on BN, BC3 and BC2N have also been reported (Chopra et al. 1995; Miyamoto et al. 1994). These nanotubes potentially possess large third-order optical non-linearity and other unusual properties (Xie and Jiang 1998). Metallic nanofibers synthesised by carbon-nanotube-template techniques are useful in the design of infrared absorption materials. The carbon nanotubes can now be catalytically produced in large quantities and have been used for reinforcement of nanostructural composite materials and concrete (Peigney et al. 1997). The elegant complexity of biological materials represents the achievement of structural order over many length scales, with the full structure developed from the "nested levels of structural hierarchy" (Aksay et al. 1992), in which self-assembled organic materials can form templates or scaffolding for inorganic components. These notions of a multilevel material structure with strong interactions among levels and an interplay of perfection and imperfection forming the final material was discussed earlier by Cyril Stanley Smith (1981). Along with characteristic length scales, there are characteristic relaxation times of the material, bringing in the consideration of the temporal stability of the structured materials (Zener 1948). A more detailed discussion of the role of biological materials as both paradigm and tool for the fabrication of nanostructured materials is given. It is interesting how many of the synthesis and assembly approaches seek out and adapt two key features of biogenic fabrication: that of "self-assembly, " and the use of natural nanoscaled templates.

Natural Templates or "Scaffolds"

Biogenic systems employ natural templates or scaffolding in the construction of nanoscaled materials, where the templates and scaffolding can help set the proper, critical dimensions that are characteristic of the final material. The templates can be formed "artificially, " through the use of lithographically defined patterning and processing of the substrate material in order to achieve selective nucleation and growth of the nanostructured materials. However, there are also a host of "natural" templating materials that researchers can avail themselves of. The zeolites have for a long time been utilised as the basis of high surface area materials that enable catalysis. Recent work has taken these natural cage structures and formed large cage structures with varying pore sizes (Estermann et al. 1991; Ying and Sun 1997; Bu et al. 1998). When zeolite cages are "loaded" with various materials, the controlled proximity enforced by the cage structure can result in various magnetically active or nonlinear optical materials.

More recently, the C-based fullerenes and C-nanotubes have provided a tremendously rich basis for nanostructure research; the impact of such materials is further described. The by now "easy" formation of such structures and ready availability of such materials provide a rich source of nanostructures with nearly immediate applications for electronic devices, storage, and enhancement of structural materials. Further, these materials can serve as the nanostructured precursors of nanostructure synthesis approaches (such as the starting material for cluster or aerosol deposition), or can provide an avenue for the synthesis of templates of varying sizes. Such templates can be electronically "doped" and filled with other materials. Recent work reported the synthesis of "nanorods" of GaN through confined reactions in carbon nanotubes (Han et al. 1998).

Outstanding Issues: Opportunities and Challenges

The broad explorations of this WTEC study have shown a great diversity of impressive work on nanostructured materials. One could argue that many aspects of the work on nanostructured materials have been long-established efforts, with well-developed techniques that have been brought forth into the manufacturing arena. How do we then explain the current excitement and interest in nanostructured materials as a promising new endeavor? Part of the answer lies in the recognition of common scientific issues and common enabling technologies that link this group of researchers together. Ready availability of ever more sophisticated characterisation methods that allow us to visualize and probe materials at the nanoscale has accelerated the pace of activities in this field; at the same time, recognition of the common critical issues of control over nanostructure size and placement motivates sharing of solutions over the boundaries of conventional disciplines.

Thus, some of the most exciting findings of this study manifest the crossfertilisation of techniques and ideas; for example, aerosol particles are being integrated with more traditionally fabricated electronic nanostructures, with placement achieved through the manipulation of STM tips, in order to explore ideas of electronic device enhancement at the nanostructure level (Junno et al. 1998). The ordered assembly of diblock copolymers define three-dimensional nanostructures, and those structures are transferred into electronic (semiconductor) substrates through high resolution pattern transfer processes (Möller 1998). Recognising the need for effective utilisation of ideas across disciplines brings the responsibility of establishing the educational infrastructure that will adequately train young scientists to more fully develop the concepts and applications of nanostructured materials in the future. Such an infrastructure will require researchers and educators who are familiar with the properties of a broad range of materials, including polymers, biomolecular materials, metals, ceramics, and semiconductors.

It will benefit from an informed perspective on critical applications, and must provide access to a wide range of synthesis and assembly techniques and characterisation methods that currently reside separately in the disciplines of physics, chemistry, electrical engineering, biology, etc. Finally, a critical enabler for the future of this field and for its educational infrastructure is further development of computational tools that encompass the full range of atomistic calculations to macroscopic materials properties. This will require a "systems approach" (Olson 1997) that will span a variety of different computational models, addressing the different critical length scales, starting with solutions of the Schrodinger and Poisson equations, solving interatomic forces, and scaling all the way up to simulations of macroscopic properties and behaviour (Goddard 1998). Increased appreciation of and access to the diverse means of nanoparticle synthesis and assembly that have been developed within many different disciplines, and a common development of enabling tools and technologies, will enhance the pace of accomplishments in this new area of nanoscale synthesis and assembly.

Current-induced Local Oxidation

A novel route is introduced for oxidising thin metal films with nanometer-scale resolution in air. By locally subjecting Ti films to high in-plane current densities, metal-oxide tunneling barriers are formed in a self-limiting fashion. The oxidation is triggered by current-induced atomic rearrangements and local heating. At the final stages of the barrier formation, when only atomic-scale channels remain unoxidised, the oxidation rate decreases drastically while the conductance drops in steps of about the conductance quantum. This behaviour gives evidence of ballistic transport and a superior stability of such metallic nanowires against current-induced forces compared with the bulk metal. We used the current-induced local oxidation process to fabricate a single-electron transistor, which exhibits a Coulomb staircase at room temperature.

The CILO process is based on thin Ti films as illustrated in the sketch and the AFM picture above. Subsequently, oxide notches are via AFM oxidation such that a narrow metallic gap is formed.

Upon applying a constant bias voltage across the gap under ambient conditions (in air), the measured current starts to decrease rapidly as a function of time, stabilising seconds later at a two orders of magnitude lower value. In the meantime, the gap has been closed by a current-induced nanometer-scale oxide barrier. After the barrier formation the current-voltage curve became strongly nonlinear (red, top right), while it was linear before (blue). The CILO process occurs at current densities of 10^7 A/cm^2 and is triggered by current-induced atomic rearrangements and local heating.

Future Application of Nanostructured Materials

Such mechanical property improvements have resulted in major interest in nanocomposite materials in numerous automotive and general/industrial applications. These include potential for utilisation as mirror housings on various vehicle types, door handles, engine covers and intake manifolds and timing belt covers. More general applications currently being considered include usage as impellers and blades for vacuum cleaners, power tool housings, mower hoods and covers for portable electronic equipment such as mobile phones, pagers etc.

The gaseous barrier property improvement that can result from incorporation of relatively small quantities of nanoclay materials is shown to be substantial. Data provided from various sources indicates oxygen transmission rates for polyamide-organoclay composites which are usually less than half that of the unmodified polymer. Further data reveals the extent to which both the amount of clay incorporated in the polymer, and the aspect ratio of the filler contributes to overall barrier performance. In particular, aspect ratio is shown to have a major effect, with high ratios (and hence tendencies towards filler incorporation at the nano-level) quite dramatically enhancing gaseous barrier properties. Such excellent barrier characteristics have resulted in considerable interest in nanoclay composites in food packaging applications, both flexible and rigid. Specific examples include packaging for processed meats, cheese, confectionery, cereals and boil-in-the-bag foods, also extrusion-coating applications in association with paperboard for fruit juice and dairy products, together with co-extrusion processes for the manufacture of beer and carbonated drinks bottles. The use of nanocomposite formulations would be expected to enhance considerably the shelf life of many types of food.

Honeywell have also been active in developing a combined active/passive oxygen barrier system for polyamide-6 materials. Passive barrier characteristics are provided by nanoclay particles incorporated via melt processing techniques whilst the active contribution comes from an oxygen scavenging ingredient (undisclosed). Oxygen transmission results reveal substantial benefits provided by nanoclay incorporation in comparison to the base polymer (rates approximately 15-20 per cent of the bulk polymer value, with further benefits provided by the combined active/passive system). Akkapeddi suggests that the increased tortuosity provided by the nanoclay particles essentially slows transmission of oxygen through the composite and drives molecules to the active scavenging species resulting in near zero oxygen transmission for a considerable period of time.

Triton Systems and the US Army are conducting further work on barrier performance in a joint investigation. The requirement here is for a non-refrigerated packaging system capable of maintaining food freshness for three years. Nanoclay polymer composites are currently showing considerable promise for this application. It is likely that excellent gaseous barrier properties exhibited by nanocomposite polymer systems will result in their substantial use as packaging materials in future years. A somewhat more esoteric possibility arising from enhanced barrier performance recently suggested has been blown–films for artificial intestines!

The ability of nanoclay incorporation to reduce solvent transmission through polymers such as polyamides has been demonstrated. Data provided by De Bievre and Nakamura of UBE Industries reveals significant reductions in fuel transmission through polyamide–6/66 polymers by incorporation of a nanoclay filler. As a result, considerable interest is now being shown in these materials as both fuel tank and fuel line components for cars. Of further interest for this type of application, the reduced fuel transmission characteristics are accompanied by significant material cost reductions.

The presence of filler incorporation at nano-levels has also been shown to have significant effects on the transparency and haze characteristics of films. In comparison to conventionally filled polymers, nanoclay incorporation has been shown to significantly enhance transparency and reduce haze. With polyamide based composites, this effect has been shown to be due to modifications in the crystallisation behaviour brought about by the nanoclay particles; spherilitic domain dimensions being considerably smaller. Similarly, nanomodified polymers have been shown, when employed to coat polymeric transparency materials, to enhance both toughness and hardness of these materials without interfering with light transmission characteristics. An ability to resist high velocity impact combined with substantially improved abrasion resistance was demonstrated by Haghighat of Triton Systems.

Water laden atmospheres have long been regarded as one of the most damaging environments which polymeric materials can encounter. Thus an ability to minimise the extent to which water is absorbed can be a major advantage. Data provided by Beall from Missouri Baptist College indicates the significant extent to which nanoclay incorporation can reduce the extent of water absorption in a polymer. Similar effects have been observed by van Es of DSM with polyamide based nanocomposites. In addition, van Es noted a significant effect of nanoclay aspect ratio on water diffusion characteristics in a polyimide nanocomposite.

Specifically, increasing aspect ratio was found to diminish substantially the amount of water absorbed, thus indicating the beneficial effects likely from nanoparticle incorporation in comparison to conventional microparticle loading. Hydrophobic enhancement would clearly promote both improved nanocomposite properties and diminish the extent to which water would be transmitted through to an underlying substrate. Thus, applications in which contact with water or moist environments is likely could clearly benefit from materials incorporating nanoclay particles.

The ability of nanoclay incorporation to reduce the flammability of polymeric materials was a major theme of the paper presented by Gilman of the National Institute of Standards and Technology in the US. In his work Gilman demonstrated the extent to which flammability behaviour could be restricted in polymers such as polypropylene with as little as 2 per cent nanoclay loading. In particular heat release rates, as obtained from cone calorimetry experiments, were found to diminish substantially by nanoclay incorporation.

Although conventional microparticle filler incorporation, together with the use of flame retardant and intumescent agents would also minimise flammability behaviour, this is usually accompanied by reductions in various other important properties. With the nanoclay approach, this is usually achieved whilst maintaining or enhancing other properties and characteristics.

Nanostructured Materials—Properties and Applications

Over the past decade nanostructured materials have been the subject of enormous interest, with the potential for wide-ranging industrial, biomedical, and electronic applications. These can include metals, ceramics, polymeric materials, or nanocomposites. The nanomaterials field has seen a huge increase in funding from private enterprises and government, and academic researchers within the field have formed many partnerships.

The Discovery and The History of Nanomaterials

Nanomaterials have actually been produced and used by humans for hundreds of years - the ruby red colour of some stained glass is due to gold nanoparticles trapped in the glass matrix. The decorative glaze known as lustre, found on some medieval pottery, contains metallic spherical nanoparticles dispersed in a complex way in the glaze, which gives rise to its special optical properties.

The Range of Nanomaterials and Their Applications

The variety of nanomaterials is great, and their range of properties and possible applications enormous, from miniature batteries, to biomedical implants, as packaging films, super absorbants, components of army, and automobile parts.

The Background To Nanocrystalline Materials

Nanocrystalline materials are polycrystalline with grain sizes in the 1-100nm range that are found to exhibit improved hardness, increased ductility and toughness, superior magnetic and optoelectronic properties. Nanocrystals are aggregates of atoms from a few hundred to tens of thousands that combine into a crystalline form of matter known as a 'cluster'. Typically around 10nm in diameter, nanocrystals are larger than molecules but smaller than bulk solids and therefore frequently exhibit physical and chemical properties somewhere in-between.

Properties of Nanocrystals

Given that a nanocrystal is virtually all surface and no interior, its properties can vary considerably as the crystal grows in size. Nanoscale crystals are often harder, stronger and more wear resistant than the same materials in bulk form. Nanocrystals might be used to make super-strong and long-lasting metal parts. The crystals also might be added to plastics and other metals to make new types of composite structures for everything from cars to electronics.

The Background To Nanoparticles

These are particles that are less than 100nm in size. At this level the percentage of exposed area in proportion to its total volume changes dramatically when compared to bulk materials. In particles of around three to 5nm in diameter, about a third of the atoms are actually surface atoms; this means they behave a lot more like gas particles rather than solids. This gives them new properties, they float where bigger particles would sink, and dissolve more easily than the same material in bulk form. The extra exposed surface area also changes depending on how the elements and particles interact with each other.

Applications of Clay Nanoparticles

Recently clay nanoparticles have made their way into composites in cars and packaging materials, where they offer transparency, reduction in permeability and increased strength, the latter being already attractive to the automotive industry, but potentially much more so to the aerospace industry. Sunscreens utilize nanoparticulate zinc oxide. Nanoparticles are also being used in antiseptics, as abrasives, in paints, in new coatings for spectacles (making them scratchproof and unbreakable), for tiles, and in electrochromic or self-cleaning coatings for windows.

Future Products Containing Nanoparticles

Nanoparticles are the basis for anti-graffiti coatings for walls and improved ski waxes and ceramic coatings for solar cells. Glues containing nanoparticles have optical properties that give rise to uses in optoelectronics. Casings, containing nanoparticles used in for electronic devices, such as computers, offer improved shielding against electromagnetic interference.

Single-charge Electronics and Quantum Dot Lasers

Even though the study of single-electron charging effects with granular metallic systems dates back to the 1950s, it was the research of Likharev and coworkers almost 10 years ago that laid much of the groundwork for understanding single-charge transport in nanoscale tunnel junctions (Likharev 1988; Averin and Likharev 1991, Chap. 6). The concept was developed of a Coulomb gap that can be exploited to control the transfer of single charges. Since then, many research groups have made use of the Coulomb blockade effect to develop SED technology. The myriad approaches to developing SEDs and representative laboratories pursuing the various SED concepts. Some of the more recent results are discussed below. The group at Hitachi Europe uses a side-gated constriction in a deltadoped GaAs structure to fabricate a magnetic tunnel junction device in which a series of small islands separated by tunnel barriers are formed (Nakazato et al. 1992). At ~ 2 K, the Coulomb gap voltage oscillates as a function of the side-gate voltage. Using the MTJ device as a building block, both memory and logic (inverter, NOR) functions have been demonstrated (Nakazato 1996, 65). The fabrication procedure makes use of standard semiconductor processing techniques and does not rely on lithography to define the nanoscale islands, since these are created by disorder in the delta-doped layer. Other groups have utilised fine-line lithography to fabricate SEDs. At IBM, a flash memory SED was demonstrated by fabricating a sub-50 nm Si quantum dot (QD) on top of a MOSFET channel using a silicon-on-insulator (SOI) substrate, with the QD acting as a floating gate (Wesler et al. 1997). Single-electron charging was observed up to 90 K, while large threshold voltage shifts of up to 0.75 V were measured at 290 K. The University of Minnesota and Fujitsu have also reported similar structures (Guo et al. 1977; Nakajima et al. 1997). To overcome the lithography limitation on the QD size, the Toshiba group used a Si edge quantum wire approach (Ohata and Toriumi 1996). An inversion layer was formed at the 15 nm high Si sidewall of a SOI structure by growing a gate oxide and depositing a poly-Si gate there. Conductance oscillations were clearly seen at 4.2 K in this edgechannel MOSFET. More recently, the Toshiba group has reverted back to a more planar device configuration, with a 50 nm wide Si quantum wire defined by e-beam lithography and oxidation of the surrounding SiO2 (Koga et al. 1997, 79). One method to form semiconductor QDs without depending on fine-line lithography is to make use of the self-organising nature inherent in the Stranski-Krastanow thin film growth mode. In the initial stages of the heteroepitaxial growth of lattice-mismatched materials, strain-induced coherent relaxation occurs and dislocation-free islands are formed that are in the tens of nanometers range in size. There has been considerable research in these self-organised quantum dots (SOQDs) in the past few years, though much of the work has been of a fundamental nature (see, for example, Petroff and Demmester 1995; Nötzel 1996). More recently, the University of Tokyo has proposed the embedding of InAs SOQDs in AlGaAs/GaAs heterojunction field effect transistors (HFETs) to form a flash-memory SED (Sakaki et al. 1995).

Depletion layer control of 2DEG area (Hitachi, University of Glasgow, University of Tokyo). This concept was further demonstrated by the Sony group, which reported observing threshold voltage shifts at 300 K (Taira et al. 1997, 53). Fujitsu has also proposed the use of InGaAs QDs in a similar manner (Futatsugi et al. 1997, 46). The one difference in this case is that the QDs are formed at the bottom of tetrahedral-shaped recesses formed by substrate patterning. Since SOQDs form in a somewhat random manner on a planar surface, this approach provides positioning control.

Tetrahedral shaped recess, TSR (Fujitsu). Al2O3. Double barrier metallic SET patterned by e-beam (NEC). Of course, QDs can be formed with materials other than semiconductors. In fact, some of the earliest work in single-electron charging was done with metallic tunnel junctions. With modern fabrication tools and techniques, some groups have investigated the formation of nanoscale Au particles between metallic contacts. The University of Cambridge group used focused ion beam deposition to place Au dots between electrodes spaced 30 - 40 nm apart (Woodham and Ahmed

1997, 73). At Lund University in Sweden, atomic force microscopy (AFM) is utilised to move a 50 nm Au particle in between contacts formed by e-beam lithography (Carlsson et al. 1997, 128). Researchers observed conductance plateaus stable for several minutes at 300 K. The group at Cambridge University/Hitachi Europe used a colloidal process to form a chain of insulated Au particles between source, drain, and gate electrodes (Tsukagoshi et al. 1997, 67). At 4.2 K, the chain exhibited a Coulomb staircase and periodic conduction oscillations in I-V measurements. For a very different approach, a molecular embodiment of a QD-based system can be realised by connecting a single molecule between metallic contacts. At Yale University a single molecule of benzene-1, 4-dithiol was self-assembled from solution onto two electrodes of a mechanically controllable break junction (Reed et al. 1997). The spacing between the electrodes is ~ 0.8 nm, and I-V measurements at room temperature showed a gap ~ 0.7 V wide, which is attributed to a Coulomb gap. The Delft University of Technology in the Netherlands is also working on transport through oligomers.

A single molecule connecting metallic contacts (Yale University, University of South Carolina, Delft University, Karlsruhe University). Concerning the architecture in which SEDs are utilised, a number of approaches have been proposed. One of the more novel ideas is that of quantum cellular automata (QCA), based on some earlier work at Texas Instruments and developed at the University of Notre Dame (Lent et al. 1993; Tougaw and Lent 1994). The basic QCA cell is made up of a group of capacitively coupled QDs. Each cell holds two electrons, resulting in two polarisation configurations that can represent the logic "0" and "1" states, and each cell interacts via Coulombic forces with neighboring cells. An array of cells can then be used to transmit binary information, which eliminates the need for physical interconnects between devices and represents a paradigm shift for ultralarge-scale integration (ULSI). Basic Boolean operations (AND, OR, etc.) can be implemented using QCA, and more complex functions have been simulated. Most recently, the Notre Dame group has demonstrated a nonlinear, bistable response of a QCA cell, albeit at a very low temperature of less than 20 mK (Snider et al. 1997, 233). The QCA approach is not without its challenges (and critics). Circuit fabrication will be difficult because stringent control in QD positioning is required. Others have pointed out that bistability is only a necessary but not sufficient condition for the operation of Boolean logic circuits, because isolation is needed between the input and output, while background charge fluctuations will hamper logic implementation (Roychowdhury et al. 1966; Barker et al. 1997, 233). Thus, there is also considerable research in the use of SEDs with more conventional architectures. The group at Hitachi Europe uses its MTJ devices in binary decision diagram logic that is commonly used in large-scale integrated (LSI) circuits (Tsukagoshi et al. 1997, 67). The Toshiba group is combining its SED with a MOSFET to compensate for the lack of gain in the former, as are, presumably, other groups working on QDbased flash memory SEDs (Koga et al. 1997, 79).

In summary, while significant progress has been made in nanofabrication techniques, the field of single-charge electronics is still limited in scope by the lack of a suitable architecture that fully utilizes the unique aspects of single-electron charging. Current approaches require many SED elements to achieve conventional functions such as adders, exclusive NORs, etc. Simulations of such circuits predict slow operating speed. The field seems to be stuck on applying conventional electronics to SEDs. Either a new architecture will be discovered or SEDs may find a niche home in those applications where measurements of single charges are needed. For the fabrication of nanoscale electronic devices, the self-organising technique appears to be the most promising. The field of self-organised semiconductor QDs is quite active, but aside from optical emitters, very few practical electronic functions have been proposed. Of those, the singleelectron flash memory is attracting attention, but there has been no serious proposal as to how the device could operate under normal integrated circuit performance conditions and reliability specifications. Using the SED as a floating gate to a MOSFET has the same kinds of problems as applying conventional approaches to SED architectures; until someone comes up with a better idea, the future of these approaches remains to be determined.

The discovery in 1988 of GMR in structures of alternating magnetic and nonmagnetic thin layers (Baibich et al. 1988) was the accumulation of several decades of intensive research in thin film magnetism (Shinjo and Takada 1987) and improvements in epitaxial growth techniques developed mainly in semiconductor materials. Not surprisingly, the first GMR structure was fabricated using molecular beam epitaxy (Baibich et al. 1988). The high quality magnetic and nonmagnetic metallic films provide electrons with a mean free path exceeding 100 Å; on the other hand, the epitaxial growth allows for each constituent layer of the structure to be as thin as a few atomic layers. The greatly enhanced spin-dependent scattering in these multilayered structures provides magnetoresistance changes as high as 50%. The institutions involved in GMR research and development, based on various publications, patents, and WTEC visits. Two subsequent major developments from IBM enabled the application of GMR materials to hard disk heads, RAM, and sensors. The first development was the demonstration of equally good or better GMR materials using high throughput and production-worthy magnetron sputtering systems (Parkin et al. 1990). The other development was the invention of magnetically soft spin-valve structures, which allow low field and low power operation (Dieny et al. 1991a; Dieny et al. 1991b).

The major U.S. players are IBM, Seagate, Quantum, ReadRite, and Applied Magnetics. In Japan, all the semiconductor companies are involved, in addition to strong magnetic media powerhouses such as TDK and Yamaha. Korea's Samsung is also actively involved in the GMR race. In Europe, Thomson CSF, Philips, and Siemens seem to have fallen behind. All in all, IBM is in a commanding position to reap the benefits of the GMR phenomenon. In November 1997, it announced the volume production of the first generation of GMR read heads. In 1995, a different class of high magnetoresistive materials was discovered in which the nonmagnetic layer separating the two ferromagnetic layers is made with an ultrathin insulating material, such as an aluminum oxide layer < 20 Å thick (Miyazaki and Tezuka 1995; Moodera et al. 1995). With the switching of magnetisation of the two magnetic layers between parallel and antiparallel states, the differences in the tunneling coefficient of the junction and thus the magnetoresistance ratio have been demonstrated to be more than 25%. A distinctive feature of this MTJ class of materials is its high impedance (> 100 kW-μm2), which allows for large signal outputs. The gradual improvement of GMR and MTJ materials have made them attractive for nonvolatile magnetic random access memory (MRAM) applications. The potential to make MRAM a high density, high speed, and low power, general purpose memory prompted the Defense Advanced Reseat Projects Agency to fund three MRAM consortia beginning in 1995, led by IBM, Motorola, and Honeywell, respectively. Other companies engaged in MRAM research are Hewlett-Packard, Matsushita, NEC, Fujitsu, Toshiba, Hitachi, and Siemens.

The key for a competitive MRAM technology is the fabrication of deep submicron-dimension magnetic cells. The further development of lithography tools utilising e-beam and deep ultraviolet radiation is essential. Magnetic storage elements as small as 0.25 μm have been demonstrated by both Motorola and IBM (Tehrani et al. 1996; Chen et al. 1997; Gallagher et al. 1997). Among the steps of MRAM fabrication that are not yet compatible with semiconductor processing is the ion milling of the magnetic cells. The possibility of dry etching the magnetic materials has, nevertheless, been demonstrated (Jung et al. 1997). The major directions in memory R&D. Granular GMR—Co, Fe (Nagoya University, Tohoku University, CNRSThomson, UCSB, UCSD). The ability to fabricate submicron magnetic elements has opened a very rich and fascinating area of micromagnetics research. Characterisation techniques having nanoscale resolution have been utilised and improved to measure and image the complex magnetisation patterns in order to understand the magnetisation switching characteristics. Examples of such techniques include the following:

- superconducting quantum interference device (SQUID) magnetometry (Zhu et al. 1997)
- magnetic force microscopy (MFM) (Ohkubo et al. 1991)
- scanning electron microscopy with polarisation analysis (SEMPA) (Scheinfein et al. 1990)

- magnetic near-field microscopy (Betzig and Trautmann 1992)
- electron holography (Mankos et al. 1995)

Another essential tool is micromagnetics modeling, which is used to predict complex magnetic domain configurations in patterned magnetic elements and to generate transient pictures that demonstrate the process of forming complex domain configurations (Zheng and Zhu 1997). By combining MFM, SQUID magnetometry, SEMPA, and micromagnetics modeling, researchers at Motorola have conducted a systematic study of the switching characteristics of single-layer and multilayer submicron magnetic structures. Three different phases of the magnetisation phase diagram have been identified with regards to material composition, dimension, shape, thickness, and other structural parameters: (1) the quasi-single domain phase can be well described by the coherent rotation model (Sakaki et al. 1995); (2) the end-domain phase is dominated by the nonuniform regions of magnetisation at the two ends of the element— the magnetisation switching process can either be rotational or through domain wall nucleation and propagation (Shi et al. 1998); and (3) the trapped magnetisation vortice (TMV) phase, which is characterised by the presence of magnetisation vortices. Nucleation from TMV sites requires lower reversal fields than coherent rotation, but a high field is needed to drive out TMVs in the element. When the driving field is not high enough, TMVs can cause unusually large fluctuations in the switching field (Shi et al. n.d.). As the size of magnetic elements scales below 20 nm, a superparamagnetic phase emerges in which the room temperature thermal energy overcomes the magnetostatic energy well of the element, resulting in zero hysteresis (Hylton 1993). In other words, although the element itself is a single-domain ferromagnet, the ability of an individual magnetic "dot" to store magnetisation orientation information is lost when its dimension is below a threshold. On the other hand, suitably prepared alloys of immiscible ferromagnetic and nonmagnetic metals that contain single-domain ferromagnetic grains in a nonmagnetic matrix have been shown to exhibit GMR characteristics.

The moments of the magnetic grains are aligned at high fields and random at the coercive field, leading to GMR characteristics. In such "granular" metals, GMR has been reported for sputtered alloy films of CoCu (Berkowitz et al. 1992; Xiao et al. 1992), FeCu (Xiao et al. 1992), NiFe/Ag (Jiang et al. 1992), and CoAg (Carey et al. 1992; Tsoukatos et al. 1992). GMR values as high as 55% at 4.2 K and 20% at room temperature have been observed. The granular films require magnetic fields of the order of 10 kOe to achieve such a change in electrical resistance. A very exciting consequence of ultrascaled magnetic particles is quantum tunneling of the magnetisation direction of a collection of spins. There is no simple Schrödinger equation that describes this process, since it is not an elementary particle that is tunneling but a collective coordinate. Below its "blocking temperature, " at which thermally assisted hopping between magnetic orientations becomes frozen out, magnetic particles of

TbCeFe at sizes around 15 nm have been observed to behave independent of temperature and with no freeze-out magnetic relaxation (Barbara et al. 1993). Because of the coherent tunneling of the magnetisation orientation between the symmetric double-well potential, a resonance line in the magnetic susceptibility and noise spectra has been observed at temperatures below 200 mK in zero applied magnetic field (Awschalom et al. 1992). This work has stimulated a number of theoretical investigations into the effects of dissipation and the feasibility of producing quantum effects in larger magnetic structures (Prokofev and Stamp 1993; Gaarg 1994; Braun and Loss 1994). Another interesting type of nanomagnetic structure is nanometer ferromagnetic wires fabricated using conventional nanolithography (Adeyeye et al. 1997; Chou 1997), nanoimprint lithography (Chou et al. 1995), AFM/MFM direct writing (Kong et al. 1997), groove deposition (Hong and Giordano 1995), and electrodeposition into pores of template polymer membranes (Piraux et al. 1994; Blondel et al. 1994). Such nanowires of either single layer or multilayers may provide new approaches to very small magnetoresistive sensors, ultrahigh-density hard disks (Chou et al. 1994), and other extensions of conventional applications. Another intriguing possibility is the suggestion to use heterostructure nanowires to investigate single electron tunneling (Cavicchi and Silsbee 1984; Kumzerov and Poborchii 1994). Recently, molecular magnetism has received much attention with the development of a variety of synthesis techniques largely adapted from biology and chemistry (Kahn 1993). Natural and artificial ferritin proteins are examples of systems obtained using these methods (Gatteschi et al. 1994). The ability to add one magnetic ion at a time has resulted in nanoscale magnets precisely defined by atomic weight. Ferritin consists of a segmented protein shell in the shape of a hollow sphere, with an outer diameter of 12.5 nm and an inner diameter of 7.5 nm. In vivo, the inner space is normally filled with a crystal of an iron oxide that is antiferromagnetic below 240 K. The empty protein shells can also serve as vessels for the synthesis of ferrimagnetic magnetite and maghemite. Thus, there exists a system in which its size as well as the nature of its magnetic interactions can be varied. Another example of molecular magnetism is a cobalt-iron-cyanide-based Prussian blue analog (Sato et al. 1996a). In the ground state the Fe+2 and Co+3 ions are low-spin and diamagnetic, and there is no interaction between them. Red light excitation transfers one electron from an iron site to a cobalt site, resulting in high-spin Fe+3 and Co+2 ions and magnetic interactions between them. The application of a blue light causes a transition back to the initial state and switches off the Fe+3-Co+2 interactions. In addition to this kind of photochemically controllable magnets, electrochemically controllable magnets have also been reported (Sato et al. 1996b).

Quantum Dot Lasers

Semiconductor lasers are key components in a host of widely used technological products, including compact disk players and laser printers, and they will play critical roles in optical communication schemes. The basis of laser operation depends on the creation of nonequilibrium populations of electrons and holes, and coupling of electrons and holes to an optical field, which will stimulate radiative emission. Calculations carried out in the early 1970s by C. Henry (Dingle and Henry 1976) predicted the advantages of using quantum wells as the active layer in such lasers: the carrier confinement and nature of the electronic density of states should result in more efficient devices operating at lower threshold currents than lasers with "bulk" active layers. In addition, the use of a quantum well, with discrete transition energy levels dependent on the quantum well dimensions (thickness), provides a means of "tuning" the resulting wavelength of the material. The critical feature size—in this case, the thickness of the quantum well—depends on the desired spacing between energy levels. For energy levels of greater than a few tens of millielectron volts (meV, to be compared with room temperature thermal energy of 25 meV), the critical dimension is approximately a few hundred angstroms. Although the first quantum well laser, demonstrated in 1975, was many times less efficient than a conventional laser (van der Ziel et al. 1975), the situation was reversed by 1981 through the use of new materials growth capabilities (molecular beam epitaxy), and optimisation of the heterostructure laser design (Tsang 1982).

Even greater benefits have been predicted for lasers with quantum dot active layers. Arakawa and Sakaki (1982) predicted in the early 1980s that quantum dot lasers should exhibit performance that is less temperaturedependent than existing semiconductor lasers, and that will in particular not degrade at elevated temperatures. Other benefits of quantum dot active layers include further reduction in threshold currents and an increase in differential gain—that is, more efficient laser operation (Asada et al. 1986). The key concepts in the laser operation. Stimulated recombination of electron-hole pairs takes place in the GaAs quantum well region, where the confinement of carriers and of the optical mode enhance the interaction between carriers and radiation. In particular, note the change in the electronic density of states, as a function of the "dimensionality" of the active layer.

The population inversion (creation of electrons and holes) necessary for lasing occurs more efficiently as the active layer material is scaled down from bulk (3-dimensional) to quantum dots (0-dimensional). However, the advantages in operation depend not only on the absolute size of the nanostructures in the active region, but also on the uniformity of size. A broad distribution of sizes "smears" the density of states, producing behaviour similar to that of bulk material. Thus, the challenge in realising quantum dot lasers with operation superior to that shown by quantum well lasers is that of forming high quality, uniform quantum dots in the active layer. Initially, the most widely followed approach to forming quantum dots was through electron beam lithography of suitably small featured patterns (~300 Å) and subsequent dryetch transfer of dots into the substrate material. The problem that plagued these quantum dot arrays was their exceedingly low optical efficiency: high surface-to-volume ratios of these nanostructures and associated high surface recombination rates, together with damage introduced during the fabrication itself, precluded the successful formation of a quantum dot laser.

Density of electronic states as a function of structure size. With the demonstration of the high optical efficiency self-assembled formation of quantum dots, formed without need of external processing and having the natural overgrowth of cladding material (which addressed issues of surface recombination), there ensued a marked increase in quantum dot laser research. The first demonstration of a quantum dot laser with high threshold density was reported by Ledentsov and colleagues in 1994. Bimberg et al. (1996) achieved improved operation by increasing the density of the quantum dot structures, stacking successive, strain-aligned rows of quantum dots and therefore achieving vertical as well as lateral coupling of the quantum dots. In addition to utilising their quantum size effects in edge-emitting lasers, self-assembled quantum dots have also been incorporated within vertical cavity surface-emitting lasers. A partial summary of the work and achievements in quantum dot lasers. As with the demonstration of the advantages of the quantum well laser that preceded it, the full promise of the quantum dot laser must await advances in the understanding of the materials growth and optimisation of the laser structure. Although the self-assembled dots have provided an enormous stimulus to work in this field, there remain a number of critical issues involving their growth and formation: greater uniformity of size, controllable achievement of higher quantum dot density, and closer dot-todot interaction range will further improve laser performance. Better understanding of carrier confinement dynamics and capture times, and better evaluation of loss mechanisms, will further improve device characteristics. It should be noted that the spatial localisation of carriers brought about by the quantum dot confinement may play a role in the "anomalous" optical efficiency of the GaN-based materials, which is exceptional in light of the high concentration of threading dislocations (~ 108 - 1010 cm-2) that currently plague this material system. The localisation imposed by the perhaps natural nanostructure of the GaN materials may make the dislocation largely irrelevant to the purely optical (but not to the electrical) behaviour of the material.

The first synthesis and characterisation of carbon nanotubes were reported by Iijima from NEC in late 1991. The initial theoretical study of their electronic structure was soon followed with the work by Dresselhaus and coworkers at MIT (Dresselhaus et al. 1992; Saito et al. 1992a; Saito et al. 1992b). Since then, the fabrication of nanotubes has been improved by several groups, and methods other than arc discharge have been explored. The main issues are to separate the nanotubes from other forms of carbon also produced in the fabrication process and to increase the yield of singlewalled nanotubes (SWNT) for potential applications. Following

Iijima's work, macroscopic quantities of MWNT were produced with an improved arc discharge method by Ebbesen and coworkers at NEC (Tsukuba) (Ebbesen and Ajayan 1992).

In the following year, that same group produced what is considered to be among the best SWNT material generated so far; over 70% of the volume of material was nanotubes bundled together into crystalline ropes of metallic character (Thess et al. 1996). Also in 1996, a group from the Chinese Academy of Science used chemical vapor deposition to produce a 50 mm thick film of nanotubes that were highly aligned perpendicular to the surface (Li et al. 1996). Progress in recent years leads one to predict that it will indeed be possible to produce high quality carbon nanotubes in macroscopic quantities needed for many of the applications outlined below. Nanotube bundles form a low density material and are expected to have high stiffness and axial strength as a result of their seamless cylindrical graphitic structure.

It is therefore predicted that they can be used to fabricate a material with better mechanical properties than the present carbon fiber materials. Information about the mechanical properties of nanotubes has been gathered recently by a study of the thermal vibrations of a single attached to a substrate (Treacy et al. 1996). Ebbesen's group at the Princeton NEC Research Institute found that nanotubes have an exceptionally high Young's modulus (~ 2 x 109 Pa) (Treacy et al. 1996). In order to reach a better understanding of the mechanical properties and intrinsic limitations of nanotubes, Bernholc's group from North Carolina State University theoretically studied the behaviour of nanotubes beyond the linear Hooke's law and the nature of the defects leading to dislocations and fractures (Yakobson et al. 1996; Nardelli et al. 1998). Nanotubes are highly polarizable nanoscale straws, a property that confers on them the capacity to ingest inorganic elements by nanocapillarity (Pederson and Broughton 1992). As a result, it has been conjectured that they could be used as minute molds to shape nanometer-sized quantum wires and as miniature test tubes. Ajayan and coworkers at NEC (Tsukuba) have first shown that lead can be introduced into carbon nanotubes (Ajayan and Iijima 1993).

The efficiency of their process is low, and prior removal of the caps from the ends of the nanotubes is expected to improve the situation (Tsang et al. 1993). More information about the mechanism of NT filling was obtained by Pascard and coworkers from the École Polytechnique in France by studying the propensity to form nanowires for 15 encapsulated metal elements (Guerret-Plécourt et al. 1994). Finally, external decoration of nanotubes with metal atoms has been demonstrated and is predicted to have applications in catalysis (Satishkumar et al. 1996). The primary methods of nanotube fabrication and the institutions engaged in specific methods of nanotube fabrication. Early theoretical studies already showed that the electronic properties of nanotubes strongly depend on their diameter and their chirality leading to metallic or semiconducting structures (Saito et al. 1992c). It was conjectured that these properties can be used to construct nanoscale electronic devices. While theoretical studies were promptly published, it was only in 1996 that Ebbesen and coworkers (1996) at the Princeton NEC Research Institute presented reliable four-point probe conductivity measurements on MWNT, confirming the theoretical predictions. In 1997, two groups, one at Lawrence Berkeley National Laboratory (Bockrath et al. 1997) and the second at Delft University (Tans et al. 1997) in the Netherlands showed that conductivity through nanotubes is controlled by low dimensional effects such as resonant tunneling and single-electron charging effects. Hall effect measurements at the École Polytechnique Fédérale de Lausanne (EPFL) in Switzerland have shown that hole transport is predominant in electronic conductance (Baumgartner et al. 1997). Despite these and other very recent and encouraging efforts such as those studying the mean free path of carriers in nanotubes, the conduction mechanism is still only partially understood (Petit et al. 1997).

For this reason and possibly others related to quantum confinement effects, nanotubes are expected to form outstanding field-emitting materials. In 1995, the Rice University group showed that nanotubes emit electrons very efficiently when immersed in an electric field and irradiated by a laser to remove their cap (Rinzler et al. 1995). They attribute their observation to the unraveling of an atomic wire of carbon atoms. Efficient field emission was also obtained from carefully aligned nanotubes by de Heer and coworkers (1995) at EPFL, whereas Collins and coworkers from the University of California at Berkeley (Collins and Zettl 1996, 1997), have used randomly oriented nanotubes with similar results. Efficient field-emitting material is highly desirable for the production of field-emission displays and microwave tubes. Recently, a group from Mie University in Japan has built a cathode ray tube (CRT) using nanotube field emitters (Saito et al. n.d.). In this work, the layers of nanotubes were cut out from the soot produced in an arc discharge chamber. This fabrication method is presently not compatible with industrial production requirements, and more progress must be made before this effort can be translated into an industrial product. The electrical and field emission properties of nanotubes, with the representative institutions pursuing these studies. In summary, macroscopic amounts of good quality nanotubes can presently be fabricated by several groups around the world, and the theoretical understanding of the electronic structure and related properties of nanotubes has reached a very good level. However, despite the fact that many potential applications are mentioned in the literature over and over again, only the outstanding field emission properties of nanotubes have achieved realisation in practical devices. One of the main obstacles simply remains the controlled manipulation of nanoscale objects. In this respect it seems that the generation of self-aligned structures is a path to explore further, especially after the encouraging successes reported in the literature in 1997-98.

REFERENCES

Becker, M.F., J.R. Brock, H. Cai, N. Chaudhary, D. Henneke, L. Hilsz, J.W. Keto, J. Lee, W.T. Nichols, and H.D. Glicksman. 1997. Nanoparticles generated by laser ablation. In *Proc. of the Joint NSF-NIST Conf. on Nanoparticles*.

Berger, R., C. Gerber, H.P. Lang, and J.K. Gimzewski. 1996. Micromechanics: a toolbox for femtoscale science: "Towards a laboratory on a tip." *Microelectronic Engineering*, 35:373-9. (International Conference on Micro- and Nanofabrication, Glasgow, U.K., 22- 25 Sept. 1996).

Berkowitz, A.E., J.R. Mitchell, M.J. Carey, A.P. Young, S. Zhang, F.E. Spada, F.T. Parker, A. Hutten, and G. Thomas. 1992. *Phys. Rev. Lett.* 68:3745.

Berndt, C.C., J. Karthikeyan, T. Chraska, and A.H. King. 1997. Plasma spray synthesis of nanozirconia powder. In *Proc. of the Joint NSF-NIST Conf. on Nanoparticles*.

Brotzman, R. 1998. Nanoparticle Dispersions. In *R&D status and trends*, ed. Siegel et al. Bu, X., P. Feng, and G.D. Stucky. 1998. Large-cage zeolite structures with multidimensional 12-ring channels. *Science* 278:2080-2085.

Brotzman, R. 1998. Nanoparticle dispersions. In *R&D status and trends*, ed. Siegel et al. Brus, L.E. 1996. Theoretical metastability of semiconductor crystallites in high pressure phases with applications to beta tin structures of silicon. *J. Am. Chem. Soc.* 118:4834-38.

Brus, L. 1996. Semiconductor colloids: Individual nanocrystals, opals and porous silicon. *Current Opinion in Colloid & Interface Science* 1:197-201. *Chemical Engineering News*. 1997. Particulate matter health studies to be reanalyzed (August 18):33.

Chang, K. "Smaller Computer Chips Built Using DNA as Template." New York Times, November 21, 2003.

Chantrell. 1994. *J. Appl. Phys.* 76:6811. Erb, U., G. Palumbo, R. Zugic, and K.T. Aust. 1996. In *Processing and properties of nanocrystalline materials*, ed. C. Suryanarayana, J. Singh, and F.H. Froes. Warrendale, PA: TMS. Gertsman, V.Y., M. Hoffman, H. Gleiter and R. Birringer. 1994. *Acta Metall. Mater*. 42:3539-3544.

Chopra, N.G., R.J. Luyken, K. Cherrey, H. Crespi, M.L. Cohen, S.G. Louie, and A. Zettl. 1995. *Science* 269:966.

Clasen, R. 1990. *Int. Journal of Glass and Science Technology* 63: 291. Crandall, B.C. and J. Lewis, (eds.). 1992. *Nanotechnology: Research and perspectives*. Cambridge: MIT Press.

Czekai, D., et al. 1994. *Use of smaller milling media to prepare nanoparticulate dispersions*. U.S. Patent application. Docket 69802(02) Filed 2/25/94.

Drexler, E., Phoenix, C.: 2004, 'Self-Replication in Nanotechnology: Feasible, Potentially Safe, and Unnecessary' (unpublished paper, in preparation).

Drexler, E.: 1986, *Engines of Creation: The Coming Era of Nanotechnology*, Anchor Books, New York (expanded edition with a new afterword, 1990).

Dupuy, J.P. "Two Temporalities, Two Rationalities: A New Look at Newcomb's Paradox." in Bourgine, P.&Walliser, B. (eds.), Economics and Cognitive Science. New York: Pergamon. 1992, 191-220.

Fennema, O. R. (1973) in *Low-Temperature Preservation of Foods and Living Matter*, eds. Fennema, O. R., Powrie, W. D. & Marth, E. H. (Dekker, New York), pp. 436-475.

Feyerabend, P.: 1981, *Realism, Rationalism and Scientific Method* (Philosophical Papers, volume 1), Cambridge University Press, Cambridge.

Feynman R. "There's Plenty of Room At the Bottom." Talk given on at the annual meeting of the American Physical Society at the California Institute of Technology, 1959.

Feynman, R.: 1960, 'There's Plenty of Room at the Bottom', *Engineering and Science*, 23, 22-36.

Fleming, D. "The Economics of Taking Care: An Evaluation of the Precautionary Principle." in Freestone, D.&Hey, E. (eds.), The Precautionary Principle and International Law. La Haye: Kluwer Law International, 1996.

Foerster, H.v.: 1962, 'Bio-Logic', in: E.E. Bernard & M.R. Kare (eds.), *Biological prototypes and synthetic systems*, Plenum Press, New York, vol. 1., pp. 1-12

Friedlander, S.K. 1993. Controlled synthesis of nanosized particles by aerosol processes. *Aerosol Sci. Technol.* 19:527.

Friedlander, S.K., H.D. Jang, and K.H. Ryu. 1998. *Appl. Phy. Lett.* 72(2):173.

Goddard, W.A. 1998. Nanoscale theory and simulation. In *R&D status and trends*, ed. R. Siegel et al.

Godsell, D.: 2003, *Living Machinery. Bionanotechnology: Lessons from Nature*, Wiley-Liss, New-York.

Guarnieri, F., M. Fliss, and C. Bancroft. 1996. Making DNA Add. *Science* 273:220-223. Hanes, J., J.L. Cleland, and R. Langer. 1997.

Held, R., T. Heinzel, A.P. Studerus, K. Ensslin, and M. Holland. 1997. Semiconductor quantum point contact fabricated by lithography with an atomic force microscope. *Appl. Phys. Lett.* 71:2689-91.

Hellemans, A. 1998. X-rays find new ways to shine. *Science* 277:1214-15.

Hoch, H.C., L.W. Jelinski, and H.G. Craighead, eds. 1996. *Nanofabrication and biosystems*. New York: Cambridge University Press.

Howland, H.: 1962, Structural, hydraulic and economic aspects of leaf venation and shape', in: E.E. Bernard & M.R. Kare (eds.), *Biological prototypes and synthetic systems*, Plenum Press, New York, vol. 1, pp. 183-192 .

Hoyningen-Huene, P.: 1993, *Reconstructing Scientific Revolutions: Thomas S. Kuhn's Philosophy of Science* (trans. by A. Levin), University of Chicago Press, Chicago.

Huang, J.Y., Y.K. Wu, and H.Q. Ye. 1996. *Acta Mater.* 44:1211. Inoue, A. 1997. Private communication. Inturi, R.B., and Z. Szklavska-Smialowska. 1992. *Corrosion* 48:398. Karch, J., R. Birringer, and H. Gleiter. 1987. *Nature* 330:556.

Jena, P., S.N. Khanna, and B.K. Rao. 1996. In *Science and technology of atomically engineered materials*, ed. P. Jena. River Edge, NJ: World Scientific.

Jonas, H. The Imperative of Responsibility. In Search of an Ethics for the Technological Age. Chicago: University of Chicago Press, 1985.

Karak, N., and S. Maiti. 1997. Dendritic polymers: A class of novel material. *J. Polym. Mater.* 14:105.

Karger, J., and D.M. Ruthven. 1992. *Diffusion in zeolites*. New York: J. Wiley. Krätschmer, W., L.D. Lamb, K. Fostiropoulos, and D.R. Huffman. 1990. *Nature* 347:354.

Kirschvink, J.L., A. Koyayashi-Kirschvink, and B.J. Woodford. 1992. Magnetite biomineralization in the human brain. *Proc. Nat'l. Acad. Sci.* USA 89:7683-7687.

Kishida, M., T. Fujita, K. Umakoshi, J. Ishiyama, H. Nagata, and K. Wakabayashi. 1995. *Chem. Commun.* 763.

Klein, J.D., et al. 1993. *Chem. Mater.* 5:902. Koch, C.C. 1989. Materials synthesis by mechanical alloying. *Annual Review of Mater. Sci.* 19:121-143.

Kourilsky, Ph.&Viney, G. Le Principe de précaution. Report to the Prime Minister, Paris, Éditions Odile Jacob, 2000.

Krejchi, M.T., E.D.T. Atkins, A.J. Waddon, M.J. Fournier, T.L. Mason, and D.A. Tirrell. 1994. Chemical sequence control of beta-sheet assembly in macromolecular crystals of periodic proteins. *Science* 265:1427-1432.

Krejchi, M.T., S.J. Cooper, Y. Deguchi, E.D.T. Atkins, M.J. Fournier, T.L. Mason, and D.A. Tirrell. 1997. Crystal structures of chain-folded antiparallel beta-sheet assemblies from sequence-designed periodic polypeptides. *Macromolecules* 30:5012-5024.

Kurzweil, R.: 1998, *The Age of Spiritual Machines, How We Will Live, Work and Think in the New Age of Intelligent Machines*, Phoenix, New York.

Kyprianidou-Leodidou, T., W. Caseri, and V. Suter. 1994. *J. Phys. Chem.* 98:8992. Kung, H.H., and E.I. Ko, 1996. *Chem. Eng. J.* 64:203.

Leonard, D., M. Krishnamurthy, C.M. Reaves, S.P. Denbaars, and P.M. Petroff. 1993. Direct formation of quantum-sized dots from uniform coherent islands of InGaAs on GaAs surfaces. *Appl. Phys. Lett.* 63:3203-5.

Liang, G., Z. Li, and E. Wang. 1996. *J. Mater. Sci.* Makino, A., A. Inoue, T. Hatanai, and T. Bitoh. 1997. *Materials Science Forum* 235-238: 723.

Mao, C., W. Sun, and N.C. Seeman. 1997. Construction of Borromean rings from DNA. *Nature* 386:137-138.

Martin, T.P., N. Malinowski, U. Zimmerman, U. Naher, and H. Schaber. 1993. *J. Chem. Phys.* 99:4210.

Martin, T.P., U. Naher, H. Schaber, U. Zimmerman. 1993. *Phys. Rev. Lett.* 70:3079. Prigogine, I., and S. Rice. 1988. *Advances in chemical physics*, Vol. 70, Parts 1 & 2. New York: J. Wiley.

McCammon, J. A., Gelin, B. R. & Karplus, M. (1977) *Nature (London)* 267, 585-590.

McCulloch, W.S.: 1962, 'The imitation of one form of life by another – Biomimesis', in: E.E. Bernard & M.R. Kare (eds.), *Biological prototypes and synthetic systems*, Plenum Press, New York, vol. 1., p. 393-97.

Mehl, R.F., and R.W. Cahn. 1983. Historical development. In *Physical metallurgy*. North Holland. Milligan, W.W., S.A. Hackney, M. Ke, and E.C. Aifantis. 1993. *Nanostructured Materials* 2:267.

Minsky, M.: 1995, 'Virtual Molecular Reality', in: Krummenacker, M. & Lewis, J. (eds.), *Prospects in Nanotechology. Proceedings of the 1st general conference on nanotechnology: developments, applications, and opportunities, November 11-14, 1992, Palo-Alto*, John Wiley & Sons, New York, pp. 187-205.

Mishra, R.S., and A.K. Mukherjee. 1997. Oral presentation at TMS meeting, Indianapolis, Indiana, 16-18 September 1997, to be published in proceedings of Symp. "Mechanical Behavior of Bulk Nano-Materials."

Mishra, R.S., R.Z. Valiev, and A.K. Mukherjee. 1997. *Nanostructured Materials* 9:473. Morris, D.G., and M.A. Morris. 1991. *Acta Metall. Mater.* 39:1763-1779.

Moore, J.C., H.M. Jin, O. Kuchner, and F.H. Arnold. 1997. Strategies for the in vitro evolution of protein function: Enzyme evolution by random recombination of improved sequences. *J. Mol. Biol*. 272:336-347.

Morris, D.G., and M.A. Morris. 1997. *Materials Science Forum* 235-238:861. Nagpal, P., and I. Baker. 1990. *Scripta Metall. Mater*. 24:2381.

Nieman, G.W., J.R. Weertman, and R.W. Siegel. 1991a. *Mater. Res. Soc. Symp. Proc*. 206:581-586.

Nomura, M. & Held, W. (1974) in *Ribosomes,* eds. Nomura, M., Tissiers, A. & Lengyel, P. (Cold Spring Harbor Laboratory, Cold Spring Harbor, NY), pp. 193-203.

Nordmann, A. (rapp.) Converging Technologies—Shaping the Future of European Societies, European Commission report, 2004.

Parker, J.C., et al. 1995. U.S. Patent 5,460,701. POST (Parliamentary Office of Science and Technology). 1995.

Phoenix, C.; Drexler, E.: 2004, 'Safe Exponential Manufacturing', *Nanotechnology*, 15, 869-72.

Rademann, K., B. Kaiser, U. Even, F. Hensel. 1987. *Phys. Rev. Lett*. 59:2319. Rao, C.N.R., B.C. Satishkumar, and A. Govindaraj. 1997. *Chem. Commun*. 1581.

Ramanan, V.R. 1998. Nanocrystalline soft magnetic alloys for applications in electrical and electronic devices. In *R&D Status and Trends*, ed. Siegel et al.

Rao, M.B., and S. Sircar. 1993. *Gas Separation and Purification* 7:279. Reetz, M.T., et al. 1995. *Science* 267:367.

Rofagha, R., R. Langer, A.M. El-Sherik, U. Erb, G. Palumbo, and K.T. Aust. 1991. *Scr. Metall. Mater*. 25:2867.

Roher, H. 1993. *Jpn. J. Appl. Phys*. 32:1335. Rohlfing, E.A., D.M. Cox, and A. Kaldor. 1984. *J. Chem. Phys*. 81:3846.

Romanov, A.E., V.I. Vladimirov. 1992. In *Dislocations in solids*, ed. F.R.N. Nabarro, Vol. 9. Amsterdam: North-Holland. Salishekev, G.A., O.R. Valiakhmetov, V.A. Valitov, and S.K. Mukhtarov. 1994. *Materials Science Forum*. 170-172:121.

Ruthven, D.M., S. Farooq, K.S. Knaebel. 1994. *Pressure swing adsorption*. New York: VCH Publishers. Sayari, A. 1996. *Chem. Mater*. 8:1840.

S. P. Ho and W. F. DeGrado, "Design of a 4-Helix bundle protein: Synthesis of peptides which self-associate into a helical protein." *J. Am. Chem. Soc.,* 109: 6751-6758 (1987).

Sankey, H.: 1994, *The Incommensurability Thesis*, Avebury, Aldershot.

Sarikaya, M.; Aksay, I. (eds.), 1995, *Biomimetics. Design and Processing of Materials*, AIP Press, Woodbury, New York.

Scanlan, R.M., W.A. Fietz, and E.F. Koch. 1975. *J. Appl. Phys*. 46:2244.

Schiefsky, M.J.: forthcoming, 'Art and Nature in Ancient Mechanics', in: W.R. Newman & B. Bensaude-Vincent (eds.), *The Artificial and the Natural: An Ancient Debate and its Modern Descendants*, MIT Press, Cambridge, MA.

Schultz, L., J. Wecker, and E. Hellstern. 1987. *J. Appl. Phys*. 61:3583.

Shen, T.D., C.C. Koch, T.Y. Tsui, and G.M. Pharr. 1995. *J. Mater. Res*. 10: 2892.

Shull, R.D., R.D. McMichael, and J.J. Ritter. 1993. *Nanostructured Mater*. 2:205.

4

Nanoscale Interfacial Materials and Molecular Manufacturing: Understanding the Dimensions and Processes

Nanoscale Interfacial Materials

In the development of nanometer-scale materials, ever since the early 1980s, scientists began discovering ultra-small crystal grains. The important point is that small is not only beautiful but also eminently useful. Therefore, the virtues of working in the nanodomain are increasingly recognised by the scientific community, the technological world, and the popular press. These small groups of atoms go by several different names: nanoparticles, nanocrystals, atomic clusters, and quantum dots. Much recent research has gone into finding ways to makesmall and uniform-size clusters that possess unusual optical, electrical, magnetic, and mechanical properties.

This effort has already paid several commercial dividends, for example, ceramics and chemical catalysts with increased efficiency due to high surface-to-volume ratios. Moreover, advances with semi- conducting nanoclusters are bringing several commercial applications into view as well. However, in order to develop new kinds of high-level functional materials, one should not only consider the proper- ties of a single nanoparticle (e.g., high surface ratios, quantum confinement effect), but also the cooperative effect among nanostructures and the macroscopic properties of their organised states. In this report, we attempt to introduce a new concept for the design and creation of new types of interfacial materials, i.e., binary cooperative complementary *nanoscale* materials.

In fact, from ancient times in China, the binary cooperative complementary concept has been used as "yin" and "yang" to explain the phenomenon and catalog the materials of nature and the universe. This initial philosophical principle has had a great influence on classical and modern science. Following the development of science, people have realised that binary cooperative complementary character is a universal feature of the material world. For example, the constitution of the chemical elements and the universe, as well as the physical and chemical properties of materials, are all involved in this binary concept.

In the development of materials science, people have also used the binary cooperative complementary effect to create unique functional materials. So far, binary atomic materials (e.g., binary alloys) and binary molecular materials have been developed that demonstrate tremendous potential in both basic research and practical applications. Well-known examples of binary atomic materials exist, such as the combination of Fe and Cr, which creates a type of stainless steel (also, cobalt is added to iron to make cobalt steels, among the hardest alloys known). Furthermore, many other binary alloys show a number of unique properties. Binary molecular, amphiphilic molecules, with both hydrophilic and hydrophobic groups on the same molecule, have been used in different fields, such as surfactants, in which a huge variety of industrial products have been developed. When electron donors and acceptors are incorporated into a single molecule, nonlinear optical materials can result. Also, binary molecular solids provide another good example. Through the synthesis of donor-acceptor complexes, organic conductors and even superconductors may be generated. The quintessential functional binary molecules (multi-binary molecules) are DNA and RNA, which precisely control the heredity of all living things, animal and plant. Still many unknowns remain, waiting to be elucidated in further research.

Above all, binary cooperative complementary atomic and molecular systems have shown us unique material words. Here we extend this concept from the molecular level to the nanostructural level and introduce a new material system, i.e., binary nanoscale materials. Before going into this topic, we would first like to start from the binary interfacial material, since understanding interfacial materials is key to the design of high-level functional materials. The motivation to develop binary cooperative complementary nanoscale interfacial materials is the desire to create unique functional macroscopic properties. By applying the binary coordinating complementary concept on the nanoscale on a solid surface, we expect to generate nanostructures with mutually compensating properties, e.g., hydrophilic and hydrophobic; conducting and insulating; convex and concave; p-type and n-type; oxidising and reducing; ferromagnetic and antiferromagnetic, and so on. Whenever an interface (or a surface) is formed by the mixture of nanostructures with a pair of these mutually complementary properties, under certain cooperative conditions, unusual interfacial properties can be created, i.e., binary cooperative complementary nanoscale interfacial materials. Our recent results indicate that such a concept is useful for designing new types of interfacial materials. In the following sections we will introduce several examples for the unique character of binary cooperative complementary nanoscale interfacial materials.

Super-amphiphilic Surfaces

Binary complementary hydrophilic and hydrophobic nanoscale structures on the surface of TiO thin films and single crystals appear to create super-amphiphilic properties, with contact angles of 0° for both water and oil. AFM images on the face of rutile TiO_2 single crystals and suggests the binary TiO_2 surface can be induced by UV light illumination, heating at 300 °C, or argon ion sputtering, respectively. The key transformation involves loss of oxygen atoms from Ti_{4+} sites to Ti_{3+} sites, which can adsorb hydroxyl groups and thus produce hydrophilic regions. In all of the images, hydrophilic domains appear as rectangular shapes, aligned along the direction of the face which was confirmed by friction force microscopy (FFM). This unique amphiphilic TiO surface character is ascribed to a composite of hydrophilic and hydrophobic (oleophilic) phases on the scale of several tens of nanometers, with the hydrophilic phases formed, for dissociative water adsorption. Microscopically, these areas may be hydrophilic and oleophilic; however macroscopically, the surface is highly amphiphilic. Furthermore, no visible water drops can form on the surface, and the material appears to have an antifogging property. The polycrystalline TiO_2 thin film, which has same character, can be coated on various substrates, e.g., glass, ceramics, plastics and metals. This highly amphiphilic surface has self-cleaning, antifogging properties. Furthermore, this basic idea can be also used to design and synthesize other types of super-amphiphilic materials, e.g., super-amphiphilic polymers.

Experimental

These kinds of surfaces or interfaces can be generated by outside stimuli, e.g., light, electric fields, magnetic fields, heat, plasma etching, or chemical synthesis on the surface of the solid. The surface nanostructures were characterised by atomic force microscopy (AFM) in air (SPI300 Seiko Instruments) or in high vacuum (SPI300HV Seiko Instruments).

Results and Discussion

Recent research indicates that the binary cooperative complementary concept is truly useful for the design and creation of *nanoscale* functional interfacial materials. Here, we will simply mention these as follows.

High-efficiency Catalysis

Pt nanograins on the surface of TiO_2 thin films, i.e., a binary (with both oxidising and reducing sites) coordinating complementary catalyst surface, can be created through synthesis processes. The effect is useful for the prevention of icing on windows and electrical cables in winter. The important advantage of this technique is that the anti-icing process is carried out through a photo-electrochemical reaction, but not a physical heating process. The necessary reaction for removing ice blocks on the substrate is just the decomposition of single molecular layer of H_2O contacting with the substrate. Furthermore, this kind of structure can be also applied to other catalysis systems for various industrial applications.

Super-"amphiphobic"surface (convex and concave binary cooperative complementary surface)

Through a phase-separation process, polyporous with hydrophobic convex and hydrophilic concave surfaces are generated in copolymer film which appears as an "amphiphobic" surface (i.e., the surface appears to have both hydrophobic and oileophobic properties). It was known that the roughness can enhance the wettability of a surface. However, a pure hydrophilic or hydrophobic surface of polyporous may only enhance the hydrophilic or hydrophobic property of a solid surface, but "amphiphobic" behaviour can not be created. Interestingly, as-grown copolymer films show an "amphiphobic" property.

Understanding Bulk Behaviour of Nanostructured Materials

Bulk nanostructured materials are defined as bulk solids with nanoscale or partly nanoscale microstructures. This category of nanostructured materials has historical roots going back many decades but has a relatively recent focus due to new discoveries of unique properties of some nanoscale materials.

Early in the century, when "microstructures" were revealed primarily with the optical microscope, it was recognised that refined microstructures, for example, small grain sizes, often provided attractive properties such as increased strength and toughness in structural materials. A classic example of property enhancement due to a refined microstructure—with features too small to resolve with the optical microscope—was age hardening of aluminum alloys. The phenomenon, discovered by Alfred Wilm in 1906, was essentially explained by Merica, Waltenberg, and Scott in 1919 (Mehl and Cahn 1983, 18), and the microstructural features responsible were first inferred by the X-ray studies of Guinier and Preston in 1938. With the advent of transmission electron microscopy (TEM) and sophisticated X-ray diffraction methods it is now known that the fine precipitates responsible for age hardening, in Al-4% Cu alloys, for example, are clusters of Cu atoms— Guinier-Preston (GP) zones—and the metastable partially coherent q' precipitate (Silcock et al. 1953-54; Cohen 1992). Maximum hardness is observed with a mixture of GPII (or q") (coarsened GP zones) and q' with the dimensions of the q' plates, typically about 10 nm in thickness by 100 nm in diameter.

Therefore, the important microstructural feature of age-hardened aluminum alloys is nanoscale. There are a number of other examples of nanoscale microstructures providing optimised properties. The critical current density JC of commercial superconducting Nb3Sn is controlled by grain size and is inversely proportional to grain size, with grain sizes of 50-80 nm providing high values of JC (Scanlan et al. 1975). The field of nanocrystalline (or nanostructured, or nanophase) materials as a major identifiable activity in modern materials science results to a large degree from the work in the 1980s of Gleiter and coworkers (Gleiter 1990), who synthesised ultrafine-grained materials by the in situ consolidation of nanoscale atomic clusters. The ultrasmall size (< 100 nm) of the grains in these nanocrystalline materials can result in dramatically improved—or different—properties from conventional grain-size (> 1 μm) polycrystalline or single crystal materials of the same chemical composition. This is the stimulus for the tremendous appeal of these materials.

While there are a number of bulk properties that may be dramatically changed when the microstructure is nanoscale, this chapter focuses on those for which the recent work with nanostructured materials has been most extensive. These are (1) the mechanical properties of nanostructured materials for a variety of potential structural applications, and (2) ferromagnetic materials with nanoscale microstructures for potential applications as soft magnetic materials and permanent magnet materials, and for other special applications such as information storage, magnetoresistance spin valves, and magnetic nanocomposite refrigerants. Other bulk applications such as hydrogen storage are discussed briefly.

Understanding Mechanical Behaviour: Strcutural Nanostructured Materials

The great interest in the mechanical behaviour of nanostructured materials originates from the unique mechanical properties first observed and/or predicted for the materials prepared by the gas condensation method. Among these early observations/predictions were the following:

- lower elastic moduli than for conventional grain size materials—by as much as 30 - 50%
- very high hardness and strength—hardness values for nanocrystalline pure metals (~ 10 nm grain size) are 2 to 7 times higher than those of larger grained (>1 mm) metals
- a negative Hall-Petch slope, i.e., decreasing hardness with decreasing grain size in the nanoscale grain size regime
- ductility—perhaps superplastic behaviour—at low homologous temperatures in brittle ceramics or intermetallics with nanoscale grain sizes, believed due to diffusional deformation mechanisms

While some of these early observations have been verified by subsequent studies, some have been found to be due to high porosity in the early bulk samples or to otherartifacts introduced by the processing procedures. The following summarizes the author's understanding of the state of the art of the mechanical behaviour of nanostructured materials, as determined from the literature, presentations at the U.S. workshop (Siegel et al. 1998), and the WTEC panel's site visits in Japan and Europe.

Elastic Properties

Early measurements of the elastic constants on nanocrystalline (nc) materials prepared by the inert gas condensation method gave values, for example for Young's Modulus, *E*, that were significantly lower than values for conventional grain size materials. While various reasons were given for the lower values of *E*, it was suggested by Krstic and coworkers (1993) that the presence of extrinsic defects—pores and cracks, for example—was responsible for the low values of *E* in nc materials compacted from powders. This conclusion was based on the observation that nc NiP produced by electroplating with negligible porosity levels had an *E* value comparable to fully dense conventional grain size Ni (Wong et al. 1994, 85). Subsequent work on porosity-free materials has supported these conclusions, and it is now believed that the intrinsic elastic moduli of nanostructured materials are essentially the same as those for conventional grain size materials until the grain size becomes very small, e.g., < 5 nm, such that the number of atoms associated with the grain boundaries and triple junctions becomes very large. For nanocrystalline Fe prepared by mechanical attrition and measured by a nano-indentation technique. Thus, for most nanostructured materials (grain size > 10 nm), the elastic moduli are not unique properties and not a "negative."

Hardness and Strength

Hardness and strength of conventional grain size materials (grain diameter, d > 1 mm) is a function of grain size. For ductile polycrystalline materials the empirical Hall-Petch equation has been found to express the grain-size dependence of flow stress at any plastic strain out to ductile fracture. In terms of yield stress, this expression is so = si + kd-1/2, where so = yield stress, si = friction stress opposing dislocation motion, k = constant, and d = grain diameter. Similar results are obtained for hardness, with Ho = Hi + kd-1/2. To explain these empirical observations, several models have been proposed, which involve either dislocation pileups at grain boundaries or grain boundary dislocation networks as dislocation sources. In all cases the Hall-Petch effect is due to dislocation motion/generation in materials that exhibit plastic deformation.

Ratio of the Young's (*E*) and shear (G) moduli of nanocrystalline materials to those of conventional grain size materials as a function of grain size. The dashed and solid curves correspond to a grain boundary thickness of 0.5 and 1 nm, respectively

(Shen et al. 1995). Most of the mechanical property data on nc materials have pertained to hardness, although some tensile test data are becoming available. Several recent reviews have summarised the mechanical behaviour of these materials (Siegel and Fougere 1994, 233–261; Siegel 1997; Morris and Morris 1997; Weertman and Averback 1996, 323–345). It is clear that as grain size is reduced through the nanoscale regime (< 100 nm), hardness typically increases with decreasing grain size and can be factors of 2 to 7 times harder for pure nc metals (10 nm grain size) than for large-grained (> 1 μm) metals.

The experimental results of hardness measurements, summarised previously, show different behaviour for dependence on grain size at the smallest nc grains (< 20 nm), including (a) a positive slope ("normal" Hall-Petch behaviour), (b) essentially no dependence (~ zero slope), and (c) in some cases, a negative slope (Siegel and Fougere 1994, 233–261; Siegel 1997; Morris and Morris 1997; Weertman and Averback 1996, 323–345). Most data that exhibit the negative Hall-Petch effect at the smallest grain sizes have resulted from nc samples that have been annealed to increase their grain size. It is suggested that thermally treating nanophase samples in the as-produced condition may result in such changes in structure as densification, stress relief, phase transformations, or grain boundary structure, all of which may lead to the observed negative Hall-Petch behaviour (Siegel and Fougere 1994, 233–261). Only a few cases of negative Hall-Petch behaviour have been reported for as-produced nanocrystalline samples with a range of grain sizes. These include electrodeposited nc alloys and devitrified nc alloys (Erb et al 1996, 93-110; Alves et al. 1996). Nanocrystalline thin films with grain sizes £ 6 nm are also observed to exhibit a negative Hall-Petch effect (Veprek 1998). While it seems likely that in many cases the observed negative Hall-Petch slopes are due to artifacts of the specimen preparation methods, it is also likely that conventional dislocation-based deformation is not operable in nanocrystalline materials at the smallest grain sizes (< ~30 nm). At these grain sizes, theoretically, mobile dislocations are unlikely to occur; nor have they been observed in *in situ* TEM deformation experiments (Siegel and Fougere 1994, 233–261; Milligan et al 1993; Ke et al. 1995). Thus, the hardness, strength, and deformation behaviour of nanocrystalline materials is unique and not yet well understood.

Ductility and Toughness

It is well known that grain size has a strong effect on the ductility and toughness of conventional grain size (> 1 mm) materials. For example, the ductile/brittle transition temperature in mild steel can be lowered about 40°C by reducing the grain size by a factor of 5. On a very basic level, mechanical failure, which limits ductility, is an interplay or competition between dislocations and cracks (Thomson 1996, 2208–2291). Nucleation and propagation of cracks can be used as the explanation for the fracture stress dependence on grain size (Nagpal and Baker 1990). Grain size refinement can make crack propagation more difficult and therefore, in conventional grain size materials, increase the apparent fracture toughness.

However, the large increases in yield stress (hardness) observed in nc materials suggest that fracture stress can be lower than yield stress and therefore result in reduced ductility. The results of ductility measurements on nc metals are mixed and are sensitive to flaws and porosity, surface finish, and method of testing (e.g., tension or compression testing). In tension, for grain sizes < 30 nm, essentially brittle behaviour has been observed for pure nanocrystalline metals that exhibit significant ductility when the grain size is conventional. Elongation to failure in tension vs. grain size for some nanocrystalline metals and alloys. In some metals, Cu for example, ductile behaviour is observed in compression, along with yield strengths about twice those observed in tension. While it is likely that the flaws and porosity present in many nc samples seriously affect the results of mechanical tests and may be partly responsible for the asymmetry of results in compression compared to tension tests, the nature of the deformation process in terms of shear banding may also be important.

The above behaviour is presumably due to the inability of usual dislocation generation and motion to occur at these smallest nc grain sizes. An intriguing suggestion based on early observations of ductile behaviour of brittle nc ceramics at low temperatures is that brittle ceramics or intermetallics might exhibit ductility with nc grain structures (Karch et al. 1987; Bohn et al. 1991). Karch and colleagues (1987) observed apparent plastic behaviour in compression in nc CaF2 at 80°C and nc TiO2 at 180°C. These observations were attributed to enhanced diffusional creep providing the plasticity at these temperatures, where conventional grain-size materials would fail in the elastic regime. It was assumed that diffusional creep was responsible for the plasticity; observations were rationalised, with boundary diffusion dominating the behaviour such that the strain (creep) rate is defined as de dt = BsWDDb d3kT where s is the applied stress, W the atomic volume, d the grain size, k the Boltzmann constant, T the temperature, B a constant, and Db the grain boundary diffusion coefficients. Going from a grain size of 1 μm to 10 nm should increase de/dt by 106 or more if Db is significantly larger for nc materials. However, these results on nc CaF2 and nc TiO2 have not been reproduced, and it is believed that the porous nature of these samples was responsible for the apparent ductile behaviour. In addition, the idea of unusually high creep rates at low temperatures has been refuted. Recent creep measurements of nc Cu, Pd, and Al-Zr at moderate temperatures by Sanders et al. (1997) find creep rates comparable to or lower than corresponding coarse-grain rates. The creep curves at low and moderate homologous temperatures (.24 – .48 TM) could be fit by the equation for exhaustion (logarithmic) creep. One explanation is that the observed low creep rates are caused by the high fraction of low energy grain boundaries in conjunction with the limitation on dislocation activity by the small grain sizes. In sum, the predicted ductility due to diffusional creep in nc brittle ceramics or intermetallics at temperatures significantly less than 0.5 TM has not been realised.

Superplastic Behaviour

Superplasticity is the capability of some polycrystalline materials to exhibit very large tensile deformations without necking or fracture. Typically, elongations of 100% to > 1000% are considered the defining features of this phenomenon. As grain size is decreased it is found that the temperature is lowered at which superplasticity occurs, and the strain rate for its occurrence is increased. As discussed previously, Equation 1 suggests that creep rates might be enhanced by many orders of magnitude and superplastic behaviour might be observed in nc materials at temperatures much lower than 0.5 TM. As mentioned above, actual creep experiments have not borne out this prediction, but instead have shown creep rates (Equation 1) comparable to or lower than those in coarse-grained samples of the same material. This is presumably why little enhancement in ductility or superplastic behaviour has been observed for nc materials at temperatures < 0.5 TM. However, there is evidence of enhancement of superplastic behaviour in nc materials at temperatures > 0.5 TM. Superplasticity has been observed at somewhat lower temperatures and at higher strain rates in nc materials. The evidence for tensile superplasticity is limited and observed typically at temperatures greater than 0.5 TM and in materials that exhibit superplasticity in coarser grain sizes (1–10 μm). For example, Mishra et al. (1997) observed superplastic behaviour in nc Pb-62%Sn at 0.64 TM and nc Zn–22%Al at 0.52 to 0.60 TM. However, Salishekev et al (1994) observed superplastic behaviour in submicron—200 nm—Ti and several Ti and Ni base alloys. Here, superplasticity (190% elongation, m = 0.32) was observed in Ti at 0.42 TM. This was at a temperature 50°C lower than for 10 μm grain size Ti. The flow stress for the 200 nm Ti at 550°C was 90 MPa, compared to 120 MPa for 10 μm Ti at 600°C. Very recently, Mishra and Mukherjee (1997) have observed superplastic behaviour in Ni3Al with a 50 nm grain size at temperatures of 0.56 to 0.60 TM to strains of 300 - 600%, but with unusual stress-strain behaviour and significant apparent strain-hardening. These new results suggest very different mechanisms may be causing superplastic behaviour in these nc materials.

Unique Mechanical Properties of Nanocrystalline Materials

While there are still only limited data on the mechanical behaviour— especially tensile properties—of nc materials, some generalisations may be made regarding the deformation mechanisms. It is likely that for the larger end of the nanoscale grain sizes, about 50 - 100 nm, dislocation activity dominates for test temperatures < 0.5 TM. As grain size decreases, dislocation activity apparently decreases. The essential lack of dislocations at grain sizes below 50 nm is presumably the result of the image forces that act on dislocations near surfaces or interfaces. The lack of dislocations in small, confined spaces such as single-crystal whiskers has been known for many years (Darken 1961). Creation of new dislocations is also made difficult as the grain size reaches the lower end of the nanoscale (< 10 nm). Stresses needed to activate dislocation sources, such as the Frank-Read source, are inversely proportional to the distance between dislocation pinning points. Since nanoscale grains will limit the distance between such pinning points, the stresses to activate dislocation sources can reach the theoretical shear stress of a dislocation-free crystal at the smallest grain sizes (~ 2 nm). Thus, at the smallest grain sizes we may have new phenomena controlling deformation behaviour. It has been suggested that such phenomena may involve grain boundary sliding and/or grain rotation accompanied by short-range diffusion-assisted healing events (Siegel 1997). Several examples of deformation by shear banding have been reported for nc materials. Carsley et al (1997, 183-192) have studied nc Fe–10% Cu alloys with grain sizes ranging from 45 to 1, 680 nm. In all cases, deformation in compression proceeds by intense localised shear banding. The stress-strain curves exhibited essentially elastic, perfectly plastic behaviour; that is, no measurable strain hardening was observed. Shear banding is also the deformation mode observed in amorphous metallic alloys and amorphous polymers. The deformation shear banding in nc Fe–10% Cu was compared to that for metallic glasses, amorphous polymers, and coarsegrained polycrystalline metals after significant plasticity and work hardening had taken place. While this suggests a close similarity between deformation in nc materials and amorphous materials, not all tensile data on nc materials exhibit a lack of strain hardening. The Fe–10% Cu samples of Carsley et al. (1997, 183-192) showed shear bands even in their larger grained specimens (i.e., about 1, 000 nm).

Theoretical Needs

Central to all of the above discussion is the lack of understanding of the microscopic deformation and fracture mechanisms in nc materials. Clearly, a stronger theoretical effort is needed to guide critical experiments and point the direction for optimising properties. There has been limited work in this area, especially in Russia, in applying disclination theory to grain rotation (Romanov and Vladimirov 1992, 191), for example. However, a much larger theoretical effort is required. Another alternate deformation mechanism that may be important in nc materials is mechanical twinning (Huang et al. 1996). Little theoretical work has been carried out to address this possibility. Another important potential approach to understanding the deformation mechanisms in nc materials is to explore the rich older literature on shear banding—mechanical instabilities that have been observed in, for example, mild steels as grain size decreases; the approach to very low strain hardening (perfectly plastic behaviour) is observed. Amorphous materials exhibit many of the phenomenological characteristics of deformation in nc materials, that is, shear banding, asymmetry between tensile and compressive behaviour, and perfectly plastic behaviour. However, deformation mechanisms are not well understood in amorphous materials either. Recent work on bulk metallic glasses may help clarify this.

Applications

Of the present or near-term applications for nc materials, the hard material WC/Co is an example of several important trends. Nanostructured WC/Co composites have been prepared that can have the following characteristics:

- nanoscale grain sizes after consolidation of powder
- hardness values about twice that of conventional micrograined WC/Co
- enhanced wear resistance and cutting performance

While this material has increased hardness and strength, preliminary reports from Stevens Institute of Technology in Hoboken, NJ, and the Royal Institute of Technology in Stockholm, Sweden, point to similar or increased fracture toughness values for nanostructured WC/Co. As noted earlier, single-phase nanostructured materials studied to date have exhibited high strength and hardness but brittle behaviour at low homologous temperatures (< 0.5 TM). The results of these studies of WC/Co two-phase nanostructured materials suggest that the combination of high hardness/strength and toughness/ductility may be possible in multiphase nanostructured materials. Other examples that point to this possibility come from the work of Professor A. Inoue at Tohoku University in Japan, whose lab the WTEC panel visited. Inoue and coworkers have synthesised a variety of multiphase Al, Mg, and Ni-base alloys with nanoscale microstructures.

Many of these alloys consist of nanocrystallites in an amorphous matrix. Some Al-rich alloys contain nanoscale quasi-crystalline particles surrounded by crystalline face-centered cubic Al. The fascinating properties of these multiphase nanostructured alloys include extremely high strength coupled with some ductility. Ductility is high in compression, but uniform elongation in tension is limited. Again, this behaviour is analogous to that exhibited by ductile amorphous alloys. These results again suggest the possibility for development of nanostructured multiphase composites that combine extremely high hardness and strength with toughness and ductility. Such materials could have many applications as unique structural materials. Multiphase ceramic nanocomposites are the focus of several efforts. One significant effort is underway at the Research Center for Intermaterials of the Institute of Scientific and Industrial Research at Osaka University in Japan. Professor Koichi Niihara has a large effort studying micro-nano composites such as Al2O3/SiC that have enhanced toughness and also a variety of hard matrix/soft dispersion or soft matrix/hard dispersion nanocomposites. The enhanced toughness in some ceramic nanocomposites observed by Prof. Niihara has been verified by parallel studies of Dr. Steve Roberts at University of Oxford in England. While "low temperature" superplasticity has not been realised in nanostructured materials, enhanced superplastic behaviour has been observed at elevated temperatures in terms of a somewhat lower temperature range for superplasticity and, perhaps more significant, a higher strain rate regime. While much needs to be understood about superplastic behaviour of nanostructured materials, the possibility for using more conventional strain rates for forming has major industrial implications. This is of particular interest for near net shape forming of ceramics.

Outstanding Issues: Opportunities and Challenges

Perhaps the overriding issue for defining the mechanical behaviour of nanostructured materials is the lack of understanding of the mechanisms for plastic deformation and fracture in these materials. An extensive experimental and theoretical research effort should be applied to solve these questions. Only with such understanding can the intelligent design of nanostructured materials with optimum mechanical properties be realised. Since many of the synthesis routes for nanostructured materials involve using powder or particulate products, compaction of such powders is needed to form bulk parts without coarsening the nanoscale microstructure. While there has been ongoing work on compaction and the thermal stability of nanocrystalline microstructures, more basic work is needed in this critical area. If some unique properties are limited to the finest grain sizes, methods must be found to stabilize the grain size while attaining theoretical density and complete particulate bonding. Furthermore, conventional processing methods are desirable if realistic scaleup and economy are to be realised. Near-term opportunities for structural applications of nanostructured materials are in the form of coatings prepared by, for example, thermal spray deposition or electrodeposition. There is a significant effort in the United States on thermal spray deposition of nc materials. Several groups in Canada lead in deposition of nc materials by electrodeposition methods. A spin-off company, Nanometals, formed under the auspices of Parteq Research and Development Innovations, has adopted a commercial electrodeposition method to produce nanostructured metals. Structural applications of nanostructured materials may be viewed as a focused approach to a long standing, well known design of materials with optimum mechanical properties. Use of nanoscale precipitates or dispersoids has been known for decades to improve mechanical behaviour. The "new" emphasis on using nanostructured materials comes from 2 significant factors: 1. the special processing methods to push microstructures to the limits of the nanoscale 2. the unique properties and phenomena—as yet not well understood— when this limit is approached

Ferromagnetic Nanostructured Materials

Soft Magnetic Nanocrystalline Alloys

The discovery of nanocrystalline Fe-based soft magnetic materials is less than ten years old. The first class of such materials was the melt-spun Fe-Si-B alloys containing small amounts of Nb and Cu (Yoshizawa et al. 1988). The Fe-Si-B-Nb-Cu amorphous phase transforms to a body-centered cubic (bcc) Fe-Si solid solution with grain sizes of about 10 nm during annealing at temperatures above the crystallisation temperature. The presence of small amounts of Cu helps increase the nucleation rate of the bcc phase while Nb retards the grain growth. These "Finemet" alloys provide low core losses (even lower than amorphous soft magnetic alloys such as Co-Fe-Si-B), exhibit saturation induction of about 1.2 T, and exhibit very good properties at high frequencies, comparable to the best Co-based amorphous alloys. These were first developed in Japan and have stimulated a large amount of research and development worldwide to optimize the magnetic properties. There has been relatively little work in the United States in this area, however. While many of the soft magnetic properties of Finemet-type nanocrystalline alloys are superior, they exhibit lower saturation inductions than Fe-metalloid amorphous alloys, mainly because of the lower Fe content to attain amorphisation and because of the addition of Nb and Cu (or other elements to control the nucleation and growth kinetics). In order to remedy this problem, another class of Fe-based nanocrystalline alloys was developed by Inoue and coworkers at Tohoku University (Makino et al. 1997), which is commercialised by Alps Electric Co., Ltd., of Nagaoka, Japan. These "Nanoperm" alloys are based on the Fe-Zr-B system; they contain larger concentrations of Fe (83-89 at.%) compared to the Finemet alloys (~ 74 at.% Fe) and have higher values of saturation induction (~ 1.6-1.7 T). The Nanoperm nc alloys have very low energy losses at power frequencies (60 Hz), making them potentially interesting for electrical power distribution transformers. The issues of composition modification, processing, and the brittle mechanical behaviour of these nanocrystalline/amorphous alloys are discussed by V.R. Ramanan in the first volume of this WTEC study, the proceedings of the May 8-9, 1997 panel workshop on the status of nanostructure science and technology in the United States (Ramanan 1998, 113-116).

Effective permeability, μe , vs. saturation magnetic flux density, Bs , for soft ferromagnetic materials (after A. Inoue 1997). While there has been extensive research on these alloys, particularly in Japan and Europe, most of the development has been carried out in Japan. The Finemet family of alloys is marketed by Hitachi Special Metals. Vacuumschmelze GmbH (Germany) and Impky (France) also market similar alloys. The Nanoperm alloys are being commercialised by Alps Electric Co. (Japan). No extensive research nor any commercialisation of these materials has been carried out in the United States. The small single-domain nanocrystalline Fe particles in the amorphous matrix gives these alloys their unique magnetic behaviour, the most dramatic being the lowest energy losses (narrowest B/H hysteresis loop) of any known materials, along with very high permeabilities. These alloys can also exhibit nearly or exactly zero magnetostriction. To date, these materials have been made by crystallisation of rapidly solidified amorphous ribbons. Other methods that might provide geometrically desirable products should be explored or developed. Electrodeposition is one such method that requires further work. Electrodeposited nc Fe-Ni soft magnetic alloys are being developed in Canada. The brittle nature of these materials is a problem for scaleup and transformer manufacture. The brittleness problem must be solved by finding less brittle materials or applying the handling and processing knowledge that exists for embrittled (after annealing) metallic glasses.

Permanent Magnet Materials

The first attempts to produce nanoscale microstructures to enhance the magnetic properties of the Nb-Fe-B permanent magnetic materials used mechanical alloying of blended elemental powders followed by heat treatment (Schultz et al. 1987). Since the grain structure so obtained does not exhibit any crystallographic texture—and limits the energy product— special processing methods such as die-upsetting were used by Schultz and coworkers (1989) to provide the crystallographic anisotropy. While the coercivities of these nanocrystalline alloys are high, the remanent magnetisation is decreased. Recent approaches to increasing the magnetic induction have utilised exchange coupling in magnetically hard and soft phases. The Fe-rich compositions (e.g., Fe90Nd7B3) result in a mixture of the hard Fe14Nd2B phase and soft aFe phase. The nanoscale two-phase mixtures of a hard magnetic phase and a soft magnetic phase can exhibit values of remanent magnetisation, Mr, significantly greater than the isotropic value of 0.5 Ms. This "remanence enhancement" is associated with exchange coupling between the hard and soft phases, which forces the magnetisation vector of the soft phase to be rotated to that of the hard phase (Smith et al. 1996). Two important requirements for alloys to exhibit remanence enhancement are a nanocrystalline grain size and a degree of coherence across interphase boundaries sufficient to enable adjacent phases to be exchange-coupled. The significant feature of the exchange coupling is that it allows crystallographically isotropic materials to exhibit remanence values approaching those achieved after full alignment. Such two-phase nanoscale ferromagnetic alloys have been prepared by nonequilibrium methods such as melt-spinning, mechanical alloying, and sputter deposition. Besides the high reduced remanence, the material cost is reduced by reduction in the content of the expensive hard rare earth-containing magnetic phase. The theoretical understanding of remanence enhancement appears to be developed to a degree enabling prediction of magnet performance; however, this performance, while a significant improvement over single-phase isotropic magnets, does not reach predicted values. Work is required on optimising the orientation relationships between the hard and soft phases and the interphase properties (coherency) between them. Research on nanocrystalline hard magnetic alloys has received attention worldwide. The U.S. efforts are summarised in the article by G.C. Hadjipanayis (1998, 107-112). While less research seems to be carried out in the world on these materials compared to the nanocrystalline soft magnetic alloys, some efforts exist in most countries. Notable programmes are those of L. Schultz and coworkers at the Institut für Festköper und Werkstofforschung (IFW) in Dresden and P.G. McCormick and coworkers at the University of Western Australia.

While the very low losses of the nc soft magnetic materials (Finemet or Nanoperm) are dependent on grain size for their properties, the hard magnetic nc alloys with remanence enhancement provide flexibility in processing, especially with powder materials. These remanence-enhanced nc hard magnetic alloys may find many applications as permanent magnet components.

Giant Magnetoresistance (GMR)

The phenomenon of giant magnetoresistance (GMR)—the decrease of electrical resistance of materials when exposed to a magnetic field—was first reported in a number of multilayer ferromagnetic/nonferromagnetic thin film systems (Baibich et al. 1988). More recently, GMR was observed in equiaxed granular nanocrystalline materials (Berkowitz et al. 1992). In particular, GMR systems with low saturation fields offer a wide area for application in magnetoresistive devices. GMR sensors have a higher output than conventional anisotropic magnetoresistive sensors or Hall effect sensors. They can operate at higher magnetic fields than conventional magnetoresistive sensors. In multilayer systems the antiferromagnetic alignment of the ferromagnetic layers in zero field becomes ferromagnetic as the field is applied and causes a decrease in resistance. Granular materials that show GMR consist of small ferromagnetic single-domain particles with randomly oriented magnetic axes in a nonmagnetic matrix. An external field rotates the magnetic axes of all magnetic particles. The rotation towards complete alignment of all magnetic axes again reduces the resistance in a similar way as for multilayers. The GMR in granular systems is isotropic. The explanation for the GMR is spin-dependent scattering of the conduction electrons at the ferromagnetic/nonmagnetic interfaces and, to a lesser extent, within the magnetic grains. The GMR scales inversely with the average particle diameter. There is worldwide research on the GMR effect. U.S. programmes are reviewed by R. Shull and G.C. Hadjipanayis in the proceedings of the WTEC U.S. nanotechnologies workshop (Shull 1998, 43-58; Hadjipanayis 1998, 107-112). The NIST work described by Shull has provided material with the largest GMR values for the smallest switching fields. Japanese research on GMR includes studies in Prof. Fujimori's group at Tohoku University. While the theory for GMR of spin-dependent scattering referred to above has been used as an explanation, other explanations taking into account interaction between magnetic regions have been proposed (El-Hilo et al. 1994). Combined theoretical and experimental studies should help to clarify the mechanism for this effect.

Ferromagnetic Nanocrystalline Materials

Magnetic nanocomposite refrigerants, which have four times the magnetocaloric effects of the best low temperature magnetic refrigerant, were developed by NIST and described by R. Shull (1998, 43-58). The entropy change at a given (low) temperature for a system of magnetic spins is enhanced when the isolated spins are clustered. Shull et al. (1993) have shown that the nanocomposite Gd3Ga5–xFexO12 gives superior magnetocaloric effects, which increase with x up to x = 2.5 and can be extended to higher temperatures than conventional materials. Magnetostrictive materials such as Terfonol-D (Tb0.3Dy0.7Fe2) have been of scientific and technological interest in recent years. It is suggested by G.C. Hadjipanayis (1998, 107-112) that nanostructured magnetostrictive materials can have improved properties, such as lower saturation fields, with reduced anisotropy and in multilayers with alternate layers of magnetostrictive and soft magnetic materials that are exchange-coupled. Hadjipanayis states that most of the research in this area is carried out in Japan and Europe.

Opportunities and Challenges

Nanocrystalline magnetic materials offer perhaps the nearest-term prospect of significant applications of bulk nanostructured materials. The remanence enhancement in two-phase hard/soft magnetic materials can result in excellent energy products—comparable to those from rapid solidification processing routes—in powder composites. The flexibility in manufacturing this allows should provide many possible permanent magnet applications. Nanocrystalline soft magnetic materials have the lowest energy losses of any material. While problems such as mechanical brittleness remain, these materials promise to replace existing transformer core materials for power applications. The U.S. research and industrial effort in these materials lags the efforts in Japan.

Nanocrystalline Hydrogen Storage Materials

R.B. Schwarz (1998, 93-95) has pointed out that nanostructured materials offer several potential advantages for hydrogen storage materials. Rapid kinetics of absorption/desorption can be aided by refining the microstructure to the nanoscale. For example, nanoscale inclusions of Mg2Ni in Mg catalyze the decomposition of the molecular hydrogen, increasing the hydrogen absorption/desorption kinetics. Another advantage of the nanoscale microstructure is that the alloy powder does not comminute on repeated charging/discharging with hydrogen. This is not strictly a bulk material, since powder agglomerates or green compacts can be used, thus obviating the need for compaction to theoretical density.

Nanocrystalline Corrosion-Resistant Materials

The limited work to date on corrosion resistance of nanocrystalline materials indicates that no generalisations can be made. Superior localised corrosion resistance in HCl was observed for nanocrystalline 304 stainless steel (Fe–18%Cr–8%Ni) prepared by sputter deposition (Inturi and Szklavska-Smialowska 1991 and 1992). This was attributed to the fine grain size and homogeneity of the nc material. However, the average dissolution rate of nc Ni was found to be higher than that for coarse-grained material (Rofagha et al. 1991).

Polymers and Polyelefin

There has been enormous interest in the commercialisation of nanocomposites for a variety of applications, and several of these applications will be successful in the near future. Mineral fillers, metals, and fibers have been added to thermoplastics and thermosets for decades to form composites. Compared to neat resins, these composites have a number of improved properties including tensile strength, heat distortion temperature, and modulus. Thus, for structural applications, composites have become very popular and are sold in billion pound quantities. These filled thermoplastics are sold in even larger volumes than neat thermoplastics. Furthermore, the volume of fillers sold is roughly equal to the volume of thermoplastic resin sold. Clearly, the idea of adding fillers to thermoplastics and thermosets to improve properties, and in some cases decrease costs, has been very successful for many years. More recently, with advances in synthetic techniques and the ability to readily characterize materials on an atomic scale has lead to interest in nanometer-size materials. Since nanometer-size grains, fibers and plates have dramatically increased surface area compared to their conventional-size materials, the chemistry of these nanosised materials is altered compared to conventional materials. Polymer nanocomposites combine these two concepts, i.e., composites and nanometer-size materials. Thermoplastics filled with nanometer-size materials have different properties than thermoplastics filled with conventional materials. Some of the properties of nanocomposites, such as increased tensile strength, may be achieved by using higher conventional filler loading at the expense of increased weight and decreased gloss. Other properties of nanocomposites such as clarity or improved barrier properties cannot be duplicated by filled resins at any loading. Polymer nanocomposites were developed in the late 1980s in both commercial research organisations and academic laboratories. The first company to commercialize these nanocomposites was Toyota, which used nanocomposite parts in one of its popular car models for several years. Following Toyota's lead, a number of other companies also began investigating nanocomposites. Most commercial interest in nanocomposites has focused on thermoplastics. Thermoplastics can be broken into two groups: less expensive commodity resins and more expensive (and higher performance) engineering resins. One of the goals of nanocomposites was to allow substitution of more expensive engineering resins with a less expensive commodity resin nanocomposite. Substituting a nanocomposite commodity resin with equivalent performance as a more expensive engineering resin should yield overall cost savings.

By a strict definition of nanocomposites, i.e., any filler submicron in size, there already are significant volumes of nanocomposites being produced. These amount to more than 20 million pounds. However, since these fillers are on the upper end of the nanocomposite size range, most sources have excluded them from consideration. This report covers these products but does not add their volumes into aggregate numbers of nanocomposite markets or production volumes to comply with the established convention. At this point in time, there has been much less commercial interest in thermoset nanocomposites compared to thermoplastics. This neglect may not continue much longer since thermoset nanocomposites have some distinct advantages over neat thermoset resins. This report covers some possible markets for thermoset nanocomposites. There are two types of recognised polymer nanocomposites:

- Thermoplastic nanocomposites: these materials are broken into two major categories, i.e., commodity resins and engineering resins. The potential of polymer nanocomposite commodity resin is covered for each resin. Engineering resin nanocomposites are restricted to resins that have been under development.
- Thermoset nanocomposites: thermoset nanocomposites have received less commercial interest in their development than thermoplastic nanocomposites, but these materials may be relatively straightforward to bring into production. Furthermore, thermoset nanocomposites can offer some significant advantages over conventional thermosets.

Polymer nanocomposite (PNC) is a polymer or copolymer having dispersed in it nanoparticles. These may be of different shape (e.g., platelets, fibers, spheroids), but at least one dimension must be ca. 1 to 50 nm (diameter of pencil lead is about 1 mm, or 1, 000, 000 nm). PNC with all three types of nanoparticles have been prepared (e.g., polycarbonate with carbon nanotubes, polyamide with iron oxide spheres), but only PNC with clay platelets are on the market for the structural (high volume) applications. These PNC's belong to the category of multi-phase systems (MPS, viz. blends, composites, and foams) that consume ca. 95 per cent of plastics production. The key to industrial success of MPS is the desired and stable morphology. Thus, these systems require controlled mixing/compounding, stabilisation of the achieved dispersion, orientation of the dispersed phase, and optimisation of interactions in the finished product. In spite of differences in the nature of the dispersed phase, the compounding strategies for all MPS, including PNC, are similar.

Mixing is the most important operation in polymer processing. Uniformity of the molecular weight and its distribution, entanglement, composition, as well as the stress history and temperature, is the key to good performance. Dispersive (or intensive) and distributive (or extensive) mixing of molten polymers is carried out within the laminar flow regime. The dispersive action is responsible for the breaking of original domains of the minor component into the desired size, while the distributive provides for their homogeneous distribution within the matrix. Microrheology may be used to describe the dispersive process in the shear or extensional flow field, while distributive mixing may be treated in terms of the laminar flow theory. Both models - microrheology and laminar mixing - indicate superiority of the extensional flow field. The presence of the vorticity component in the shear tensor is responsible for rotation of the flow elements that cause excessive energy consumption, polymer degradation and attrition of anisometric solid particles - shear mixing is inefficient. Mixing in the extensional flow field may be carried out by means of the convergent-divergent flow geometry. A motionless extensional flow mixer (EFM) and its newer, dynamic version (DEFM) are fully adjustable, general-purpose mixers used for homogenisation of reaction products, removal of "fish eyes" from films, alloying, blending or dispersing of nanoparticles[4]. In the search for synergism the industry moves toward progressively complex MPS, e.g., foaming of nanoclay reinforced polymer blends. Here, the difficulty is the presence of several phases with potentially different thermodynamic and rheological character as well as reinforcing and foaming the appropriate blend components. However, once the difficulties are over, the advantages translate into superior performance.

The clay-containing polymeric nanocomposites (CPNC) constitute new type of polymeric systems, with a great variety of molecular structures and processing difficulties. The only type of nanoparticles used in commercial production of CPNC is the lamellar, expandable, smectite clay. For the best performance of CPNC the clay platelets (ca. 1 nm thick and 300 nm diameter) must be homogeneously dispersed as individual crystalline lamellae (known as exfoliation). To accomplish this task, a balance between the thermodynamics and mechanical mixing must be achieved. Owing to the large aspect ratio, at a concentration of clay exceeding ca. 1 wt per cent the platelets start to interact, creating 3D structures. In commercial CPNC the clay content is about twice as large, hence there is local ordering. Generation of the most suitable orientation is the key for the optimisation of the performance. Three recent developments in the CPNC science and technology are worth noting.

The use of an elongational flow field greatly accelerates dispersion and orientation, facilitating the processing. Several CPNC types (e.g., with polyethylene terephthalate,

PET, polystyrene, PS, or polypropylene, PP, matrix) have been prepared using commercial EFM attached either to single-screw extruder (SSE) or twin-screw extruder (TSE). The results demonstrated significant beneficial role even of the standard EFM (developed for melt homogenisation and polymer blending). For example, PET was extruded with organoclay using a TSE and the same machine, but with attached to it EFM. The presence of EFM increased the PNC properties over that from TSE by (data from EFM Inc.):

(i) Tensile modulus by ca. 67 per cent.

(ii) Tensile strength by ca. 60 per cent.

(iii) Impact strength 80 per cent.

In Union Chemical Laboratory, Industrial Technology Research Institute (UCL/ITRI) in Taiwan, PP was compounded with organoclay and a compatibilizer (maleated polypropylene, PP-MA)[5]. Three processing methods were used: #1: TSE; #2: TSE + EFM; and #3: SSE + EFM. The method #1 resulted in mixtures with clay dispersed as micron-sized aggregates. By contrast, process #2 produced well-intercalated CPNC with short stacks of organoclay uniformly dispersed in the molten PP. However, the best dispersion was obtained from #3, i.e., using SSE + EFM. This unexpected result was most likely due to higher thermo-mechanical degradation in TSE.

The ammonium-ion intercalated organoclay is the main type used for the production of CPNC. However, these materials start to degrade at temperatures above 180°C by the Hofmann elimination mechanism, leaving the clay surface bare. Numerical simulations were carried out to evaluate the effects of independent parameters on the degree of CPNC dispersion. The most dramatic worsening of the thermodynamic conditions for exfoliation was obtained in the presence of bare clay platelets - even as small amount as 1 or 2 per cent of clay surface precludes exfoliation.

Kato et al. from the Toyota Central R&D Lab proposed a revolutionary method of CPNC production. The method uses un-treated clay, e.g., sodium montmorillonite (Na-MMT), which becomes exfoliated and the morphology stabilised during the mixing/compounding step. The process uses a long (L/D = 77) TSE, divided by a pair of seal rings into three zones. In the first zone PP + PP-MA + Na-MMT were melt mixed together. In zone #2 ca. 10 per cent water was injected into the melt and the pressure of the water vapor was well controlled. Adsorption of water by clay platelets, followed by reaction with compatibilizer (PP-MA), resulted in exfoliation and stabilisation of the dispersion. Finally, in zone #3 devolatilisation took place. The resulting CPNC had almost the same properties as the conventionally prepared one, but at a fraction of cost. The method bypasses the need for the costly intercalation of Na-MMT with ammonium ions, avoids the introduction of readily degradable organoclay to molten polymer (which in turn gets degraded by combination of organoclay degradation and air) and at the same time eliminates the need of costly in time and energy pre-drying step. Evidently, the method is in an early stage of development. Its extension to other polymers (e.g., polyesters or polyamides), optimisation of the compatibilizer, processing equipment and process parameters are required.

The main advantages of CPNC stem from the improved stiffness (by a factor of 2 at 5 wt per cent loading), reduction of gas and vapor permeability (by a factor of 100 at 10 wt per cent loading), and reduced flammability. Since these enhancements are achieved with only a slight increase of density, the transport, and packaging industries are expect to be the first to profit from this new technology. Up to now two principal obstacles hindered progress of CPNC into the market: the cost and reproducibility of performance. The demonstrated by Kato et al. suitability of Na-MMT reduces the cost of the nano-filler from ca. $7 to $2 per kilogram, and at the same time reduces the probability of thermo-mechanical degradation during processing and forming. Many questions remain the foremost being whether the achieved improvement of technology is sufficient to lower the cost and improve stability to satisfy the industry standards. Evidently, the future of CPNC technology hinges on the intelligent application of the available compounding and processing technologies, adopted for the specific needs of the nano-scale mixing.

The subject of hybrids based on layered inorganic compounds such as clays has been studied for a considerable time, but the area is enjoying a resurgence of interest and activity as a result of the exceptional properties which can be realised from such nanocomposites. Materials variables which can be controlled and which can have a profound influence on the nature and properties of the final nanocomposite include the type of clay, the choice of clay pre-treatment, the selection of polymer component and the way in which the polymer is incorporated into the nanocomposite. The last of these may be dictated by the available processing methods and whether the prospective user is an integrated polymer manufacturer or a specialist processor.

Common clays are naturally occurring minerals and are thus subject to natural variability in their constitution. The purity of the clay can affect final nanocomposite properties. Many types of clay are aluminosilicates, which have a sheet-like (layered) structure, and consist of silica SiO_4 tetrahedra bonded to alumina AlO_6 octahedra in a variety of ways. A 2:1 ratio of the tetrahedra to the octahedra results in smectite clays, the most common of which is montmorillonite. Other metals such as magnesium may replace the aluminium in the crystal structure. Depending on the precise chemical composition of the clay, the sheets bear a charge on the surface and edges, this charge being balanced by counter-ions, which reside in part in the inter-layer spacing of the clay. The thickness of the layers (platelets) is of the order of 1 nm and aspect ratios are high, typically 100-1500. The clay platelets are truly nanoparticulate. In the context of nanocomposites, it is important to note that the molecular weight of the platelets (ca. 1.3×10^8) is considerably greater than that of typical commercial polymers, a feature which is often misrepresented in schematic diagrams of clay-based nanocomposites. In addition, platelets are not totally rigid, but have a degree of flexibility. The clays often have very high surface areas, up to hundreds of m^2 per gram. The clays are also characterised by their ion (e.g. cation) exchange capacities, which can vary widely. One important consequence of the charged nature of the clays is that they are generally highly hydrophilic species and therefore naturally incompatible with a wide range of polymer types. A necessary prerequisite for successful formation of polymer-clay nanocomposites is therefore alteration of the clay polarity to make the clay 'organophilic'. An organophilic clay can be produced from a normally hydrophilic clay by ion exchange with an organic cation such as an alkylammonium ion. For example, in montmorillonite, the sodium ions in the clay can be exchanged for an amino acid such as 12-aminododecanoic acid (ADA):

Na^+-CLAY + HO_2C-R-$NH_3^+Cl^-$ à.HO_2C-R-NH_3^+-CLAY + NaCl

The way in which this is done has a major effect on the formation of particular nanocomposite product forms and this is discussed further below. Although the organic pre-treatment adds to the cost of the clay, the clays are nonetheless relatively cheap feedstocks with minimal limitation on supply. Montmorillonite is the most common type of clay used for nanocomposite formation; however, other types of clay can also be used depending on the precise properties required from the product. These clays include hectorites (magnesiosilicates), which contain very small platelets, and synthetic clays (e.g. hydrotalcite), which can be produced in a very pure form and can carry a positive charge on the platelets, in contrast to the negative charge found in montmorillonites.

The synthetic route of choice for making a nanocomposite depends on whether the final material is required in the form of an intercalated or exfoliated hybrid. In the case of an intercalate, the organic component is inserted between the layers of the clay such that the inter-layer spacing is expanded, but the layers still bear a well-defined spatial relationship to each other. In an exfoliated structure, the layers of the clay have been completely separated and the individual layers are distributed throughout the organic matrix. A third alternative is dispersion of complete clay particles (tactoids) within the polymer matrix, but this simply represents use of the clay as conventional filler.

In recent years, there has been extensive study of the factors which control whether a particular organo-clay hybrid can be synthesised as an intercalated or exfoliated structure. Since clay nanocomposites can produce dramatic improvements in a variety of properties, it is important to understand the factors which affect delamination of the clay. These factors include the exchange capacity of the clay, the polarity of the reaction medium and the chemical nature of the interlayer cations (e.g. onium ions). By modifying the surface polarity of the clay, onium ions allow thermodynamically favourable penetration of polymer precursors into the interlayer region. The ability of the onium ion to assist in delamination of the clay depends on its chemical nature such as its polarity. The loading of the onium ion on the clay is also crucial for success and it should be borne in mind that a commercial organo-clay may not have the optimum loading for a particular application. For positively charged clays such as hydrotalcite, the onium salt modification is replaced by use of a cheaper anionic surfactant. Other types of clay modification can be used depending on the choice of polymer, including ion-dipole interactions, use of silane coupling agents and use of block copolymers. An example of ion-dipole interactions is the intercalation of a small molecule such as dodecylpyrrolidone into the clay. Entropically-driven displacement of the small molecules then provides a route to introducing polymer molecules. Unfavourable interactions of clay edges with polymers can be overcome by use of silane coupling agents to modify the edges. These can be used in conjunction with the onium ion treated organo-clay. An alternative approach to compatibilising clays with polymers has been introduced by TNO (Netherlands), based on use of block or graft copolymers where one component of the copolymer is compatible with the clay and the other with the polymer matrix. This is similar in concept to compatibilisation of polymer blends. A typical block copolymer would consist of a clay-compatible hydrophilic block and a polymer-compatible hydrophobic block. The block length must be controlled and must not be too long. High degrees of exfoliation are claimed using this approach.

The correct selection of modified clay is essential to ensure effective penetration of the polymer or its precursor into the interlayer spacing of the clay and result in the desired exfoliated or intercalated product. Indeed, further development of compatibiliser chemistry is undoubtedly the key to expansion of this nanocomposite technology beyond the systems where success has been achieved to date. Polymer can be incorporated either as the polymeric species itself or via the monomer, which is polymerised in situ to give the corresponding polymer-clay nanocomposite. The second of these is the most successful approach to date, although it probably limits the ultimate applicability of these systems. Polymers can be introduced either by melt blending, for example extrusion, or by solution blending. Melt blending (compounding) depends on shear to help delaminate the clay and can be less effective than in situ polymerisation in producing an exfoliated nanocomposite. Both thermo sets and thermoplastics have been incorporated into nanocomposites, including:

- Nylons
- Polyolefins, e.g. polypropylene
- Polystyrene

- Ethylene-vinyl acetate (EVA) copolymer
- Epoxy resins
- Polyurethanes
- Polyimides
- Poly(ethylene terephthalate) (PET)

Polyelefin

Polyolefin nanocomposites are useful for a wide variety of applications in packaging, engineering and fire retardant markets. These nanocomposites are near molecular blends of montmorillonite nanoclay and a polyolefin matrix. Concentrate trade name. Here performance properties are outlined for typical polyolefins with explanations of options for creating nanocomposites.

Polyolefins represent a wide range of homo and copolymers, principally containing propylene and ethylene. Also included are polyolefin/rubber alloys usually called TPO's and polyolefin/EVA blends. Because propylene and ethylene contents as well as polydispersities vary widely, some Nanomer Concentrate screening is required to determine the optimum match between the matrix and nanoclay surface treatment.

Polyolefin nanocomposites are produced via extrusion compounding using Concentrates. Refer to data for concentrate specifications and grade selection. Film or bottle grade products require twin screw compounders or intensive mixers. Most engineering and fire retardant applications can be accommodated using single screw technology. A specially designed screw, called a NanoMixer™, must be used. Compounding guidelines are available on Tech Data Sheets P-803 through P-806.

Nanocomposite performance depends on the degree of nanoclay de-agglomeration and dispersion within the polyolefin matrix. The process of de-agglomeration/dispersion is commonly referred to as exfoliation. Three factors influence exfoliation: the nanoclay surface treatment, nanocomposite loading, and the amount of dispersive energy used. The optimum nanoclay surface treatment is readily determined through screening, as described earlier. Nanomer loading will not hinder exfoliation within the 2-8 per cent wt/wt range and is optimised at 6 per cent wt/wt. The proper choice of compounding equipment and use of the guidelines assists in achieving the dispersive energy requirement. However, another factor impacting dispersive energy is matrix resin melt flow. Resin melt flow influences the level of shear resulting from dispersive energy input. The lower the melt flow, the greater the shear on Nanomer particles. It follows that exfoliation is maximised in lower melt flow resins, where shear is the highest.

Polyolefin nanocomposites exhibit mechanical properties similar to highly filled systems, but with the added advantage of low density. The density (and resulting part weight) is just slightly higher than unfilled resin. Nanocompositing also provides a high degree of reinforcement without significant loss of impact. In most cases notched Izod impact remains the same as neat (unfilled) resin.

Polyolefin nanocomposites have specific gravities and melt flows nearly the same as neat resin, so they process like neat resin. Part output rates will be significantly higher than conventionally filled composites and machine wear will also improve noticeably. Nanocomposites can be used in conjunction with conventional fillers, i.e. glass fiber or talc, to create synergistic materials with customised property sets. For example a 6 per cent wt/wt nanocomposite containing 10-12 per cent wt/wt glass fiber will deliver a property set similar to 30 per cent wt/wt straight glass fiber, but lower specific gravity and part weight.

Polyolefin nanocomposites are potent char formers. In fire retardant (FR) applications they function by insulating the resin substrate from the air-surface interface, starving the flame by reducing the fuel needed to continue the burning process. The result is a dramatically lower peak heat release rate (HRR). Nanocomposites alone do not conform to a UL V0 rating. But in conjunction with reduced levels of conventional FR additives they act synergistically to achieve V0. Often traditional FR additives can be reduced by 50 per cent wt/wt. With less FR additive dilution and the enhanced reinforcing effect of nanocompositing, mechanical properties are restored, usually at lower cost than would be the case without nanocompositing. Another benefit of nanocomposite/FR additive synergy is lower density and part weight, since a lot of high density additive has been removed from the formulation.

As polyolefins take an increased role in packaging applications, gas barrier becomes an important requirement. Nanocomposites improve oxygen transmission rates (OTR) between 1.2-2X. Carbon dioxide transmission rates (CO2TR) improve 1.4-1.8X. There is also some improvement in water vapor transmission (WVTR), around 10-15 per cent. Improved gas barrier follows the tortuous path model and is valid for cast, blown and molded products. Present performance data for a variety of polyolefins. Predictably, nanocomposite films possess greater tensile strength and modulus, although elongation-atbreak is reduced. Simultaneous improvement in tensile properties and gas barrier provides the formulator with opportunities to downguage films and containers. Tear strengths in both the machine and transverse directions remain the same as neat polyolefin films of the same mil thickness. Tear strength of downguaged nanocomposite film drops in proportion to the reduction in film thickness.

The earliest example of the in situ polymerisation method was work by Toyota on synthesis of clay-nylon nanocomposites and this remains probably the most studied system, including work by Bayer and Ube. In a typical synthesis ADA-modified clay is dispersed in the monomer caprolactam, which is polymerised to form the nylon-6-clay hybrid as an exfoliated composite. Complete exfoliation may be preceded by intercalation of the monomer in the clay. Generally, low concentration of clay (a few per cent) are incorporated in these nanocomposites, partly because this is often sufficient to modify the desired properties significantly, but also because higher levels of clay can adversely increase the system viscosity leading to poor processability, although the viscosity increase is shear rate dependent. Other nylons and copolyamides (e.g. nylon-6/6, 6) have also been incorporated in clay nanocomposites. Functionality such as hydroxyl groups can be introduced into the onium salt modifiers to improve compatibility with the nylon via hydrogen bonding – this can lead to an enhancement of desirable nanocomposite properties.

Similar modification of the chemistry may be required for successful exfoliation with other types of polymer system. In the case of ethylene-vinyl alcohol (EVOH) copolymers, use of hydroxylated quaternary ammonium ions improves compatibility between the clay and the EVOH by introducing favourable hydroxyl group interactions. In polypropylene (PP) nanocomposites, maleic anhydride grafted PP is used as a compatibiliser. Polymerisation initiators can be anchored to the clay platelet surface and this approach has been extended to living free radical polymerisation of styrene, where an initiating species bonded to the TEMPO free radical is attached to the surface of the clay platelets via ion exchange.

For preparation of nanocomposites from high temperature engineering thermoplastics, a major limitation in the use of conventional onium ion modified clays is the thermal instability of the alkylammonium species during processing. The block copolymer route developed by TNO (vide supra) offers one potential solution to this problem. Imidazolium salts such as that depicted. A further alternative is the use of phosphonium salts in place of ammonium salts—this can lead to an increase in the degradation temperature of the organo-clay from 200-300°C to >300°C. Wholly synthetic organo-clays clays, on the other hand, can exhibit thermal stability to beyond 400°C. By using these last two approaches, Triton Systems has succeeded in producing nanocomposites from high temperature resins such as polyetherimide (PEI). Forming the nanocomposite provides a potential route to producing 'speciality' engineering resin performance from cheaper 'standard' engineering resins.

Future goals include the extension of this nanotechnology to additional types of polymer system, where the development of new compatibilisation strategies is likely to be a prerequisite. Production of PVC-based systems is still some way off and challenges remain to be solved in PET nanocomposites. Additional reinforcement of clay nanocomposites by glass fibre is currently being investigated. There is also interest in the development of electrically conducting clay nanocomposites.

Polymer-Silicate and Nanoparticles

Polymer nanocomposites, particularly polymer-silicate nanocomposites, represent an emerging class of specialty plastic materials. The melt-processing route would be a powerful approach in the manufacture of these compounds. Techniques such as transmission electron microscopy and x-ray diffraction were used to determine that nylon-6 nanocomposites could be produced via melt compounding. As a result of the "nano"-sized dispersion, reinforcement efficiency is realised at low loading levels (3-5 per cent) which is a small cost in terms of specific gravity. Dramatic enhancements are observed in heat distortion temperature and stiffness. Besides mechanical properties, nanocomposites also have much lower gas permeability than the neat resin. Other potential benefits include good surface appearance, dimensional stability, and lower heat release. Nanocomposites, particularly polymer-silicate nanocomposites, are materials that have been given much attention recently and represent an emerging class of specialty plastic compounds.

One of the reasons for this attention is that due to the "nano"-scale dispersion, which results in high aspect ratio and surface area, the reinforcement efficiency of polymer-silicate nanocomposites can be significantly better than conventional mineral fillers. The potential benefits of polymersilicate nanocomposites include higher heat resistance, increased flexural modulus, lower gas permeation rate, easy flow, dimensional stability, good surface appearance, and lower heat release rate. Typically, the type of silicate used in polymersilicate nanocomposites is montmorillonite which is a 2:1 layered smectite clay. This clay has a platy structure with individual platelets having thicknesses of one nanometer and surface lengths on the order of 100 to 1000 nanometers. Montmorillonite is hydrophillic which makes proper exfoliation and dispersion into conventional polymers difficult. Thus, montmorillonite is usually modified through substitution of sodium ions with organic onium ions. The result is an organo-clay complex.

The intercalation procedure expands the gallery and improves the compatibility of the complex with the polymer so that individual platelets can be more easily separated. Possible morphologies of polymers and organo-clay complexes.There are several potential approaches for the production of the polymer-silicate nanocomposite. One procedure involves the synthesis of the compound in the polymerisation stage with the complex *in-situ* with the liquid monomer. Also, nanocomposites could be prepared from solvents in which both polymer and complex are dissolved. Another method involves the preparation of polymer-silicate nanocomposites via melt processing. In this process, the application of shear during compounding assists exfoliation and dispersion. The melt-compounding route is a powerful approach in the production of polymersilicate nanocomposites since existing technologies and equipment could be utilised and easily scaled to commercial quantities. Additionally, this approach allows specialty materials such as those with custom colours, static dissipation, or wear resistance to be readily manufactured.

Polymer-silicate compounds were prepared by compounding various organo-clay complexes at different loading levels with commercially available nylon-6 using a twin-screw extruder. Neat nylon-6 and the compounds were dried at 80°C under vacuum prior to compounding, molding, and rheometry. The extent of intercalation, exfoliation, and dispersion of the complexes in the compounds was determined. This was accomplished through wideangle x-ray diffraction (XRD) for preliminary determinations. Transmission electron microscopy (TEM) was used to confirm the morphology development established by XRD. Test specimens were prepared by injection molding. The mechanical and electrical properties of the specimens were evaluated according to ASTM test methods: D-638 for tensile, D-790 flexural, D-256 for Izod impact, D-648 for deflection temperature, D-955 for shrinkage, and D-570 for water absorption. The barrier capability of these materials was explored by measuring oxygen transmission rates (OTR) with a Mocan apparatus. OTR measurements were performed at constant temperature and humidity for 200 hours on specimens thermoformed from injection molded plaques. Compound melt flow behaviour was analysed with a capillary rheometer. The effect of shear rate at 235°C and melt flow stability at 235°C and 1000 sec^{-1} were determined.

Suitable morphology development in a nanocomposite via the compounding route consists of wetout, intercalation, exfoliation, and dispersion. Therefore, the selection of an appropriate complex for the matrix polymer plays a critical role in determining the potential degree of exfoliation and dispersion in a polymer-silicate compound. One procedure that can be used to assess the extent of intercalation and exfoliation is to analyze x-ray diffraction patterns. XRD patterns for nylon-6 compounds containing two different types of complexes. Finally, not only does this show that XRD patterns can be used to evaluate nanocomposite development but also that nylon-6 nanocomposites are indeed possible via melt compounding. The properties of a typical nylon-6 nanocomposite are given in Table and compared to the unfilled matrix nylon-6. At almost no cost in specific gravity, significant improvements in certain properties are realised. For instance, flexural modulus is increased by almost 50 per cent and deflection temperature raised by over 35°C. Also, tensile strength is increased by over a third. For comparison purposes, the properties of typical mineral and glass fiber filled nylon-6 compounds are given. Reinforcement efficiencies of nanocomposites compared to conventional mineral fillers on a weight basis.

Besides mechanical and heat resistance property improvements, nanocomposite compounds can exhibit dramatically enhanced barrier properties as represented by lower oxygen transmission rates. Nylon-6 nanocomposites can achieve an OTR almost four times lower than the unfilled nylon-6. The reduction in OTR is typically attributed to the increase in effective diffusion length as solutes must travel a tortuous path around well-dispersed platelets of high aspect ratio. It is also important to point out that as a result of the nanometer length scale that excellent transparency is retained in sheets and films formed from nylon-6 nanocomposites. There has been some concern about the ability to process nanocomposites due to the high surface area of the nanoclay upon exfoliation. For instance, the effect of other high surface area additives such as high structure carbon black is to dramatically increase the viscosity. The viscosity of a typical nylon-6 nanocomposite is compared to the viscosity of the neat polymer as a function of shear rate. Contrary to the concerns, the viscosity is observed to be lower than the neat resin at the shear rates of most interest to potential processors. As a result, nylon-6 nanocomposites may be processed as easily as the neat resin. There are several potential explanations for the shear thinning at higher shear rates. One possible reason could be due to orientation of the platelets upon shear. Another reason could be that the polymer has been degraded during compounding with the nanoclay. However, degradation was not obvious in the nylon-6 nanocomposite samples. The melt flow stability of a nylon-6 nanocomposite was determined. The viscosity was found to not significantly decrease with time. Additionally, the viscosity is equivalent to the viscosity measured under similar conditions during prior testing of the effect of shear rate. Therefore, the degradation if present, would not be expected to continue on further processing. The dimensional stability of nylon-6 nanocomposites is compared. The unfilled resin and other typical reinforced compounds. As seen, the water absorption of the nanocomposite is dramatically reduced by as much as 55 per cent. This is comparable to a filled compound which contains much higher content. This is due to the barrier properties since resin replacement cannot entirely account for all of the change as is the case for conventional mineral and glass compounds. Also observed in table is that the shrinkage of nylon-6 nanocomposites is more similar to the base resin due to the lower filler content. Additionally, the shrinkage in the transverse direction does not differ significantly from the flow direction. This gives a warpage factor much closer to one as shown.

Polymer-clay nanocomposites have been developed via the compounding route in nylon-6. XRD and TEM analysis confirmed that good exfoliation and dispersion was attained. Notable property improvements were observed in tensile strength, flexural modulus, heat deflection temperature, and oxygen transmission rate. Proposed application areas include those requiring structural properties. For structural applications, nylon-6 nanocomposites exhibit equivalent or better properties than typical 20-30 wt per cent mineral filled compounds depending on the property being investigated. It is important to note that these gains come at almost no cost in specific gravity from the unfilled nylon-6. However, the properties of nylon-6 nanocomposites have yet to approach those of glass fiber reinforced compounds.

Even so nylon-6 nanocomposites are excellent candidates when improved heat deflection temperature, stiffness, or strength is required especially when glass fiber reinforcement cannot be considered either due to specific gravity, mold shrinkage/ warpage, or wear. Another potential application area is barrier. A significant four-fold improvement was observed in the oxygen transmission rate. This may allow nylon-6 nanocomposites to be utilised in certain packaging and film applications.

Nanoscience has taken scientists around the world by storm. It claims to revolutionize the world we live in with radical breakthroughs in areas such as materials and manufacturing, electronics, medicine and healthcare, environment and energy, chemical and pharmaceutical, biotechnology and agriculture, computation and information technology.

Nanoparticles are small clusters of atoms about 1 to 100 nanometers long. 'Nano' derives from the Greek word "*nanos*", which means dwarf or extremely small. It can be used as a prefix for any unit like a second or a liter to mean a billionth of that unit. A nanosecond is a billionth of a second. A nanoliter is a billionth of a liter. And therefore a nanometer is a billionth of a meter or 10^{-9} m.

- If a baseball is the size of Earth, a nanoparticle would be the size of an apple.
- We can also compare it with things in the natural world.
- The smallest synthetic nanotechnology made so far are the individual components of an Intel Pentium III microprocessor and they span about 200 nanometers.

Nanoparticles are larger than individual atoms and molecules but are smaller than bulk solid. Hence, they obey neither absolute quantum chemistry nor laws of classical physics and have properties that differ markedly from those expected. There are two major phenomenons that are responsible for these differences. First is the high dispersity of nanocrystalline systems. As the size of a crystal is reduced, the number of atoms at the surface of the crystal compared to the number of atoms in the crystal itself, increases. For example, a 4 nm diameter CdS nanoparticle has about 1500 atoms, of which about a third are on the surface. Properties, which are usually determined by the molecular structure of the bulk lattice, now become increasingly dominated by the defect structure of the surface. The second phenomenon occurs noticeably only in metals and semiconductors. It is called size quantisation and arises because the size of a nanoparticle is comparable to the de Broglie wavelength of its charge carriers (*i.e.* electrons and holes). Due to the spatial confinement of the charge carriers, the edge of the valance and conduction bands split into discrete, quantised, electronic levels. These electronic levels are similar to those in atoms and molecules. The spacing of the electronic levels and the bandgap increases with decreasing particle size. This is because the electron hole pairs are now much closer together and the Coulombic interaction between them can no longer be neglected giving an overall higher kinetic energy. This increase in bandgap can be observed experimentally by the blue-shift in the absorption spectrum or sometimes even visually by the colour of the samples. A larger bandgap means that more energy is required to excite an electron from the valance band to the conduction band and hence light of a higher frequency and lower wavelength would be absorbed. The increase in molar absorption coefficient shows that the oscillator strength also increases as particle size decreases. This is due to the strong overlapping of the wave functions of confined charge carriers. Experiments carried out in the past have found the absorption spectrum to be potential dependent, with decreasing wavelengths as applied potential decreases. This change is fully reversible before the threshold potential.

The first generation of nanofillers is emerging from R&D into commercial application in composites. Federal funding for nanotechnology research and development in USA has increased six-fold, from $116 million (USD) in 1997 to an estimated $961 million in 2004. In the U.S. alone, the FY 2005 budget request that President George W. Bush has sent to Congress calls for a total National Nanotechnology Initiative (NNI) budget of $982 million, which represents a 2 per cent increase. With all this research, what has come out of the R&D pipeline ready for commercial application? As it turns out, plenty. The following are new and, surprisingly, not-so-new applications of two types of nanotechnology: ceramic nanoparticles and carbon nanotubes used as highly functional fillers/additives in polymer products of interest to composites manufacturers. Introduction of these materials can result in the improvement of the polymer's mechanical properties in molded parts and coatings on a wide variety of fronts, including increased part strength, ductility and dimensional stability as well as resistance to damage by abrasion and exposure to caustic chemicals and/or extreme heat.

Nanotechnology involves the control and manipulation of materials at the nanoscale—particle sizes from 1 to 100 nanometers (nm), where 1 nm equals 1 *billionth* of a meter— to create new materials and structures that have novel properties due to their small size. Conventional multifunctional materials, such as fiber-reinforced or filled polymers, require a trade-off between various mechanical and physical properties, cost and processability. However, combinations of properties derived at the nanoscale provide opportunities to circumvent these traditional trade-offs. The transition from microparticles to nanoparticles leads to a number of changes in physical properties. One of the major factors is the increase in the ratio of surface area to volume. As the surface area of a particle increases, the portion of its constituent atoms at or near the surface increases exponentially, creating more sites for bonding, catalysis or reaction with surrounding materials, resulting in improved properties such as increased strength or chemical or heat resistance. Additionally, the fact that nanoparticles have dimensions below the critical wavelength of light can render them transparent, which has implications for part aesthetics. One of the most common groups of the new generation of commercial nanotechnologies is ceramics. Ceramics can be categorised into two subgroups: silicates (or silicon oxide), generally in the form of nanoscale flakes of clay, and metal oxide ceramics such as titanium, zinc, aluminum and iron oxides.

Nanoclays—The most common type is montmorillonite (MMT), a layered aluminosilicate in the smectite family of clays. MMT is a component of bentonite clay, formed when volcanic ash settled into ancient brine seas, and now mined in Wyoming, western Europe, China, Australia, Iran and elsewhere worldwide. Unlike clay minerals such as talc and mica that have been used as fillers for years, MMT can be delaminated and dispersed into individual layers only one nanometer thick by about 70 nm to 150 nm across. The result is a radical increase in the surface area-to-volume ratio: With a surface area of 750 m2/g, 20g of MMT platelets could cover a football field! Southern Clay Products Inc. (Austin, Texas), a manufacturer of functional clay-based additives, began MMT product development in the early 1990s. The company's Cloisite additives, commercially available since 1999, consist of organically modified, nanometer-sized MMT platelets. Unseparated MMT particles are often referred to as tactoids. When tactoids are partially separated in a polymer, they are referred to as being intercalated. When thoroughly separated, individual platelets are said to be exfoliated. To form a nanocomposite, the clay must first be modified with an appropriate organic surface treatment to enable complete dispersion into and miscibility with the polymer matrix. Then the tactoids must be exfoliated into platelets either through the shear stresses exerted during melt compounding or via chemical processes during polymerisation of the matrix. As Karl Kamena, Southern Clay's commercial manager of Cloisite nanoclays, describes it, "The incorporation of single-digit percentages of high surface-area nanoclays into host polymer systems improves the performance of the specific polymer matrix without significant deterioration of properties." The numerous benefits that Cloisite can confer on the host polymer include increased mechanical properties, such as modulus, melt strength and scratch resistance, as well as increased heat resistance, dimensional stability and flame retardation.

The extent of enhancement is related directly to the degree of exfoliation. Currently, about 40 per cent of the MMT platelets can be fully exfoliated, with the remainder forming tactoids of 2 to 10 platelets. Southern Clay continues to improve exfoliation methods, anticipating that, eventually, remaining tactoids will contain as few as one to three platelets. In 1998, Southern Clay began working with General Motors (GM, Detroit, Mich.) and resin manufacturer Basell (Hoofddorp, The Netherlands) to develop a nanoclay-reinforced thermoplastic olefin (TPO) that would outperform the talc-filled TPO then in use. An 8-micron Cloisite clay, each particle of which contains one million platelets, was selected as the filler. Work carried out jointly by the companies allowed the new material to be formulated to different shrinkage rates and to match colour with conventionally filled molded-in-colour parts. Early on, the 8-micron particles had a tendency during the extrusion stage (when the clay was mixed into the TPO) to aggregate into 200-micron clumps rather than exfoliate, causing surface defects and painting problems in the injection molded parts. The team then proved that this issue could be resolved by making adjustments to the extrusion temperature, the extruder screw rotation speed, the material feed rate and other variables during the nanocomposite manufacturing process. GM first introduced a 3 per cent nanoclay-filled TPO on the step assists for the 2002 GMC *Safari* and Chevrolet *Astro* vans, which had been using a 20 per cent talc-filled TPO. The same material moved into a higher volume application in the body side trim moldings for the 2004 and 2005 Chevrolet *Impala*, molded at Blackhawk Automotive Plastics (Troy, Mich.). Will Rodgers, GM staff research scientist, says GM saw weight savings of 3 to 21 per cent on these redesigned parts. With increased confidence in the material, GM now is using 7 lb of molded-in-colour nanocomposite parts to make this year's *Hummer H2 SUT* cargo bed trim (the center bridge, the sail panels and the box rail protectors), all molded by SportRack Automotive (Sterling Heights, Mich.).

These are the first parts that GM has designed as nanocomposites from conception, choosing the material primarily because of its low coefficient of thermal expansion (CTE) and improved appearance. For example, by varying the surface finish of the tool, three levels of gloss were molded into the box rail protectors to match adjoining parts, something that was never possible with talc-filled TPO. GM also reports better scratch/mar performance, increased cohesive strength between the paint and the part, greater ductility at low temperatures, lower flammability, a larger injection molding process window, sharper feature lines, improved knit line appearance and the elimination of tiger striping, (Tiger striping is caused when a filled material groups itself into lines of higher and lower fill weight or viscosity as it flows into a mold, giving the part a striped appearance.) Although the MMT-filled TPO is about an order of magnitude more expensive than the talc-filled material (about \$3/lb vs. \$0.35/lb), the improvements in part performance and appearance apparently justify the cost.

Metal oxides—Scratch resistance is a highly sought quality in coatings, such as automotive finishes. However, use of silica or clay fillers in quantities sufficient to improve scratch resistance tends to make the coating hazy and reduce its viscosity. Alumina particles are harder than silica but also lead to haziness. This is particularly troublesome in clear coats. Nanophase Technologies Corp. (Romeoville, Ill.) and BYK-Chemie (Wesel, Germany) have teamed to address this problem with two new additives marketed under the tradenames Nanobyk-3600 and Nanobyk-3601. Both additives contain Nanophase's 37-nm diameter Nanodur aluminum oxide particles in concentrations of 30 to 50 per cent by weight, and are intended for use with aqueous (Nanobyk-3600) and nonaqueous UV-curable coatings (Nanobyk 3601). Nanophase Technologies produces its aluminum oxide nanoparticles via physical vapor synthesis: Arc energy is applied to a metallic precursor to generate a vapor at a high temperature. A reactant gas is introduced to the vapor and then the mixture is cooled at a controlled rate so that it condenses and forms nanoparticles, which then are coated with a thin polymeric shell that ensures their compatibility with various polymers and fluids. When the additives are blended into resins and coatings at very low 1.5 to 6 per cent concentrations, scratch resistance reportedly goes up dramatically.

In Nanophase tests, particle loadings of 1 per cent resulted in five- to ten-fold increases in scratch resistance, depending on the resin formulation tested, with less than 0.5 percent haze. (At greater than 1 per cent haze, or about a 2 per cent loading of Nanodur alumina particles, a resin is no longer considered "clear.") Exploration of carbon nanotubes is relatively new. Single wall nanotubes (SWNTs) are long, thin cylinders made of carbon atoms arranged in a graphitic lattice structure that may or may not have closed ends. Multiwall nanotubes (MWNTs) are concentric cylinders of carbon in a similar graphite structure, held together by weak intermolecular-level forces. SWNTs and MWNTs have length-to-diameter ratios that can range from 100:1 to 10, 000:1. Their large surface-area-to-volume ratios mean that carbon nanotubes (CNTs) affect matrix properties far beyond what would be predicted from rule-of-mixtures calculations. Nanotubes are currently being commercialised in applications where electrical conductivity and increased mechanical performance are primary design drivers.

Electrical properties. SWNTs can possess electrical conductivity of 1 billion amps/ cm2, Young's modulus of more than 1 TeraPascal and a maximum strain of 10 per cent. MWNTs have lesser properties but are easier to produce, though they also are less well understood, due to their greater complexity and variety. Orders of magnitude less costly than SWNTs, MWNTs are more attractive for commercial applications. While they have been used for the last 10 years in thermoplastics as a direct replacement for carbon black in electrical applications, recent developments are enabling their use in thermosets, and significant efforts are underway to exploit their mechanical properties as well. Carbon black and carbon nanotubes are both pure graphitic carbon and so have the same electrical properties. Carbon black particles, however, are spherical, and the rod-like CNTs have a higher aspect ratio, which means far fewer CNT particles are needed to form a conductive "bridge" through a nonconductive polymer. Typical carbon black loading in polymers is 7 to 18 percent (depending on the polymer and the level of conductivity needed), which will make the polymer brittle. CNT's achieve the same conductivity in 1 to 3 percent loadings. Moreover, the CNT's conductivity-versus-loading curve is more gradual and more easily tailored to a specific level of conductivity than carbon black's curve.

In 1984, Hyperion Catalysis (Cambridge, Mass.) was the first company to file a patent on a nanotube manufacturing method. Hyperion's MWNTs have 8 to 12 walls and are about 10 nm wide by 10 micrometers long, an aspect ratio of 1, 000:1. They are made through a process which Hyperion's sales and marketing manager Pat Collins describes as "continuous, catalysed vapor phase decomposition and recombination, using light hydrocarbon feedstock." When an atomised metallic catalyst comes in contact with a hydrocarbon gas in a reactor under appropriate time, pressure, and temperature conditions, nanotubes will grow on the catalyst. Each nanotube has a tiny remnant of catalyst inside it. The MWNTs, as formed, tend to clump together in aggregates, attracted to each other by intermolecular Van der Waals forces. The high shear forces which are necessary to disperse the CNTs into the resin are obtained by using a high-viscosity resin and a twin-screw extruder typically used to compound thermoplastics. The disentangled and dispersed nanotube/resin mixture is formulated in concentrations of up to 20 per cent. A wide variety of nanotube-filled thermoplastics can be purchased, either ready to use with low CNT loadings or as a master batch with higher nanotube concentrations, which then can be let down (diluted) to the appropriate usage level.

The ready-to-use resins are offered in antistatic, electrostatic dissipative and conductive grades, with volume resistivities ranging from 1010 ohm/cm for antistatic grades to 1 ohm/cm for conductive grades (as compared to 1.68 x 10-6 ohm/cm for copper). Because the nanotubes are in such low concentrations and are so small, their addition has little effect on the base resin's properties, allowing the material to be processed on the injection molder's existing equipment under conditions similar to those used to process the base resin. Hyperion's continuous process runs 24 hours a day, seven days a week, producing tons of pure (no byproducts, such as amorphous carbon) nanotubes per week. Automotive fuel systems have to be designed for electrostatic discharge.

The turbulent flow of fast-moving fuel creates a static charge which must be dissipated. If charge is allowed to build up it can cause pinholes to develop in the fuel system and may damage electronic sensors. On rare occasions, sparks may even discharge during refueling. The fuel system, which includes the filler neck, inlet check valves, fuel module (pump/sender unit) inside the tank, the fuel lines and the fuel filter, must all, therefore, be conductive. (The Society of Automotive Engineers (SAE) will soon issue its newly revised *Recommended Practice, SAE J1645*, which will outline how all materials that touch fast-moving fuel must be grounded.) Fuel systems static dissipation is a rapidly growing segment of Hyperion's business, according to Collins. BASF (Ludwigshafen, Germany) has already commercialised a conductive polyoxymethylene (POM) with Hyperion nanotubes, specifically for automotive fuel systems. CNT-reinforced nylon 6, manufactured by Degussa (Parsippany, N.J.), is used for the tubing in the flexible portion of the fuel line in 70 percent of cars made in North America. O-rings in the system also must be conductive or they will build unwanted capacitance. Precix (New Bedford, Mass.) recently introduced its F98 series of O-rings, made by incorporating Hyperion's multiwall nanotubes into Precix's own proprietary flouroelastomer resin system. Compared to the company's carbon black-filled O-rings, the F98 compound is said to have three to five times greater resistance to fuel vapor permeation, with significantly reduced fuel swell (the expansion of part dimensions due to absorption of the fuel) as well as a durometer of 79 versus 88, allowing for much lower insertion pressure when assembling parts. Hyperion also supplies various conductive thermoplastics to manufacturers such as GE Plastics (Pittsfield, Mass) for use as injection molding material for electrostatically painted automotive body parts, such as mirror housings, door handles and fenders.

According to a *CT* web search, the Renault *Clio* and *Mégane* currently sport fenders made from a nanotube-filled polyamide/polyphenylene ether (PA/PPE) blend. Since CNT loadings are typically 2 to 3 percent rather than the 12 to 15 per cent necessary with carbon black, the plastics do not become brittle, thus negating the need for impact modifiers which would produce the undesirable side effect of reducing the plastic's glass transition temperature (Tg). The CNT-reinforced parts need no primer and can be painted at the same temperature and on the same electrostatic line used with metal parts, reducing cost and improving colour matching. Because the CNT concentrations are low, the surface finish also is smoother than plastics with other types of conductive fillers. Collins adds that Hyperion now has branched out beyond thermoplastics into thermosets, with nanotube-filled versions of Resolution Performance Products' (Ennis, Texas) Epon 1009 and Epon 828 resins. Epon 1009 is a high molecular weight, solid pellet form of bisphenol-A, which can be compounded into liquid epoxies or vinyl esters. Hyperion sells the Epon 828 nanocomposite in paste form, which is made by grinding the Epon 1009 nanocomposite pellets into a powder and blending the powder into the 828 with a multishaft double planetary mixer in a liquid dispersion process. Collins also notes that Hyperion will introduce easier-to-use liquid master batches of other nanotube-filled epoxies in the near future, primarily for electrical applications, such as conductive epoxies, potting compounds and connectors. Although the nanotubes may add \$5/lb to \$6/lb to the final polymer mix, Collins says there are many applications where the cost is justified either through cost savings in the final product manufacturing process or because the necessary properties cannot be obtained any other way.

Mechanical properties. Zyvex Corp. (Richardson, Texas) approaches the market from a different angle, seeking to use specially modified carbon nanotubes to increase the mechanical performance in sporting goods, aerospace and defense applications. Zyvex does not make its own nanotubes, but instead uses SWNTs and MWNTs from other suppliers, and certifies them for quality and batch-to-batch consistency using a series of standard analytical tests. Zyvex uses a patented process to functionalize the surface of the nanomaterials with a non-covalently bonded "Kentera" molecule which non-covalently bonds the nanomaterial surface via pi-pi bonding and also adheres well to the host matrix. Pi-pi bonding, which involves a slightly stronger molecular force than the intermolecular forces that cause the nanotubes to clump together, enables the Kentera molecule to slip between the nanomaterials and disperse them. Zyvex offers its Nanosolve functionalised nanotubes and nanofibers as dispersions in solvents or water additives, or in solvent-free concentrates, which can be used in thermoplastics and a subset of thermosets comprising polyurethanes and epoxies.

This includes concentrates in Resolution Performance Products' Epon 826, 828 and 862 at customised weight loadings specific to an application, whether it is thermal, electrical or mechanical. Marni Rutkofsky, Nanosolve chemist and technical sales representative, is overseeing third-party ASTM physical and mechanical testing on several materials. She says preliminary results at less than 5 per cent loading of MWNTs in thermoplastic polycarbonate show increases of 80, 50 and 150 per cent, respectively, in tensile strength, elongation and Young's modulus. Results at similar loading in polyurethane show a four-fold increase in thermal conductivity. Data on epoxy is expected later this year.

One beneficiary of Zyvex research is Easton Sports (Van Nuys, Calif.) a leading manufacturer of bicycling, hockey, baseball, softball and archery equipment. Three years ago, Easton embarked on an ambitious project to explore the advantages that nanotubes could bring to bicycle components bladder molded at its production facility in Dongguan City, Guangdong Province, China. Easton drum winds its own prepreg and planned to incorporate CNTs into the resin before introducing the resin to the fiber. The challenge was to find a nanotube chemistry that would remain well dispersed in the resin during the course of Easton's manufacturing process and yield improved properties at a reasonable cost. Zyvex and Easton tried several iterations of the resin filled with the CNT/Kentera chemistry before developing a formulation that met the challenge. "We were looking more for strength increase than weight decrease, " says John Harrington, VP of Easton's Bicycle Div. "Our products are already some of the lightest on the market, and we want to be conservative with weight reductions when we are using a new material." However, Easton has shaved 10g to 20g from most of the CNT-filled prepreg components, while achieving an average 15 per cent increase in product flexural strength. Road, mountain, and triathlon handlebars, forks, seatposts and OEM seatstays and chainstays are already on the market. The CNT Technology logo distinguishes components which have incorporated nanotubes, and most of the rest of Easton's bicycle components will follow suit. Retail prices on the logoed components have gone up an average of only 10 per cent—an increase that Harrington says consumers are willing to pay for the increase in performance. While nano-sized fillers aren't an overnight revolution, it is clear that there is significant activity on many fronts, with increasing use on all. Beyond the applications highlighted here, nanofillers have been incorporated into in golf clubs and golf balls, tennis rackets and a number of other products, with many more to come. The main barriers to nanomaterials use now seem to be development cost and the end-user's ability to obtain proper dispersion in and compatibility with resins. However, all of the nanosuppliers interviewed above are focused on making materials less expensive and more accessible by scaling up production, standardising products, issuing test data and offering them in user-friendly forms.

You must be wondering by now after all the 'over-the-head' theory about how different nanoparticles are from bulk and atomic material, what this really mean to us average everyday individuals. A lot!! Due to the two unique phenomenon which occurs in nanoparticles, their properties (electrical, optical, chemical, mechanical, magnetic, etc.) can be selectively controlled by engineering the size, morphology, and composition of the particles. These new substances will have enhanced or entirely different properties from their parent materials. Nanoparticles can make metals stronger and harder, give ceramics enhanced ductility and formability, enable normally insulating materials to conduct heat or electricity, and make protective coatings transparent. Hence, industries can re-engineer many existing products and design novel new products/processes that function at unprecedented levels. Use the menu bar on the top to read about more about some examples of the possible applications.

Solar cells are photovoltaic devices, which are devices that can convert light into electricity. They are much in demand as they offer an inexhaustible and environmentally benign energy source. Most solar cells are made of amorphous silicon. The problem with this is that the silicon must be of a very high purity and have a near perfect crystal structure. This makes it very expensive to produce. The efficiency of such a cell is also very small, typically converting only 13-18 per cent of sunlight to electricity. However, low efficiency wouldn't matter if huge arrays of cells could be produced cheaply. After all, nature's solar cells, chloroplasts in plants are less than 1 per cent efficient. Most solar powered devices rely on the same principle: a photon of sunlight boosts an electron in the material into a mobile state so that it can be used to generate electricity. The problem with this simple mechanism is that the electrons are negatively charged and will leave a positive charge. These opposite charges attract one another and therefore will tend to recombine, squandering the absorbed energy as heat or as re-emitted light. Silicon solar cells use an electric field to push the negatively charged electrons and positive charges apart. While chloroplasts adopt a more subtle approach of separate charges by making a distinction between the units that generate the electron and those that transport it away. Deciding to copy nature's trick, Michael Gratzel and Brian O'Regan at the Swiss Federal Institute of Technology began research and produced the Grätzel cell in 1991. The Grätzel cell uses intensely coloured organic dye molecules to capture light energy to inject an electron from the dye into a semiconductor such as titanium dioxide (TiO_2). This remarkably efficient charge separation reaction initiates current flow and the output of electrical energy by the cell.

The ideal material used in the cell must have a high surface area for light absorption and charge separation. Nanoparticles, having a comparable surface area to volume ratio, provides for just that. Titanium dioxide nanoparticles are used to make nanoporous thin film supported upon a glass substrate. The material obtained has optical transparency, excellent stability and good electrical conductivity. The benefit of these novel photoelectrical solar cells is that they can be fabricated from cheap, low purity materials by simple and low cost procedures. Contrary to expectation, some of the new devices also have strikingly high conversion efficiency. The size-tunable bandgaps of the semiconductor nanoparticles, due to size quantisation, also means more efficient solar cells can be produced for photovoltaics (electricity production) and water splitting (hydrogen production) processes.

Electrochromism is defined as a reversible colour change of a material caused by the application of an electrical current or potential. It has potential applications in making smart windows for the modulation of incoming light, as well as displays, a field in which competition with liquid crystal displays (LCD) is possible.

There used to be only two types of electrochromic devices. In the first type, the electrodes are made of conducting glass and covered with an organic or inorganic polymer. The two materials would usually display complementary electrochromism and hence produce the same colour change when one is oxidised while the other is reduced. This type of construction is used in electrochromic windows. It is bistable, which means that once the colour change has occurred, the state of the device remains even in absence of applied voltage. The limitation of this typ e of electrochromic devicesis the slowness of the colour change, due to thelow migration rate of the counterions in the bulk polymer. It is also difficult to obtain strong colour changes or bright colours. In the second type, two complementary electrochromic molecules are dissolved in a solvent. This type of system is very simple to build, reacts very fast and can produce dark or bright colours. The drawback of this type is that an electrical current is needed to maintain the coloured state because the two types of coloured molecules diffuse through the system and react with each other to restore the bleached states. It therefore cannot be used for large area devices or for battery-powered displays.

We therefore need a system which is both bistable and changes colour rapidly. This can be done by attaching a suitable molecule that is colourless in the oxidised state and coloured in the reduced state onto the surface of a colourless semiconductor on conducting glass. When a sufficiently negative potential is applied, electrons are injected from the conducting glass into the conduction band of the semiconductor and reduce the adsorbed molecules. The reverse process occurs when a positive potential is applied and the molecules get bleached. The advantage of this system is that combines immobility of the electrochromic material with the rapidity and colouration efficiency of molecular systems. However, a single molecular monolayer does not absorb a perceptible amount of light. This is where nanoparticles come in!

By using nanocrystalline semiconductor films, it was possible to amplify the light absorption of the molecular monolayer to obtain visible colour changes. The nanocrystalline layer are highly porous and usually the real surface of a 5 μm thick layer is about 500 times the projected area. Therefore, as light passes through the layer, it crosses several hundreds of monolayers of coloured molecules giving a strong absorption.

Some commercial products have already been produced! The most successful is the automatic dimming rear-view mirror sense ambient light and glare from behind, and reduce annoying glare from bright lights. While the company NTera has produced paper quality displays. These displays are bi-stable, not requiring much power keep to a pixel coloured, and reflective, not needing power for backlighting, and therefore can operate with very little power. Coupled with the fast switching, (*i.e.* changing of colour) due to amplification with nanoparticles, this technology could in most flat panel displays.

The invention of the transistor in 1948 by John Bardeen and William Shockley, triggered a new era in electronics. The trend since then has been to create smaller and smaller product using fewer chips of greater complexity and smaller 'feature' sizes. The development of integrated circuits and storage devices have continued to progress at an exponential rate. At present it takes two or three years for each successive halving of component size. However, the technologies currently in use have fundamental limits below, which the devices no longer function in a predictable manner. For instance, oxide layers used in Complimentary Metal Oxide Semiconductors (CMOS) devices are becoming so thin that leakage currents are conducted quantum mechanically by electron tunneling. Therefore, replacement technology to advance miniaturisation even further down, eventually to the dimensions of single atoms and molecules, is keenly sought. In 1985, Dmitri Averin and Konstantin Likharev, proposed the idea of a new three-terminal device called a single electron tunneling (SET) transistor. Two years later Theodore Fulton and Gerald Dolan at Bell Labs in the US fabricated such a device and demonstrated how it operates. Since then much research has gone into this area and better single electron devices has been fabricated. These devices usual have a molecule or atom as a functional unit and operate based on the quantum effect. Due to the small dimensions of the device (a couple of nanometers and hence is a nanoparticle), there are repulsive Coulombic interactions between the electrons and energy is required to add an extra electron. This energy cost can be tuned to zero by applying a voltage to the gate electrode. The electrostatic potential at this gate will be such that an extra electron can hop from the source onto the molecule. However, due to Coulomb repulsion, a second, extra, electron cannot hop on at the same time. The first electron must leave the molecule, moving into the drain before the next electron can enter. This one-by-one electron motion, is known as the single electron quantum tunneling. The creation an ultrafast, single electron transistor would lead to the development of "quantum" computers with supercomputer powers and the size of a thumbtack. However, there is still a long road ahead before atomic or molecular transistors can be assembled into viable, dense, fast logic-circuits. Currently these single electron devices are still no competition for silicon transistors. But they will serve both scientifically, for studying electron motion through nanoscale objects, and technologically, for developing chemical techniques with which to fabricate electronic devices on single molecules.

Nanocrystals and High Surface Area Materials

Semiconductor nanocrystals (also called nanocrystals or quantum dots) can be described as a state of matter intermediate between molecular and bulk crystalline. Their small size and high optical activity make them a very interesting material for study in disciplines ranging from opto-electronics (integrating optical and electronic signals on a single substrate, i.e. crystalline Si) to fluorescence microscopy (as a bright, photostable and minimally perturbative fluorescent marker). This work will concentrate on experiments designed to optically characterize silicon nanocrystals as fluorescent beacons in solution since in biology fluorescent labels are invariably used to image structures or track/probe particles/biomolecules in the liquid state. Images of single immobilised Si nanocrystals on quartz substrates will also be presented. Just as bulk semiconductors are characterised by the structure of their conduction and valence bands as being direct or indirect, so are semiconductor nanocrystals. Direct semiconductor nanocrystals (e.g. CdSe, CdS) are characterised by having the minimum transition energy to promote an electron from the valence band to the conduction band without a change in the electron momentum (the energy of this energy separation is known as the band-gap). For indirect semiconductors however excitation at the band gap energy must be accompanied by a change in the electron's momentum (supplied by a phonon). While both types of semiconductors are efficient absorbers of light, it has only been the direct semiconductors which have been emissive enough to be used as an optical source (e.g. in semiconductor GaAs quantum well lasers, Saleh and Teich, 1991) or as fluorescent labels (CdSe) (Chan and Nie, 1998, Bruchez et al, 1998). The discovery of porous silicon as an efficient luminescent material (Canham, 1990) has created great excitement as a prospective candidate for opto-electronic circuit integration. [Integrating direct semiconductor structures on silicon has been unsuccessful due to lattice mismatch]. Though the waited for advances have not materialised, it has been the starting point for experimental and theoretical investigations on nanocrystalline silicon.

Nanocrystalline materials are build from nanoparticles. These particles are also called quantum dots or artificial atoms. These nanoparticles are usually spherical colloids of a semi-conducting material with a diameter of nanometers. Because these colloids are so small new material properties exist which are not present in bulk materials.

An important property is that the energy difference between the energy bands, the valence and conducting bands in bulk materials, increases and separate energy levels can be formed. These changes in the energy levels and the small size of the particles result in a new optical and electrical properties which can be used in the devices which were mentioned before.

Due to their novel properties and varied potential applications, nanocrystalline materials, with typical grain sizes <100 nm, are attracting increasing attention from researchers all over the world. These materials are traditionally classified as one-dimensional (1-D), or layered structures, two-dimensional (2-D), or rod- or wire-like, and three-dimensional (3-D), or equiaxed, with the 1-D layered and 2-D wire-like structures being extensively investigated for coating and electronic applications and the 3-D equiaxed structures exploited for potential structural or bulk applications. Because of the small grain size of these materials, and consequently the large volume fraction of atoms in or near the grain boundaries, these *materials* exhibit properties that are often superior, and sometimes completely new, in comparison to those of conventional coarse-grained *materials*. However, it should be realised that a full understanding of the structure, thermal stability, and properties of these novel materials is not yet available. A number of issues and concerns still need to be solved.

Nanocrystalline materials can be synthesised by a number of techniques with the starting material existing either in the vapor, liquid, or solid state. Inert-gas condensation, in which the evaporated material is quenched onto a cold substrate, was the first technique used to synthesize nanocrystalline metals and alloys. Subsequently, plasma processing and other physical and chemical vapor deposition techniques have also been used. Electrodeposition and rapid solidification processing use the liquid state as the starting material. Mechanical alloying, friction stir welding, severe plastic deformation, spark erosion, sliding wear, and repeated cold rolling produce nanocrystalline materials in the solid state. While some of these techniques are commercially in use for large-scale production purposes (e.g., inert-gas condensation, electrode-position, and mechanical alloying), many of the other techniques are essentially used for laboratory investigations. The choice of technique for the synthesis of nanocrystalline materials is determined by the ease of the process, its economic viability, its scalability, and the purity of the end product desired. Most of the techniques referred to previously produce the nanocrystalline material in the form of powder. The application of nanocrystalline materials requires that these powders are consolidated to full density with minimal or no porosity and this is not a trivial matter.

Full densification with complete inter-particle bonding requires exposure of the powder to high temperatures and pressures for extended periods of time, which is likely to coarsen the micro structural features. Retention of the material in the nanocrystalline state, however, warrants that the powder is not taken to high temperatures for long periods of time; this may not result in complete densification. Thus, successful consolidation to full density requires innovative methods of consolidation. However, since nanocrystalline materials have a large fraction of the atoms in the grain boundaries, the (effective) diffusivity of nanocrystalline materials is much higher than that of a coarse-grained material of the same composition. This will facilitate achieving full consolidation of nanocrystalline materials at temperatures 300°C to 400°C lower than those required for the coarse-grained materials. Successful consolidation of nanocrystalline powders has been achieved by electro-discharge compaction, plasma-activated sintering, shock (explosive) consolidation, hot isostatic pressing (HIP), Ceracon processing, hydrostatic extrusion, strained powder rolling, and sinter forging. Consolidation is not required if the powder could be used as-is (e.g., in slurries, as catalysts, or for coating purposes).

Membranes for Biological Separations

Supported polymeric membranes can be used to remove low molecular weight organics from aqueous solution (Ho et al. 1996). They work by filling the pores of microfiltration or ultrafiltration membranes with functional, polymeric, or oligomeric liquids that have an affinity for the compound of interest. These ideas can be extended to microemulsions and used for separations of large biological molecules. For example, non-ionic microemulsions (i.e., oil-water systems) have been used for hemoglobin extraction. These results augur well for the separation of other biological materials (Qutubuddin et al. 1994). Molecular imprinting can also be used to produce high surface area adsorbents that are enantioselective for amino acids and other biological molecules (Vidyasankar et al. 1997).

Crystalline bacterial cell surface layers (S-layers) are composed of repeating protein units. These layers self-assemble and have a high binding capacity. They have been explored as patterning elements for molecular nanotechnology. For example, they have been used to pattern cadmium sulfide superlattices (Shenton et al. 1997).

Consolidated Materials: Opportunities and Challenges

Biomolecule/inorganic interactions can be used to produce ceramics with increased toughness. Fundamental studies of biomineralisation, in which an organic substance (usually protein or peptide or lipid) interacts with an inorganic phase (e.g., calcium carbonate or hydroxyapatite) have led to the bioinspired synthesis of composite materials. Novel combinations of DNA, metal ligands, DNA templating, and proteins are being investigated for molecular wires, inductors, and switches (photo courtesy of Shionoya and coworkers, Inst. for Molecular Science).

The structure and porosity of the inorganic phase can be controlled by templating with an organic surfactant, vesicular arrays, or liquid crystalline materials. Micelle-templated synthesis can produce ceramics with 20-100 Å pore dimensions (Ying 1998). These tailored pores can be used as catalysts and absorbents, and for gas/liquid separations and thermal and acoustic insulation. Their selectivity makes them very useful for biochemical and pharmaceutical separations. Bioceramics can also be made that are more compatible with teeth and bone. An interesting example of an organic/inorganic composite is the new packaging material that has been developed to replace the polystyrene "clam-shell" for fast food products. Composed of potato starch and calcium carbonate, this foam combines the advantages of good thermal insulation properties and light weight with biodegradability (Stucky 1978).

It has been well documented that a very large number of organisms have the ability to precipitate ferrimagnetic minerals such as Fe3O4 and Fe3S4. In addition, linear chains of membrane-bound crystals of magnetite, called magnetosomes, have been found in microorganisms and fish (Kirschvink et al. 1992). For example, the Fe3O4 domain size in the organism *A. magnetotactum* is about 500 Å, and a chain of 22 of those domains has a magnetic moment of 1.3 x 10-15. It is not immediately clear how these particles could be exploited for nonbiological applications.

This section of the review describes current limits on the biological aspects of nanotechnology. It outlines selected broad areas of research opportunity, and it sets forth challenges for the intermediate and long-term future.

Enabling Technologies

We owe much of the recent progress in the biological aspects of nanotechnology and nanoparticles to important enabling technologies. Many of these enabling technologies are described in other parts of this report. Those that have had a significant impact on biological measurements include techniques and instrumentation such as optical traps, laser tweezers, and "nano-pokers" to measure femtoNewton forces (Svoboda and Block 1994), and AFM and STM (Wildoer et al. 1998).

There is also a need to move detection away from the ensemble, to the single molecule scale. This includes driving instrumentation toward singlemolecule spectroscopy; single-spin detection (Rugar et al. 1994); chemical analysis of nanoliter volumes (Hietpas et al. 1996, 139-158); nuclear magnetic resonance microcoils; nanoscale electrode arrays and chemical sensing and detecting technology (McConnell 1996, 97-102); separations technology; chemical analysis of single cells (Hietpas et al. 1996); new biological transformations; and chemical probes of nanostructures. Enhanced computational infrastructure, both hardware and algorithm development, will also be required to investigate supramolecular assemblies and to understand interactions that occur over a wide variety of time and length scales. At present, we are fairly well equipped to handle quantum mechanical calculations. At the next level of computational complexity, the molecular dynamics force fields and atomic charges can presently handle up to about a million particles and still retain their accuracy. The next scale of complexity, mesoscale modeling, currently requires the use of pseudo atoms. Much more research is required to improve the accuracy of finite element calculations and the modeling of materials applications (Goddard 1998).

Surface Interactions and the Interface Between Biomolecules and Substrates

Much of the progress in the biological aspects of nanotechnology has come from research on surface interactions. It will be important to continue to develop a better understanding of the interface between biomolecules and surfaces. Of particular interest are questions about whether surfaces cause biomolecules to denature, the optimum length of linker groups, and ways to communicate from the biomolecule, through the linker, to the substrate. Another challenge is to produce nanometer ultrathin films that have stable order (Jaworek et al. 1998).

Robustness of Biomolecules and Their Interactions in Aqueous Solution

Although much progress has been made on ways to confer additional robustness on biomolecules, it will be necessary to continue to improve the stability and reliability of biomolecular assemblies. Such research may be along the lines of understanding the thermal stability of biomolecules, perhaps by examining extremophiles and molecules produced by directed evolution. It will also be necessary to produce molecules that can work in the absence of water, or to devise ways in which aqueous solutions can be used reliably.

Assemblers or Templating

The hierarchical self-assembly of biomolecules is often touted as one of their key attributes for nanotechnology applications. However, there remains much to be learned about how to make periodic arrays of biomolecular assemblies, how to use biological templating in an efficient manner, how to mimic biological self-assembly with nonbiological molecules, and how to exploit differences between biological and nonbiological self-assembly.

Combinatorial Approaches to Nanotechnology

Combinatorial chemistry has had an enormous impact on drug discovery and development in the past several years. Ability to develop molecularly sensitive screening techniques is one of the limiting factors of combinatorial chemistry. It would be intriguing to discover screening methods that are particularly sensitive for picking up new phenomena exhibited by biological nanoparticles.

Biomimetics and Polymer-Biopolymer Chimeras

The largest impact of biological sciences on nanotechnology may prove to be through the lessons one can learn from biology, rather than from directly employing the biological molecules themselves. To this end, it is important to support research on various aspects of biomimetics, including activities in which synthetic and biologically-based molecules are combined in chimeric fashion.

Molecular Manufacturing

This more radical form of nanotechnology originated in the mind of an M.I.T. undergraduate in the mid-1970s. Eric Drexler, who went on to obtain a Ph.D. from M.I.T., realised that the biological "machinery" already responsible for the full diversity of life on Earth could be adapted to build nonliving products upon command. Molecule-sized machines, modeled after those found in nature, could be used to manufacture just about anything man wished.

Drexler, who began to develop these theories even before he'd heard of Feynman's lecture, first published his ideas in a 1981 journal article. Five years later, he brought the notion of molecular manufacturing to the general public with his book *Engines of Creation*. An astonishingly original work of futurism, *Engines of Creation* pointed out how molecular manufacturing would revolutionize other areas of science and technology-leading to breakthroughs in medicine, artificial intelligence, and the conquest of space. Drexler refutes every technical objection he can anticipate, explaining how such phenomena as quantum uncertainty and thermal vibrations don't make molecular manufacturing impossible. It was also in *Engines* that Drexler introduced the term "nanotechnology"-a reference to the nanometer, one-billionth of a meter-to describe this approach to molecular manufacturing, although the term is now also used for the more mundane applications (cosmetics, tennis balls, etc.) described above.

To shore up his technical arguments for the feasibility of his vision, Drexler further expanded on his ideas in the world's first nanotechnology textbook. *Nanosystems* (1992), a dense volume that grew out of a class he taught at Stanford, is crammed with equations and diagrams and designs for molecular machines, and it has gone far to put the theory of molecular manufacturing on sound technical footing-although scientific debate about the achievability and the best routes to developing nanotechnology has continued.

In the past decade, theorists have begun to flesh out the details of how nanotechnology might be used in manufacturing and medicine, although it is unclear how soon any of this will be possible. Some analysts have estimated that major breakthroughs in molecular manufacturing are at least three decades away; others have suggested that major progress might occur in the next five years.

Molecular Machinery

A comparison of biochemical to macroscopic components will show the possibilities of the former by analogy to the latter. With structural members, moving parts, bearings, and motive power, versatile mechanical systems can be built. Molecular assemblages of atoms can act as solid objects, occupying space and holding a definite shape. Thus, they can act as structural members and moving parts. Sigma bonds that have low steric hindrancc can serve as rotary bearings able to support ~ 10^{-9} N. A line of sigma bonds can serve as a hinge. Conformation-changing proteins (such as myosin) can serve as sources of motive power for linear motion; the reversible motor of the bacterial flagellum can serve as a source of motive power for rotary motion. The existence of this range of components in nature indicates that power-driven mechanical systems can be constructed on a molecular scale. By analogy with macroscopic devices, feasible molecular machines presumably include manipulators able to wield a variety of tools. Thermal vibrations in typical structures are a modest fraction of interatomic distances; thus, such tools can be positioned with atomic precision. As present microtechnology can lay down conductors on a molecular scale (10 nm) and molecular devices can respond to electric potentials (through conformation changes, etc.), such devices can be controlled by human operators or macroscopic machines. Further, by analogy with biological sensors, molecular scale instruments can evidently produce macroscopic signals, indicating the feasibility of feedback control in molecular manipulations.

Together, these arguments indicate the feasibility of devices able to move molecular objects, position them with atomic precision, apply forces to them to effect a change, and inspect them to verify that the change has indeed been accomplished. It would be foolish to minimize the time and effort that will be required to develop the needed components and assemble them into such complex and versatile systems. Still, given the components, the path seems clear.

Ordinary chemical synthesis relies on thermal agitation to bring reactant molecules in solution together in the correct orientation and with sufficient energy to cause the desired reaction. Enzyme-like molecular machines can hold reactants in the best relative positions as bonds are strained or polarised. Like some enzymes, they can do work on reactant molecules to drive reactions not otherwise thermodynamically favoured.

These are clearly techniques of great power, yet the synthetic capabilities of systems based on polypeptide chains might seem limited by amino acid properties. However, enzymes show that other molecular structures bound to the polypeptide (such as metal ions and complex ring structures) (11) can extend protein capabilities. The range of such tools is large and greater than found in nature. Thus, the synthetic capabilities of enzymes set only a lower bound on the capabilities of engineered protein systems. Indeed, as tool-wielding protein systems can control the chemical environment of a reaction site completely, they should be able, at a minimum, to duplicate the full range of moderate-temperature synthetic steps achieved by organic chemists. Further, where chemists must resort to complex strategies to make or break specific bonds in large molecules, molecular machines can select individual bonds on the basis of position alone. Conventional organic chemistry can synthesize not only one-, two-, and three-dimensional covalent structures but also exotic strained and fused rings. With the addition of controlled site-specific synthetic reactions, a broad range of large complex structures can doubtless be built.

Still, the synthetic abilities of protein machines will be limited by their need for a moderate temperature aqueous environment (although applied forces can sometimes replace or exceed thermal agitation as a source of activation energy and reaction sites and reactive groups can be protected from the surrounding water, as in some enzymatic active sites). These limits may be sidestepped by using the broad synthetic capabilities outlined above to build a second generation of molecular machinery whose components would not be coiled hydrated polypeptide chains but compact structures having three-dimensional covalent bonding. There is no reason why such machines cannot be designed to operate at reduced pressure or extreme temperatures; synthesis can then involve highly reactive or even free radical intermediates, as well as the use of mechanical arms wielding molecular tools to strain and polarize existing bonds while new molecular groups are positioned and forced into place. This may be done at high or low temperature as desired. The class of structures that can be synthesised by such methods is clearly very large, and one may speculate that it includes most structures that might be of technological interest.

Firmness of the argument

The development path described above should lead to advanced molecular machinery capable of general synthesis operations. As the results of this path can be shown to have consequences for the present, it is of interest to discuss the degree of confidence that should be placed in its feasibility.

It might be argued that complex protein or nonprotein machines are impossible or useless, on the grounds that, if they were possible and useful, organisms would be using them. A similar argument would, however, conclude that bone is a better structural material than graphite composite, that neurons can transmit signals faster than wires, and that technology based on the wheel is impossible or useless. Nature has been constrained less by what is physically possible than by what could be evolved in small steps. Thus, the absence of a proposed kind of molecular machinery in organisms in no way suggests its infeasibility.

To deny the feasibility of advanced molecular machinery, one must apparently maintain either (i) that design of proteins will remain infeasible indefinitely, or (ii) that complex machines cannot be made of proteins, or (iii) that protein machines cannot build second-generation machines.

In light of the expected improvements in computation, the simplified task of design engineers (compared with scientists), the possibilities offered by sheer trial-and-error modification of natural proteins, and the progress already made in protein design, the first seems difficult to maintain. Further, even if protein design were to prove intractible (because of difficulties in predicting conformations), this would in no way preclude developing an alternative polymer system with predictable coiling and using it as a basis for further development.

In light of the presence of the needed components for mechanical devices in the cell, the second seems difficult to maintain. Indeed, the cytoskeleton provides a fair counterexample.

In light of the results of synthetic organic chemistry and the ability of molecular machines to make reactions site specific, it seems difficult to maintain that nonprotein machine components cannot be built and assembled.

Each of the development steps outlined above seems closely analogous to past steps taken by nature or by technology. Each of these steps can be accomplished in many ways. To argue their infeasibility would seem to require some general principle precluding success, and it is difficult to see what such a principle might be like. Thus, the claim that advanced molecular technology can be developed seems well founded.

Although the existence of molecular machinery in cells indicates the feasibility of some sort of artificial molecular machinery, errors in assembly might limit the synthesis of structures of great complexity. In the cell, molecular machinery uses

DNA to direct the assembly of DNA and other molecules. In some eukaryotic cells, DNA directs DNA synthesis with an error rate of ~ 10^{-11} per nucleotide added (12). As engineers commonly design systems to function reliably with many more failed components than 1 in 10^{11}, such an error rate seems no barrier to the construction of quite complex devices.

The possibility of low error rates is not surprising. For synthesis systems permitting error detection and correction (such as DNA synthesis), the net error rate in assembly can be reduced to roughly the product of the raw error rate in assembly and the rate at which errors are falsely identified as correct. As no uncertainty principle prohibits accurate discrimination between objects of different kinds (such as correctly and incorrectly assembled molecular structures), no limits to the detection and correction of errors are apparent.

Applications to computation

Molecular technology has obvious application to the storage and processing of information. A crude approach would involve literal "molecular machinery" patterned on the Babbage machine. In a more subtle approach, bits could be represented by protons, bound electrons, reactive groups, or conformation changes and transferred by movement of protons or of well-localised electrons (13), excitons, or phonons. The range of plausible device speeds is suggested by the 10^{-6} -sec turnover time for a fast enzyme, by the 10^{-13} -sec scale of collisional interactions (11), and by the 10^{-16} sec taken for an electron to cross an interatomic distance at a typical Fermi velocity.

It seems highly likely that a cubic cell 0.1 micrometers on a side (containing some 10^8 optimally arranged atoms) can hold a bit or perform a logic operation and, at the same time, transmit bits through itself to provide communication from cell to cell in a lattice. If so, then computers can be built with at least 10^{15} active elements per cubic centimeter. In a well-designed computer (with elements closer to their true technological limit and not laid out in regular cubical cells), this volume estimate should prove quite conservative. Elements so small will be sensitive to radiation damage; to be reliable, systems will require a large measure of redundancy.

Concern might be raised about the cost of such intricately patterned matter, either because of labor or energy requirements. It seems clear, however, that molecular-scale production systems can be completely automated (what use is there for hands?). Thus, labor costs of production (including production of additional production equipment) can approach zero. The energy needed to produce molecularly engineered material will generally be greater than the energy needed to produce ordinary materials of similar bulk composition, but analogy suggests that the energy cost need not be vastly greater than for the production of biological materials. In many cases (e.g., advanced computers or any of a number of applications not discussed here), the unique value of the products would make such energy costs unimportant, even if energy costs remained high.

Some biological applications

Molecular devices can interact directly with the ultimate molecular components of the cell and thus serve as probes of unique value in studying processes within the cell. Further, molecular devices can characterize a frozen cell in essentially arbitrary detail by removal and characterisation of successive layers of material (atomically thin layers, if desired). Although the amount of data involved is large (a typical cell contains billions of protein molecules), the physical bulk of a device able to store and manipulate this amount of data will be quite small.

The change of temperature and water distribution during freezing modifies cell structures in several ways, primarily by physical displacement of structures by ice crystals and denaturation of proteins by concentration of solutes in the residual liquid (14). With frozen tissue, knowledge of normal structures (membrane geometries, natural protein structures) and analysis of frozen structures (position of ice crystals, position of denatured proteins) should permit quite accurate reconstruction of the nature of the tissue before freezing.

Such procedures would have special utility in analysing the structure of tissue in the brain. Unlike, say, muscle or liver tissue, the function of brain tissue depends on the detailed three-dimensional structure of intertwined cells and their interfaces. The freezing process is far too slow to stop such dynamic processes as action potentials and synaptic transmission; short-term memory, however, is suspected to involve chemical modification of the neurons, and long-term memory is believed to involve the growth and modification of neuronal structures, particularly synapses. At the modest freezing rates possible in substantial pieces of tissue, ice crystals may be expected to nucleate and grow in the intercellular fluid, displacing the cell membranes as they do so. Electron micrographs, however, show that synapses (like many intercellular junctions) involve complementary structures on both sides of the intercellular gap, which should provide information enough to reconstruct the pre-freezing configurations of the cells almost regardless of ice crystal locations.

The ability to reconstruct the prefreezing structure of tissue, when combined with the general synthetic capabilities outlined above, will make feasible the physical restoration of tissue damaged by ordinary freezing through characterisation, reconstruction, and restoration of successive segments of frozen material. Although restored to a frozen condition, such tissue would lack the characteristic damage caused by the freezing process. As many tissues can survive the gross insult of ordinary freezing, it seems likely that most could survive freezing followed by repair. The remaining mode of damage would seem to be denaturation of proteins sensitive to cold alone during the thawing process. Should cell components of some species prove sensitive to short periods of cold, they could presumably be modified to resemble those of hardier species without changing either cell function or DNA.

Implications for the present

The existence of a path to an advanced molecular technology has implications for the present. As with all technologies, long-range promise should tend to increase interest in undertaking the early steps, even beyond the interest springing from more immediate benefits. The longer the expected wait, however, the less the interest. On the other hand, molecular engineering of materials and devices can extend the capabilities of technology many fold in many areas. The implications of the feasibility of molecular technology are important to present day speculations concerning the probable behaviour (and likelihood of existence) of extraterrestrial technological civilisations. Similarly, those concerned with the long-range future of humanity must concern themselves with the opportunities and dangers arising from this technology. Finally, the eventual development of the ability to repair freezing damage (and to circumvent cold damage during thawing) has consequences for the preservation of biological materials today, provided a sufficiently long-range perspective is taken.

Development of the ability to design protein molecules will, by analogy between features of natural macromolecules and components of existing machines, make possible the construction of molecular machines. These machines can build second-generation machines able to perform extremely general synthesis of three-dimensional molecular structures, thus permitting construction of devices and materials to complex atomic specifications. This capability has implications for technology in general and in particular for computation and characterisation, manipulation, and repair of biological materials.

Molecular Mechanical Components

In order to lay a foundation for molecular manufacturing, it is necessary to create and to analyze possible designs for nanoscale mechanical components that could, in principle, be manufactured. Because these components could not yet be built in 1998, such designs could not be subjected to rigorous experimental testing and validation. Designers were forced instead to rely upon ab initio structural analysis and computer studies including molecular dynamics simulations. Noted Drexler: "Our ability to model molecular machines (systems and devices) of specific kinds, designed in part for ease of modeling, has far outrun our ability to make them. Design calculations and computational experiments enable the theoretical studies of these devices, independent of the technologies needed to implement them."

In nanoscale design, building materials do not change continuously as they are cut and shaped, but rather must be treated as being formed from discrete atoms. A nanoscale component is a supermolecule, not a finely divided solid. Any stray atoms or molecules within such a structure may act as dirt that can clog and disable the device, and the scaling of vibrations, electrical forces, thermal expansion, magnetic interaction and surface tension with size lead to dramatically different phenomena as system size shrinks from the macroscale to the nanoscale.

Molecular bearings are perhaps the most convenient class of components to design because their structure and operation is fairly straightforward. One of the simplest examples is Drexler's overlap-repulsion bearing design, shown with end views and exploded views using both ball-and-stick and space-filling representations. This bearing has exactly 206 atoms including carbon, silicon, oxygen and hydrogen, and is comprised of a small shaft that rotates within a ring sleeve measuring 2.2 nm in diameter. The atoms of the shaft are arranged in a 6-fold symmetry, while the ring has 14-fold symmetry, a combination that provides low energy barriers to shaft rotation. An exploded view of a 2808-atom strained-shell sleeve bearing designed by Drexler and Merkle using molecular mechanics force fields to ensure that bond lengths, bond angles, van der Waals distances, and strain energies are reasonable. This 4.8-nm diameter bearing features an interlocking-groove interface which derives from a modified diamond {100} surface. Ridges on the shaft interlock with ridges on the sleeve, making a very stiff structure. Attempts to bob the shaft up or down, or rock it from side to side, or displace it in any direction (except axial rotation, wherein displacement is-extremely smooth) encounter a very strong resistance. Whether these bearings would have to be assembled in unitary fashion, or instead could be assembled by inserting one part into the other without damaging either part, had not been extensively studied or modeled by 1998.

Molecular gears are another convenient component system for molecular manufacturing design-ahead. For example, Drexler and Merkle designed a 3557-atom planetary gear, shown in side, end, and exploded views. The entire assembly has twelve moving parts and is 4.3 nm in diameter and 4.4 nm in length, with a molecular weight of 51, 009.844 daltons and a molecular volume of 33.458 nm^3. An animation of the computer simulation shows the central shaft rotating rapidly and the peripheral output shaft rotating slowly. The small planetary gears rotate around the central shaft, and they are surrounded by a ring gear that holds the planets in place and ensures that all of the components move in the proper fashion. The ring gear is a strained silicon shell with sulfur atom termination; the sun gear is a structure related to an oxygen-terminated diamond {100} surface; the planet gears resemble multiple hexasterane structures with oxygen rather than CH_2 bridges between the parallel rings; and the planet carrier is adapted from a Lomer dislocation array created by R. Merkle and L. Balasubramaniam, and linked to the planet gears using C-C bonded bearings. View (c) retains the elastic deformations that are hidden in (a) — the gears are bowed. In the macroscale world, planetary gears are used in automobiles and other machines where it is necessary to transform the speeds of rotating shafts.

W. Goddard and colleagues at CalTech performed a rotational impulse dynamics study of this "first-generation" planetary gear. At the normal operational rotation rates for which this component was designed (e.g., <1 GHz for <10 m/sec interfacial velocities), the gear worked as intended and did not overheat. Started from room temperature, the gear took a few cycles to engage, then rotated thermally stably at ~400 K. However, when the gear was driven to ~100 GHz, significant instabilities appeared although the device still did not self-destruct. One run at ~80 GHz showed excess kinetic energy causing gear temperature to oscillate up to 450 K above baseline. One animation of the simulation shows that the ring gear wiggles violently because it is rather thin. In an actual nanorobot incorporating numerous mechanical components, the ring gear would be part of a larger wall that would hold it solidly in place and would eliminate these convulsive motions which, in any case, are seen in the simulation only at unrealistically high operating frequencies.

Drexler and Merkle later proposed a "second-generation" planetary gear design with 4235 atoms, a molecular weight of 72, 491.947 daltons and a molecular volume of 47.586 nm^3. This new version was indeed more stable but still had too much slip at the highest frequencies. Commenting on the ongoing design effort, Goddard suggested that an optimal configuration could have the functionality of a planetary gear but might have an appearance completely different from the macroscopic system, and offered an example: "Because a gear tooth in the xy plane cannot be atomically smooth in the z-direction, we may develop a Vee design so that the Vee shape of the gear tooth in the z-direction nestles within a Vee notch in the race to retain stability in the z-direction as the teeth contact in the xy plane. This design would make no sense for a macroscopic gear system since the gear could never be placed inside the race. However, for a molecular system one could imagine that the gear is constructed and that the race is constructed all except for a last joining unit. The parts could be assembled and then the final connections on the face made to complete the design" analogous to the ZARBI system of Rebek.

Another class of nanodevice that has been designed is a gas-powered molecular motor or pump. The pump and chamber wall segment contain 6165 atoms with a molecular weight of 88, 190.813 daltons and a molecular volume of 63.984 nm^3. The device can serve either as a pump for neon gas atoms or (if run backwards) as a motor that can convert neon gas pressure into rotary power. The helical rotor has a grooved cylindrical bearing surface at each end, supporting a screw-threaded cylindrical segment in the middle. In operation, rotation of the shaft moves a helical groove past longitudinal grooves inside the pump housing. There is room enough for small gas molecules only where facing grooves cross, and these crossing points move from one side to the other as the shaft turns, moving the neon atoms along. Goddard reported that preliminary molecular dynamics simulations of the device showed that it could indeed function as a pump, although "structural deformations of the rotor can cause instabilities at low and high rotational frequencies. The forced translations show that at very low perpendicular forces due to pump action, the total energy rises significantly and again the structure deforms." Merkle acknowledged that the pump moves neon atoms at an energy cost of 185 Kcal/mole-Angstrom (12, 900 zJ/atom-nm), which is not very energy-efficient. Further refinement of this crude design is clearly warranted.

Conveyor systems would also be useful. Employing a primitive molecular CAD software package called Crystal Sketchpad, G. Leach created a design for a 5-nanopart, ~2500-atom conveyor belt system comprised of two rollers, two axles, and a belt consisting of a strained thin-walled diamond sheet. The design has not been subjected to further computational analysis, either to minimize rotational energy barriers or to optimize rotational dynamics or operational stability.

Drexler and Merkle have also produced a preliminary design for a 2596-atom fine-motion controller. A general-purpose molecular assembler arm must be able to move its "hand" by many atomic diameters, position it with fractional-atomic-diameter accuracy, and then execute finely-controlled motions, perhaps to transfer one or a few atoms in a guided chemical reaction. Human arms use large muscles and joints for large motions and more finely-controlled finger motions for precision; the device presented here can execute precise finger-like motions over several atomic diameters with associated 90-degree rotations. The core of the device consists of a shaft linking two hexagonal endplates, sandwiching a stack of eight rings, making a modified Stewart platform. In a complete system, each ring would be rotated by a lever driven by a cam mechanism. Each ring supports a strut linked to a central platform (here shown raised, displaced, and twisted). Rotating a ring moves a strut; moving a strut moves the platform; positioning all eight rings (over)determines a platform position in x, y, z, roll, pitch, and yaw. (If the struts were rigid, six would do the job; here, two struts have been added to increase stiffness.) Notes Drexler: "The chief design problem is to enable an adequate range of motion without mechanical interference or unacceptable bond strains, and within the size constraints set by available modeling tools and patience."

Almost all current design research in molecular nanotechnology is restricted to computer simulation, which allows the design and testing of large structures or complete nanomachines, and the compilation of growing libraries of molecular designs. The work is relatively inexpensive and does not require the support of a large team. Of course, calculations of many-body systems are notoriously difficult, with many computer packages making a number of simplifying assumptions — e.g., nuclei as point masses, electrons treated as a continuous charge distribution, and 3-D potential energy functions derived semi-empirically from experimental data and treated as a classical field despite their true quantum mechanical character (for ease of computation). Notwithstanding these shortcuts, Tuzun and colleagues claim that classical simulations may overestimate the rate of energy transfer between vibrational modes even at low energies, in which case "current designs for various nanocomponents [would] actually perform better and be more stable than recent molecular dynamics simulations suggest."

Goddard notes that future nanosystem simulations may require 1-100 million atoms to be considered explicitly, demanding major improvements in molecular dynamics methodologies. By 1998, new algorithms for parallel processing of massive molecular dynamics simulations were being developed, producing methods and optimised parallelised computer programmes efficient for high capacity molecular dynamics simulations of 10, 000-1, 000, 000 atoms for finite molecular structures.

Ultimately, Computer-Aided (Molecular) Design (CAD) systems will be needed to efficiently design and analyze molecular components and their higher-order assemblies. CAD systems are commonplace in macroscale engineering and architecture, as are Computer-Aided Manufacturing (CAM) systems in macroscale manufacturing. In the molecular realm, Computer Aided Synthesis Design (CASD) has been avidly pursued by computational chemists since the 1980s, and by 1998 many university and commercial molecular modeling software packages were already well-known, including Alchemy (Tripos Inc.; http://www.tripos.com/), Cerius2 (Molecular Simulations Inc.; http://www.msi.com/), Chem3D (CambridgeSoft; http://www.camsoft.com/), Conformer (Princeton Simulations; http://www.conformer.com/), Gaussian94 (Gaussian Inc.; http://www.gaussian.com/), Hyperchem (Hypercube Inc.; http://www.hyper.com/), Molecular Operating Environment (Chemical Computing Group Inc.; http:// www.chemcomp.com/), MOPAC97 (Fujitsu Ltd.; http://www.winmopac.com/), RealMol/ CAVE/NAMD (Fraunhofer Institute for Computer Graphics; www.igd.fhg.de/www/ igda4/research/chemie), Sculpt (Interactive Simulations Inc.; www.intsim.com/products/ index.htm), and Spartan (Wavefunction Inc.; http://www.wavefun.com/).

In 1998, the capacity of such systems to design and manipulate large nanoscale mechanical components was extremely limited. A few very primitive molecular nanotechnology design packages had been attempted, including Crystal Sketchpad, DiamondCAD (www.zyvex.com/diamond.html), Molecular Assembly Sequence Software (www.carol.com/mass.shtml), Molecular Modelling Toolkit (starship.python.net/crew/ hinsen/mmtk.html), and NanoCAD (world.std.com/~wware/ncad.html). The possible design of molecules for specific purposes using genetic software techniques had also been investigated.

Molecular Assemblers

Drexler has proposed the molecular assembler, a device resembling an industrial robot which would be capable of holding and positioning reactive moieties in order to control the precise location at which chemical reactions take place. This general approach would allow the construction of large atomically precise objects by a sequence of precisely controlled chemical reactions. Much like the ribosome in biology, an assembler would build various classes of useful molecular structures following a sequence of instructions. During this process, the assembler would provide three-dimensional positional and full orientational control over the molecular component (analogous to the individual amino acid in the ribosome model) that is being added to a growing complex molecular structure (analogous to the growing polypeptide in the ribosomal model). In one approach, a molecular assembler may be capable of forming any one of several different kinds of chemical bonds (e.g., by changing tool tips), not just a single kind such as the peptide bond that the ribosome makes. In bonding atoms or molecules to one another, the assembler would provide any needed energy (especially if the reaction happens not to be energetically favoured) through physical force, thus performing mechanosynthesis (as opposed to the traditional means of chemical synthesis in solution). In another approach, a molecular assembler might be capable only of noncovalent assembly operations, wherein nanoparts are fabricated by other means and then presented to the assembler, which assembles the nanoparts into working nanomachines.

The first simple molecular assembler will almost certainly be a macroscale device, perhaps a modified SPM system as was being pursued by Zyvex in 1998. Multiple SPM heads could be equipped with a small number of nanoscale tool tips. In one scenario, nanoparts fabricated using bulk chemistry techniques would be inspected and selected by the SPM, then assembled one by one into working nanomachines (e.g., the desired useful nanoscale products). Such assembly operations will be very slow, because the placement of each new component may require simultaneous rotations and translations of large macroscale SPM components. Assembly time scales roughly linearly with assembler size because smaller assembler components moving at a given velocity need to travel less distance to accomplish a given physical operation, hence consuming less time and energy per physical operation. An important early developmental goal thus will be to design and fabricate nanoscale molecular assemblers.

At its most basic level, the simplest possible nanoscale molecular assembler may be comprised of one or more nanoscale manipulators. For example, the diamondoid telescoping manipulator arm described in has about 4 million atoms excluding the base and control and power structures; doubling the size to account for support structures gives a molecular weight per arm of ~100 megadaltons and a total molecular volume of ~140, 000 nm^3. Designed for high strength and stiffness, this robot arm should be able to hold a molecular fragment stiffly enough to make it react with a chosen end of a carbon-carbon double bond with an error rate of only 10^{-15}. The robot arm could guide chemical reactions with high reliability at room temperature in vacuo, with little need for sensing the positions of the molecules with which it is working. Alternatively, with appropriate tool tips, such an arm could grasp and manipulate individual prefabricated nanoparts in solution phase. Assembly of nanoparts without fabrication may require only a very small tool set.

The manipulator arm must be driven by a detailed sequence of control signals, just as the ribosome needs mRNA to guide its actions. However, such detailed control signals can be provided by external acoustic, electrical, or chemical signals that are received by the robot arm via an onboard sensor or power transducer, using a simple "broadcast architecture", a technique which can also be used to import power. Such transducers may be extremely small, on the order of $(\sim 10\ nm)^3$ each. Interestingly, the biological cell may be regarded as an example of a broadcast architecture: the nucleus, located external to the cytoplasm, broadcasts mRNA chemical signals to millions of spatially diverse cytoplasmic ribosomes, thereby remotely controlling the construction of cellular proteins.

Thus a two-armed mechanical molecular assembler that receives power and instructions from some external agency may have a total molecular volume of ~300, 000 nm^3 (a cube ~67 nm on an edge), containing ~16 million atoms with a molecular weight of ~200 megadaltons — very roughly the mass and scale of a medium-size virus particle, such as an adenovirus. For comparison, the average enzyme (a biochemical crimping tool) weighs about ~0.1 megadalton, while a typical ribosome (a primitive "protein assembler") weighs ~4.2 megadaltons. A simple mechanical assembler in the 10-100 megadalton range cannot be ruled out. However, such a device might have a very small set of tool tips and an extremely limited manufacturing repertoire.

Multiple nanoscale assemblers each capable of independent simultaneous actuation may require control signals more conveniently provided by an onboard nanocomputer. This programmable nanocomputer must be able to accept stored instructions which are sequentially executed to direct the manipulator arm to place the correct moiety or nanopart in the desired position and orientation, thus giving precise control over the timing and locations of chemical reactions or assembly operations. The mechanical nanocomputer analysed by Drexler requires ~16 nm^3 per logic gate and ~40 nm^3 per data register. If such components could be used to construct the equivalent of the most primitive 4-bit Intel 4004 microprocessor, then the nanocomputer could process ~10^5 bits/sec (~25, 000 ops/sec) at a ~1 KHz clock speed within a mechanism volume of ~36, 000 nm^3, again neglecting power supply, I/O linkages, and the like. An additional ~160, 000 nm^3 of rod logic registers adds ~1 kilobyte of onboard RAM memory or ~40 kilobytes of internal tape memory (conservatively assuming a tape storage density comparable to linear DNA). (A memory tape containing all bits needed for complete self-description may be considerably longer, even allowing for substantial data compression.) Doubling the total volume to account for support structure and other overhead gives a minimum mechanical nanocomputer molecular volume of ~400, 000 nm^3, roughly 70 million atoms with a molecular weight of ~800 megadaltons. Thus the smallest nanocomputer-driven nanoscale molecular assembler with two manipulator arms may have a total molecular volume of ~700, 000 nm^3 (a cube ~88 nm on an edge), containing ~86 million atoms with a molecular weight of ~1 gigadalton.

By 1998, only a small amount of research targeted at actual assembler design had begun. Following Drexler's original discussions, during 1991-1998 R. Merkle authored or coauthored a continuing series of papers discussing various operational aspects and specific components of assembler design, including mechanosynthetic positional control, general design considerations for assemblers, the broadcast architecture, convergent assembly, binding sites, positioning devices, mechanosynthetic path sets, designs for a neon pump and a fine motion controller, and possible assembler casings. J.S. Hall has considered high-level designs for a nanoscale parts-fabrication and parts-assembly nanorobot. Zyvex, founded in 1996 by James von Ehr, has set itself the task of building the first programmable nanoassembler in a 5-10 year time frame (e.g., by ~2006). W. Goddard and colleagues have proposed a series of molecular dynamics simulations of simple assemblers, although by 1998 these studies evidently had not yet begun. According to Goddard's original proposal:

"Ultimately we need a programmable synthetic system to make a real device. Even though we may not have tools for all the chemical steps and may not have designs for all the pumps, engines, and transmissions needed, we propose to study the dynamics of simplified prototype assemblers. In these studies we anticipate having (1) a reservoir or supplies area for providing the various building units (atoms and fragments) required, (2) a work area in which we construct the nanomachine device (initially we will consider assembling the structure on top of a diamond surface), and (3) a molecular scale nanohand which will extract the atoms from (1) and carry them to (2). We will then use extensions of our massive molecular dynamics programme to operate the system: moving the tip from reservoir to work area, moving it to contact the appropriate surface site, moving it to regenerate the active tip, and then moving it back to add new atoms and molecules. This will include proper temperature effects, molecular vibrations, energy release upon the various chemical steps, etc. The ground rules here are that a realistic force field be used and that all pieces be treated at the atomic level (but some might be semi-rigid). This will use the force field developed for nanosynthesis. The purpose of these simulations is to examine issues of vibration caused by chemical forces as the tool picks up and delivers atoms to the growing surface. Also we want to consider the effect of energy release in the chemical steps on the thermal fluctuations in these systems (which may cause displacements and vibrations)."

Building mega-atom or giga-atom nanoproducts one at a time would be incredibly expensive and time consuming. For example, imagine that we wish to construct a simple medical nanorobot such as the 1-micron spherical respirocyte described in which consists of ~18 billion atoms (dry structure). A factory employing a coordinated team of 100 macroscale SPM assemblers, each able to place ~1 atom/sec-SPM on a convergently-assembled workpiece achieves a pitiful manufacturing throughput rate of ~2 respirocytes per decade. Nanoscale assemblers with appendages ~10^6 times smaller than macroscale SPM assemblers might plausibly achieve net assembly rates of ~10^6 atoms/sec-nanoassembler, but even a production line employing ~300 such nanoassemblers (which the abovementioned 100-SPM factory team could build in ~1 year, assuming ~10^7 atoms/nanoassembler) can only manufacture ~1 respirocyte nanorobot per minute. At that rate, it would take ~2 million years to build the first ~1 cm^3 therapeutic dosage containing ~10^{12} respirocyte nanorobots.

The necessary solution to this mass-production bottleneck is to employ any of several massively parallel approaches to manufacturing, including such techniques as self-assembly, convergent assembly, or, most usefully, self-replication. The basic advantage of self-replication is readily illustrated. Consider a single ~10^8 atom "seed" nanoassembler (having an onboard nanocomputer) that has been painstakingly built in ~10^6 sec using the abovementioned 100-SPM macroscale assembler team, working at ~1 atom/sec-SPM. The seed nanoassembler is first programmed to build a copy of itself, which it accomplishes (working at ~10^6 atoms/sec-nanoassembler) in ~100 sec. These two nanoassemblers then each build a copy of themselves in another ~100 sec; now there are four nanoassemblers. After ~48 generations, requiring a total of ~80 minutes to complete, there are ~3 x 10^{14} nanoassemblers. These nanoassemblers are then reprogrammed for the manufacture of 18-billion-atom respirocytes and are fed the appropriate, presumably different, feedstock. This vastly expanded nanoassembler manufacturing system can now produce ~10^{12} respirocytes (~1 cm^3 therapeutic dose) per minute.

The design of machines able to make copies of themselves was first described by von Neumann. Many variations on this theme have been reviewed by Freitas and Gilbreath and Sipper, and design for self-replication in the context of nanoscale assemblers has been considered by Drexler, Merkle, and Hall. As Drexler notes: "It may seem somehow paradoxical that a machine can contain all the instructions needed to make a copy of itself, including those selfsame complex instructions, but this is easily resolved. In the simplest approach, the machine reads the instructions twice: first as commands to be obeyed, and then as data to be copied. Adding more data does not increase the complexity of the data-copying process, hence the set of instructions can be made as complex as is necessary to specify the rest of the system. By the same token, the instructions transmitted in a replication cycle can specify the construction of an indefinitely large number of other artifacts."

The estimated information content of self-replicating systems — the length of the instruction tape — can be surprisingly small. Von Neumann's original analysis concluded that perhaps twelve different kinds of units of unknown complexity could be required as building materials, and Haldane inferred that as many as ~10^5 individual parts might be needed to make a replicator. This inference was refuted just three years later with the arrival of the first in a stream of very simple but ingenious designs for mechanical self-replicating machines that were assembled and operated in the late 1950s. In the first example, a pair of joined ratchet-like blocks, when placed in a "sea" of left and right blocks and then physically agitated, replicated itself from this well-ordered input substrate, making more block-pairs until all the single blocks were used up, a process similar to chemical autocatalysis. More complex congeries of blocks, including clever four-block, eight-block, and 12-block replicators that could replicate themselves in a sea of blocks were presented by Penrose. Jacobson demonstrated a 3-unit replicator consisting of toy train engines circulating on HO model railroad tracks, and Morowitz designed a simple two-unit device with one unit comprised of about a dozen components including switches, batteries and electromagnets, that could assemble copies of itself from parts floating on the water surface of a bathtub. In 1998, Lohn and colleagues gave two designs for simple self-replicating systems that could be constructed out of wood, batteries and electromagnets, with explicit analogies to nanoscale fabrication. Several important conclusions may be drawn from these examples:

First, replication is fundamentally so simple a task that machines capable of displaying this behaviour predate most of the modern electronic computer era.

Second, assembly is an inherently simpler operation than fabrication. A complex part may embody hundreds or thousands of prior fabrication and assembly operations, yet may be installed within an assemblage in a single step. Hence nanopart assembly may be an easier candidate for early implementation in first-generation self-replicating molecular manufacturing systems than is molecular fabrication.

Third, and most important, the simplest nanoreplicator may require the ability to assemble, but not to (atomically) fabricate, in order to replicate itself. A nanomachine capable of assembly alone can replicate itself only from a very limited set of well-ordered input materials of relatively high complexity, but such a machine can be extremely simple both in structure and function. Certainly a nanomachine capable of both assembly and fabrication can replicate itself from a more diverse and disordered set of more elementary input materials but only at the expense of far greater internal structural and functional complexity.

At least two distinct models of replication have been identified in replicating systems design. The first model may be called the "unit replication" or organismic model, in which the replicator is an independent unit which employs the surrounding substrate to directly produce an identical copy of itself; both the original and the copy remain fertile and may replicate again, thus exponentiating their numbers. The second model may be called the "unit growth" or factory model, in which a population of specialist devices, each one individually incapable of self-replication, can collectively fabricate and assemble all necessary components comprising all specialist devices within the system, hence the factory is capable of expanding its size (or of manufacturing duplicate factory systems) indefinitely in an appropriate environment. The factory model is sometimes called "bootstrapping," which is any production of more productive capacity, from less. T. McKendree [personal communication, 1999] likens the situation to an ant colony: "There necessarily is the line of reproducing queens, making a sufficient number of ants feasible; [however,] most of the queen's products are 'workers' that perform useful functions, but themselves are incapable of reproduction."

It cannot be emphasised too strongly that mechanical medical nanodevices will not self-replicate inside the human body, nor will they have any need for self-replication themselves. Machines that perform medical tasks are fundamentally different from machines that manufacture other machines. As R. Merkle explains: "While self-replicating systems are the key to low cost [manufacture], there is no need (and little desire) to have such systems function in the outside world. Instead, in an artificial and controlled environment they can manufacture simpler and more rugged systems that can then be transferred to their final destination. Medical devices designed to operate in the human body don't have to self-replicate: we can manufacture them in a controlled environment and then inject them into the patient as needed. The resulting medical device will be simpler, smaller, more efficient and more precisely designed for the task at hand than a device designed to perform the same function and self-replicate. This conclusion should hold generally: optimize [product] device design for the desired function, manufacture the [product] device in an environment optimised for manufacturing, then transport the [product] device from the manufacturing environment to the environment for which it was designed. A single device able to do everything would be harder to design and less efficient."

In an effort to stimulate scientific and engineering interest in constructing the first nanoassembler, in November 1995 the Foresight Institute created the $250, 000 Feynman Grand Prize, with funding contributed by Zyvex founder James von Ehr and St. Louis venture capitalist Marc Arnold. The prize will be awarded to the individual or group that first achieves both of two significant nanotechnology breakthroughs — first, the design and construction of a functional nanometer-scale robotic arm, and second, the design and construction of a functional 8-bit adder computing nanodevice. According to Grand Prize rules, the robotic nanomanipulator must fit entirely inside a 100-nm cube, carry out actions directed by input signals of specified types, be able to move to a directed sequence of positions anywhere within a 50-nm cube, complete all directed actions with a positioning accuracy of 0.1 nanometer or better, and perform at least 1, 000 accurate, nanometer-scale positioning motions per second for at least 60 consecutive seconds. The adder must fit entirely within a 50-nm cube, and must be capable of adding accurately any pair of 8-bit binary numbers, discarding overflow, accepting input signals of specified types, and producing its output as a pattern of raised nanometer-scale bumps on anatomically precise and level surface. (J.S. Hall notes that a conventional 8-bit adder may be constructed using a total of 94 AND, OR, and NOT gates; using XOR gates, the total may be reduced to 37 gates.) Both devices may accept inputs from acoustic, electrical, optical, diffusive chemical, or mechanical means, although any mechanical driving mechanism used for input must be limited to a single linkage that either slides or rotates on a single axis. To demonstrate the capacity for mass production, at least 32 copies of each device must be provided for analysis and destructive testing by judges.

Once nanoassemblers are available, the design of far more complex nanomachines each containing tens or hundreds of billions of precisely arranged atoms will require new molecular CAD tools and techniques, including automated hierarchical design decomposition of objects (that must be built up from an array of nanoparts) and a shape description language. Nanosystem design compilers, conceptually similar to silicon compilers that generate a complex pattern of transistors and conductors from an abstract specification of the properties of a digital circuit, will also be required for efficient nanorobot design. Given a fully specified design that meets all constraints necessary to physically permit assembly, all operations of the assembly process must be specified using an assembly process compiler. Explains Drexler: "The design of structures and assembly procedures by hierarchical decomposition directly generates a tree of assembly steps. If this tree has been chosen to generate parts of the appropriate sizes and numbers, then it can be mapped onto a manufacturing layout. The vast number of manipulator motions to be specified at the finest, earliest, and most-dispersed levels of the assembly process can be planned in parallel by identical software systems running almost independently on separate processors. The result of this parallel assembly process compilation is a set of instructions that, when executed by the transportation system and manipulator controllers of a manufacturing mechanism, will result in the assembly of an object corresponding to the initial design." Scale-up to larger systems, including the control, coordination, and programming of large aggregates of cooperating nanorobots.

Since molecular manufacturing systems can be used to make more molecular manufacturing systems, it is believed that the capital costs of production can be quite low. Drexler claims that an analysis of inputs, outputs, and productivity suggests the total cost of production can be in the range familiar in agriculture and in the production of industrial chemicals — on the order of tens of cents per pound. Merkle has made similar assertions, for example: "Eventually (after amortisation of possibly quite high development costs), the price of assemblers (and of the objects they build) should be no higher than the price of other complex structures made by self-replicating systems. Potatoes — which have a staggering design complexity involving tens of thousands of different genes and different proteins directed by many megabits of genetic information — cost well under a dollar per pound." Such analogies from conventional agriculture or industrial chemistry are not strictly applicable to the medical products and pharmaceutical areas, where structural precision, product repeatability, and intensive quality control are of supreme importance, and where most products are subject to rigorous government regulation — unlike potatoes, which vary widely in size, shape, and molecular constitution, and generally do not require stringent product liability insurance, multiphase clinical trials, or FDA approval.

Still, it is likely that molecular manufacturing will eventually allow reasonably inexpensive production of large batches of designed medical nanorobots. As a very crude analogy, even in 1995 a typical mail-order bioscience vendor could manufacture 15 micromoles (~9 x 10^{18} product items) of customised oligonucleotides for $18/base up to at least 110 bases with a convenient 48-hour turnaround on orders; other online vendors offer similar or slightly higher prices up to 200-mer sequences with additional charges for necessary purifications. If a billion-base structure could be induced to self-assemble from 10 million different sets of 100-mer custom-built oligonucleotides (total ~20 billion atoms), forming a desired ~20-billion-atom nanomachine, then, optimistically assuming a ~100% yield, the cost would be ~$2 x 10^9 per nanomachine or ~$2000 per ~1 cm^3 of product volume which would include 10^{12} finished nanomachines. In 1998, many specialty biotechnology-related drug treatments were comparably expensive. For example, treatments for multiple sclerosis using two closely related ~22.5 kilodalton interferons were administered in dosages of ~1 milligram/week at a treatment cost of ~$10, 000/yr, which is equivalent to a treatment cost of ~$2 x 10^9 (2 "nanodollars") per giga-atom. The famous HeLa (Henrietta Lacks) cell line is an example of a once tiny population of unique cancerous cells that were purposely replicated in vitro (~24-hour replication time) over many decades and are now widely available worldwide at very low cost (e.g., $216 per 1-$cm^3$ ampule of the ATCC CCL-2 HeLa cell line). Swallowable therapeutic pills containing bacteria (i.e., natural biological nanomachines) are already widely available over the counter for gastrointestinal refloration, as for example Salivarex which "contains a minimum of ~2.9 billion beneficial bacteria per capsule", and

Alkadophilus which "contains 1.5 billion organisms per capsule", both at a 1998 price of ~\$0.2 x 10^9 per bacterium.

Another useful analogic approach to cost estimation is suggested by the plunging price of computer power during the 20th century. The cost, in constant 1998 U.S. dollars, of a computer processing capability equivalent to ~1 TeraFLOP (10^{12} floating-point operations per second, ~10^{14} bits/sec assuming 64-bit words). The plotted data is from a compilation of 67 historical computers by Moravec through 1989, with an additional 27 data points for other computers spanning 1973-1999 added by the author. The chart shows that the cost of 1 TFLOP has fallen from \$2 x 10^{20} in 1908 (the mechanical Hollerith Tabulator) to just \$6 x 10^6 in 1998 (the SGI/Cray T3E-1200E), a 90-year price decline of almost 14 orders of magnitude — on average, a halving every ~2 years, a century-long example of Moore's Law. This trend may give us some confidence that even if the first individual 1 micron^3 nanorobot costs ~\$100, 000 to assemble, eventually the unit price may fall at least 14 orders of magnitude to below ~\$$10^9$ per nanorobot or <\$1000 per ~1 cm^3 dosage (10^{12} nanorobots). Recycling or remanufacturing each nanomachine ~1000 times before final discard would then imply a net ~\$1/cm^3 treatment cost. Of course, in 1998 all such estimates were mere guesswork, but the ultimate expectation of relatively low-cost nanomedical treatments was certainly rational, if not provable.

Controversy and Policy

Since 2000, awareness of nanotechnology among environmental activists, regulators, and lawmakers has been on the rise. Environmental organisations have expressed fears about the potential ecological and health consequences of mainstream nanotechnology, and have called for increased research into safety of nanoparticles.

The Drexler version of advanced nanotechnology has also been the subject of public fear, largely centered on the notion that nanotechnology could spiral out of control and convert all life on Earth into "gray goo." Drexler, who originally introduced this apocalyptic prospect in *Engines of Creation*, has since repeatedly distanced himself from it-but gray goo retains its grip on the public imagination.

There are other serious reasons to be worried about the development of nanotechnology, including the risk of severe economic disruption; the possibly dehumanising effects of using nanotechnology on ourselves; and the potential criminal, military, or terrorist use of advanced nanotechnology. A few organisations are paying full-time attention to these concerns, including the Foresight Institute (established in 1986) and the Center for Responsible Nanotechnology (established in 2002).

Public policy discussions have barely begun to reflect those long-term concerns.

Although some agencies in the U.S. government have been involved in nanotechnology since the 1980s, federal funding of nanotechnology research did not begin in earnest until the late 1990s. In 2000, the National Nanotechnology Initiative was established to coordinate the government's work in nanotechnology; soon, federal spending on nanotechnology is scheduled to cross the $1 billion-per-year mark. Along with the increased funding has come a government commitment to investigate the "social, economic, health, and environmental implications" of nanotechnology. As public interest continues to grow, and as scientific progress make advanced nanotechnology seem ever more attainable, policymakers are likely to increasingly turn their attentions to the promise and peril of nanotechnology.

Molecular Manufacturing (MM)

Definition Manufacturing today works by cutting or deforming large chunks of matter, then fastening together the remaining pieces into products. Molecular manufacturing plans to be more efficient and make better products by assembling products directly from the smallest pieces: atoms and molecules. The basic idea is to develop a small set of chemical reactions that can be applied repeatedly to build large molecules, then control the sequence and/or position of the reactions by computer in order to build engineered molecular systems.

Molecular manufacturing is different from biology in that biological systems are not engineered. The functional properties of a cell, or even a protein, are complex and hard to predict. However, the process of building protein molecules from small molecular fragments is quite programmable, and scientists are developing the ability to design and synthesize proteins with desired properties. This will allow protein chemistry to be used in an engineering, rather than a biological, context. This would be one approach to molecular manufacturing. Other approaches using different kinds of chemistry may also work, producing better materials.

Eric Drexler has proposed two kinds of chemistry that might work. The first, in 1981, was protein chemistry as described above. The second, in 1992, was three-dimensional carbon lattice, similar to diamond. This would be synthesised by direct robotic operation in vacuum. Although the required chemical reactions have not yet been demonstrated, some of them have been simulated in detail. Despite this, some scientists assert that this kind of chemistry can't possibly work as planned.

Discussion of molecular manufacturing has been distorted by several factors. From the beginning, it has been associated with "gray goo" (runaway biosphere-eating self-replicators), leading to excessive fear and attempts to counteract that fear. It has also been associated with with extreme science-fictional projections-though it can be hard to tell fantasy from sober calculation, because the calculations predict some pretty amazing capabilities. In reaction to these factors, some mainstream scientists have attempted to shut down discussion entirely by declaring that molecular manufacturing is impossible or that its major proponents are not credible. Discussion has been further distorted by a variety of widespread conceptual confusions.

The position that molecular manufacturing is impossible is not supportable. Living organisms are not an example of molecular manufacturing, because they are not based on engineering but rather on interlocking complex systems. However, the biochemistry of life could be adapted almost unchanged to a molecular manufacturing system.

A few prominent scientists have nonetheless claimed that molecular manufacturing is impossible, and others have echoed them. However, study of their objections shows that the arguments are weak, based on intuition rather than calculation. Some of the arguments elevate engineering difficulties to the status of fundamental limitations. Others are built on basic misunderstandings of the proposals. See *"Molecular Manufacturing Myths"* for examples of arguments that have been advanced against the theory, and why they are incorrect.

Although there are some practical questions remaining to be answered, there is no scientific study demonstrating a limitation or problem with molecular manufacturing theory. Indeed, the chemical mechanisms of life demonstrate that machines of a sort can be constructed out of molecules and can build more machines. Thus the scientists who dismiss the entire field out of hand have not bothered to study it in any depth. An intermediate position is that some form of molecular manufacturing will be possible, but will be limited. This is usually based on the assumption that engineering cannot improve on biology. This may be true for complex problems, but not for the relatively simple tasks that are tackled by engineering.

Perhaps because of the opposition of these scientists, not many people are working on molecular manufacturing in the lab; most of the work has been theoretical study, which is easier to do without funding. For now, it remains a controversial theory. But after more than a decade, no one has managed to find a real problem with the theory, and evidence is accumulating in favour of it. Some predicted phenomena, such as exceptionally low friction in certain cases, have now been observed. A single atom has been mechanically removed from a crystal and put back in the same place. Although these were not intended as demonstrations of molecular manufacturing theory, they indicate that at least some of the predictions are valid.

Separating science from science fiction, and alarmism from serious policy concerns, will not be an easy task. Most nanotechnology researchers have not paid enough attention to molecular manufacturing to be able to give informed commentary. The section on *"Molecular Manufacturing Myths"* lists some of the common misconceptions about molecular manufacturing.

According to its proponents, molecular manufacturing could be revolutionary. Because chemistry is very precise and repeatable, the manufacturing operations should be reliable enough to allow complete automation. Because the chemical steps are simple, a single fabricator could build a wide variety of products by changing the control programme. Furthermore, machines built this way could have very precise features. This is a good thing, because building with molecules would be quite slow, and small machines can be built more quickly. In fact, if the calculations are right, a complete nanoscale manufacturing system could build a complete copy of itself in a few hours-and this remains true if many small manufacturing systems are combined into one large one, so that even human-scale factories and products can be built quite rapidly. Finally, it appears that the materials produced by the manufacturing system may be extremely high-performance, allowing products to be far more compact and powerful. The section on *"Risks and Benefits of Molecular Manufacturing"* discusses some implications of the technology.

Exponential general-purpose molecular manufacturing — that's a mouthful, but what does it mean? Let's take the phrase apart to see why it is so important.

Manufacturing: The ability to make products, in this case ranging from clothing, to electronics, to medical devices, to books, to building materials, and much more.

Molecular manufacturing: The automated building of products from the bottom up, molecule by molecule, with atomic precision. This will make products that are extremely lightweight, flexible, durable, and potentially very 'smart'.

General-purpose molecular manufacturing: A manufacturing technology that will find many applications across many segments of society. Its extreme flexibility, precision, high capacity, and low cost will cause rapid adoption almost everywhere, and therefore will have disruptive effects in many industries.

EXPONENTIAL general-purpose molecular manufacturing: The word exponential refers to the rapid pace — probably unprecedented — at which this technology may be deployed. A compact automated molecular manufacturing system will be able to make more manufacturing systems. We're talking about factories that can build duplicate factories — and probably do it in less than a single day. The math is simple: if one factory makes two, and two factories make four, then within ten days you could have one thousand factories, in ten more days a million factories, and ten days after that a billion factories. Within the span of just a few weeks, in theory, every household in the world could have one of their own, to make most of the products they need, at just the cost of raw materials.

Exponential general-purpose molecular manufacturing means a manufacturing system capable of making a wide range of technologically advanced products, far superior to what we have today, much cheaper, much faster, and able to multiply its own source of production exponentially.

From *Responsible Nanotechnology*. See *Unanswered Questions, Part 1* for details and comments.

Molecular Manufacturing Myths

Perhaps the biggest myth is that is impossible. Before we get into the alleged reasons, it helps to remember how mundane it is: it's just very small machines, making chemistry happen according to plan.

The explanations given here are accurate, but simplified. If you want more detail, please contact the author at cphoenix [at] CRNano.org. He will be happy to defend and explain these statements in as much detail as necessary.

Part 1: "The chemistry won't work."

Myth: You can't build machines with chemistry because you can't pick up individual atoms.

Fact: This is a *straw man*. No one has proposed picking up individual atoms. What can be done-it happens all the time in nature-is to transfer groups of atoms from one molecule to another.

Myth: Atoms won't stay where they're put, because of *quantum uncertainty*, *Heisenberg uncertainty*, thermal noise, etc.

Fact: That depends on how the atoms are connected to other atoms. For example, diamonds don't evaporate. Not at all-not even one atom per year. This objection is like arguing that airplanes are impossible because they can't be built out of mud.

Myth: You can't make chemical reactions happen where you want them because it's too hard to control the surrounding atoms. (Sometimes stated as: It would take too many "fingers.")

Fact: Again, this depends on how the atoms are connected. Underwater chemistry is complex, but biology manages to do it. Chemistry in other environments may be a lot simpler. There's quite a bit of evidence that points to several different ways to build programmable shapes. If someone says it's impossible, ask for their evidence, and ask them about DNA folding, organic synthesis of graphite and fullerenes, and removal and replacement of single silicon atoms.

Myth: Chemistry is too complex, and it'll be difficult or impossible to build machines with it.

Fact: The point of molecular manufacturing is to use just a few reactions over and over again to build aggregated molecules. Building a brick wall is easy because you only need one kind of brick. Similarly, building a shape out of diamond is expected to be easy if you can add the atoms a few at a time, using the same reaction in multiple programmed locations. This plan shifts the complexity from chemistry to positional control-which can be controlled by software, and software is easy to write.

Part 2: "The machines won't work."

Myth: There's no way to control such small machines.

Fact: There are lots of ways, including simply running wires to them. Whoever says this is obviously unfamiliar with the scientific literature.

Related Myth: There's no way to power such small machines.

Fact: Again, there are lots of ways, with chemicals, electricity, light, and even sound pressure. If someone says this, ask them about the "nano-elevator" which might be controlled by chemicals, electricity, or light, and which may allow "tight control over the timing of a reaction."

Myth: Brownian motion (or thermal noise) will prevent the machines from working.

Fact: Thermal noise will make things wiggle a bit. But this has been taken into account in detailed engineering calculations. Large machines also have to deal with vibration. And cellular machinery, which is built a lot more loosely, manages to work just fine.

Related Myth: Nanomachines can't work because they don't use Brownian motion like biology does.

Fact: Several nanomachine designs do use Brownian motion. Others simply work around it. Biology doesn't always use it either. There's nothing magical about Brownian motion, and no rule that it has to be used, any more than larger machines must all use electricity.

Myth: Friction will prevent the machines from working.

Fact: With precise molecular construction techniques, surfaces can be made hard and smooth enough and positioned so that the friction drops to near-zero. This is called "superlubricity" and has recently been demonstrated.

Myth: Things are weird ("quantum") at that small scale, and familiar engineering just won't work.

Fact: Cells are chock-full of molecular machines like turbines, trusses, ratchets, and so on. It's possible to build things that depend on quantum effects, but we already know of lots of molecular machines that don't.

Myth: Entropy will keep molecular machines from working.

Fact: Entropy can be held at bay by putting energy into a system. Your computer does trillions of calculations without a single error. Your body is constantly rebuilding itself. Entropy doesn't magically get worse at the nanoscale-in fact, the more precisely a machine is built, the more efficiently it can run. Entropy has been accounted for in molecular machine designs, and it turns out not to be a big deal. (On average, entropy would require less than one error per billion tons of diamond - one defect per 1.6x10^25 tons.)

Unraveling the Big Debate over Small Machines:

Nanobot naysayers argue that molecular manufacturing is impossible, but the evidence goes against them

Bailey thoroughly and systematically rebuts the arguments that have been used to retard the growth of our understanding of *molecular manufacturing*. This article sets the record straight, once and for all, and is well worth reading.

—Rocky Rawstern, Editor Nanotechnology Now. August 2004

In the spirit of "myth busting, " and presented in an article for BetterHumans, writer Patrick Bailey sets forth on a journey of explanation, into the nanorealm. In *Unraveling the Big Debate over Small Machines,* Bailey carefully lays out the history of several complex and key issues; issues that have been used to slow the pace of our understanding of molecular manufacturing.

In just over 2, 500 words, Bailey tells an intriguing tale about truths, concepts, and competing points of view, set in the last gasps of the old Millenium, fast forward to the present.

In easy to understand language, from a trained observer's point-of-view, read here a story that is sure to steer the debate.

In *Italics*, here are a few of his well-considered observations. "Arguments" here indicates a combination of myth and straw man.

Behind the public face of nanotechnology—the press releases, science fiction novels and environmental doomsday reports—a debate has been evolving and swirling for the past decade around a fundamental question: Can tiny machines build things useful to humans by moving molecules or even individual atoms?

(Dr K. Eric) *Drexler's proposal for molecular manufacturing is to use a process called mechanosynthesis or "machine-phase" chemistry. "By guiding and positioning molecules, molecular manufacturing will control which molecules react, and where, building up complex structures with atomically precise control, " he says.*

What quickly became apparent is that the debate is larger than any individual scientist and the consequences more significant than simply settling who's right and who's wrong. This is a debate over conceptions, misconceptions, theories and misrepresentations of physics and chemistry, not men.

This isn't a debate of concern only to those in the ivory tower of academia. A major goal of nanotechnologists is to create nanoscale robots—nanobots—that can perform various functions at the nanoscale. These functions include molecular manufacturing: Using nanobot designs, such as fabricators and assemblers, to build products with atomically precise control. Considering the possible applications of nanobots and molecular manufacturing in medicine, computing, industrial production and more, the debate's importance becomes apparent.

One of the arguments is *"what we know about chemistry doesn't support Drexler's nanobot and molecular manufacturing proposals. Two key arguments here are known as the "sticky fingers" and "fat fingers" problems. Both state that the fingers used by a nanobot are not able to move individual atoms with the precision needed to make molecular manufacturing feasible."*

Sticky fingers refers to the inherent chemical bonding properties of atoms, which, says (Dr. Richard E.) Smalley, will prevent accurate atomic positioning because the atoms will stick to the end of a nanobot finger. Similarly, fat fingers refer to nanobot fingers being too large to move individual atoms.

Another case for the argument: *"When you have a tiny robot, the fingers on the robot have to be comparable to the size of the object that's being picked up or smaller, I mean you can't pick up a pea using a boxing glove very well and when you get down to the scale of atoms there's nothing smaller than an atom, so that it's not quite clear how that works. And atoms have a tendency to be very reactive so they stick to the fingers and you can't pull them out in the first place...it absolutely wouldn't work."* (Professor George Whitesides)

And Bailey's counter: *"Statements about sticky and fat nanofingers, however, overlook a very important point: Grabbing individual atoms is not part of Drexler's proposal for nanobot and molecular manufacturing. "No one has seriously proposed picking up individual atoms with pincers or fingers," says Chris Phoenix, director of research at the Center for Responsible Nanotechnology (CRN) in Brooklyn, New York.*

"What can be done—it happens all the time in nature—is to transfer atoms from one molecule to another."

If fat and sticky fingers were the only opposition arguments, chances are the debate would have ended long ago. Clearly, molecular manufacturing doesn't require fingers at all. There are other arguments, however, that, if valid, cause considerable problems for molecular manufacturing.

Another of the arguments is based on *"a chemistry-based objection claiming that enzyme catalysis only works in water."*

And the counter: *"Discussing enzyme catalysis is important, not because enzyme catalysis is necessary or proposed by Drexler, but because Smalley has used it to reject mechanosynthesis—Drexler's proposal for molecular manufacturing." ... Nearly two decades ago, (Alexander) Klibanov and his then graduate student Alex Zaks demonstrated that enzyme catalysis in a nonaqueous (non-water-based) environment is possible. Their findings were first published in 1984.*

At the end of the day, it's not just understanding the technical issues, but the enormous *human implications* that matters most.

Bailey has taken a nanoScalpel to the straw men, and they are no more. Now maybe we can forge ahead and fully explore the many paths that lead to molecular manufacturing.

From *Nanobot naysayers argue that molecular manufacturing is impossible, but the evidence goes against them.* 8/16/2004.

Bailey holds a masters degree in philosophy from the University of South Florida. He specializes in the philosophy of mind, neuroscience, and theory of knowledge, and is a technical writer for the IT industry. Better Humans is an editorial production company that's dedicated to having the best information, analysis and opinion on the impact of advancing science and technology.

Part 3: "There's no point"

Myth: Mechanical chemistry will be far too slow to build anything useful.

Fact: A mechanical chemistry machine won't be too slow to build additional tiny machines. In fact, a nanoscale robot arm doing chemistry should be able to build its own mass in a few hours-maybe even minutes. Detailed plans exist for combining lots of these tiny machines into an integrated factory that could make large products. So there's no problem with productivity.

Myth: Biology is better than anything we can build.

Fact: Biology never invented the digital computer, internal combustion, or even the wheel. Try putting three hundred actual horses under the hood of your sports car! Biology is very good at surviving on its own in ecosystems. But engineering, whether large-scale or nanoscale, is better at simple powerful things that modern humans tend to want-like compact motors and fast computers.

Related Myth: Nanomachine designs are unnatural and therefore inefficient. (Sometimes stated as: nanomachines are too rigid and have to spend lots of energy to do anything.)

Fact: Biology includes some great tricks-and nanomachines can use them too. But sometimes, rigid is better (see the "friction" myth). And by the way, if someone gives diffusion as an example of biology being better, they're wrong: making something diffuse takes just as much energy as dragging it (the Einstein, Stokes, and Shannon equations cancel out).

Related Myth: Other technologies will do just as well.

Fact: Precise nanoscale machines have several special advantages. Being small, they can operate more quickly-and thus more efficiently. Being precise, they can have very low friction. And being built with atoms, which are discrete, pliant, and identical, each copy of a machine can be very predictable and reliable. All of this is basic physical law, not speculation.

Myth: It's too hard to do-it'll take many decades if it can be done at all.

Fact: We won't know for sure until it's done. But it doesn't look that hard to design a few chemical reactions, and a few molecular machines, and some larger systems, and some software to pull it together, and a nano-manipulation or enzymatic system to build the first one. Each step requires only engineering, not complex system design. The military payoff alone will probably tempt nations to spend billions on developing it. A lot of people are now saying it could take ten to fifteen years; I think it could be as little as five, especially if one or more nations are already working on it.

Myth: It'll be nearly impossible to design products for a molecular manufacturing system, so it won't have much impact.

Fact: Once a few basic molecular machines have been designed, they can be recombined in many ways to make almost any product you can think of. Software engineers do this kind of thing all the time, compiling a few instructions into large programmes-and never having to think about the physical chips. And most of today's products are a lot less complex than today's software programmes.

From *Leading nanotech experts puts 'grey goo' in perspective*.

"Contrary to previous understanding, self-replication is unnecessary for building an efficient and effective molecular manufacturing system. Instead of building lots of tiny, complex, free-floating robots to manufacture products, it will be more practical to use simple robot arms inside desktop-size factories. A robot arm removed from such a factory would be as inert as a light bulb pulled from its socket. The factory (*) as a whole would be no more mobile than a desktop printer and would require a supply of purified raw materials to build anything."

"An obsession with obsolete science-fiction images of swarms of replicating nanobugs has diverted attention from the real issues raised by the coming revolution in molecular nanotechnologies, " said Drexler. "We need to focus on the issues that matter - how to deal with these powerful new capabilities in a competitive world."

—Mike Treder, Executive Director of CRN, said, "We hope that this article will advance the discussion of the actual implications of molecular manufacturing. There is no need for panic, but there are urgent concerns that must be addressed before the technology arrives." This article is well worth a read - *link*.

Part 4: "It's all just science fiction"

Myth: Molecular manufacturing hasn't appeared in peer-reviewed scientific literature.

Fact: This is incorrect. For example, there's Eric Drexler's 1981 PNAS paper, Ralph Merkle's publications in *Nanotechnology* in the mid-1990's, and a wide range of theory and experiment papers on mechanosynthesis

Myth: Nanobots just aren't well thought out: no power, control, heat dissipation, etc.

Fact: There's a lot of fiction and "artist's conception" floating around. But anyone who tars all nanomachine work with that brush is seriously ignorant. There's a lot of serious engineering work, with nanomachines, robots, and even desktop integrated manufacturing systems worked out in detail.

Part 5: "It's all about dangerous self-replicating nanobots"

Myth: Products of molecular manufacturing must be small.

Fact: It won't be hard to integrate many nanoscale production systems into a factory that joins their output to make large products. This will be a lot more convenient and efficient than trying to keep track of zillions of tiny assemblers and nanobots.

Myth: Every product of molecular manufacturing will self-replicate.

Fact: This would be as inefficient as building an oil refinery and a machine shop into every car. In fact, even the manufacturing systems won't self-replicate, since they'll require preprocessed chemicals.

Myth: A laboratory accident could unleash a "gray goo" of nanobots on the world.

Fact: During the development as well as the use of mechanical molecular manufacturing systems, there will be no reason to build anything even close to a self-replicator. Gray goo would take a lot of hard, pointless design work. Even for military purposes, it would be quite inefficient (think of the car with the built-in oil refinery). A bigger concern appears to be deliberate production of non-replicating weapons.

We encourage debate. Please use this form to tell us what you think.

Risks and benefits of Molecular Manufacturing

The risks and benefits of molecular manufacturing depend heavily on several factors: what it is capable of, when it will be developed, and how it is used. There is substantial disagreement over the first two questions, and almost no information about the third.

As explained above, the idea that molecular manufacturing is impossible is not credible. A separate question is how difficult it will be to develop and whether it will be worth doing. Other technologies are advancing so rapidly that if molecular manufacturing is not developed by 2050 or so, it may not have a very large impact. If it is limited to biochemical materials such as protein, its products may be limited in strength (though still extremely compact compared to today's computers and MEMS).

Despite the weight of rhetoric indicating that molecular manufacturing will have little or no impact, the weight of calculation indicates that it will. In the absence of demonstrated problems with the theory, this section is written assuming product performance consistent with carbon-lattice construction-diamond, graphite, and carbon nanotubes, also called buckytubes or fullerenes.

The high speed of small systems implies that manufacturing systems based on molecular machines will probably be able to produce their own mass in an hour or so. (Bacteria can do this, and bacteria spend only a fraction of their energy on reproduction.) Building complicated shapes from simple chemistry (under direct computer control), there is no apparent reason why manufacturing systems cannot duplicate themselves. This implies that the systems would become extremely cheap, and manufacturing capacity could increase to any desired level almost immediately.

The products of molecular manufacturing would be very high performance. Even with biochemistry, computers would be millions of times more powerful than today's. With carbon lattice, motors would also be millions of times more powerful. Computation and actuation would be a miniscule (and virtually free) fraction of any human-scale product. And carbon-based materials may be hundreds of times as strong as plastics or even metals. With molecular-scale features, very low friction, and very strong materials, the main limitation on products would be the skill of the designers.

The benefits are obvious. Cheap manufacturing of advanced products would allow inefficient infrastructures to be replaced-including our entire energy infrastructure, and much that is wasteful about our agriculture. This would greatly reduce our ecological footprint. Medical research and treatment would be greatly accelerated by the ability to design medical equipment, including massive sensor arrays and tiny robotics, and have them built instantly at negligible cost. Medical research would also be significantly helped by advanced computers. Cheap local self-contained manufacturing offers the possibility of eliminating material poverty. Advanced materials, extremely compact avionics, and rapid, inexpensive manufacturing would allow remarkably cheap access to space and its resources.

However, these capabilities also have downsides. Advanced weapons could also be developed quickly. An arms race based on this technology would be far less stable than the nuclear arms race: weapons would be easier to develop and build in secret, easier to use and control, and harder to stockpile (since they would be almost immediately obsolete) than nuclear bombs. An unstable arms race leading to a devastating war is probably the largest as well as the earliest danger posed by this technology.

Criminal technologies, from weapons to spy systems to communication to smuggling, would also get a boost from the ability to fabricate advanced products as needed. It remains to be seen whether defensive or offensive technologies will be stronger. Even if defensive technologies are adequate, they will probably be used unevenly. This implies that at least some criminal and terrorist acts using the technology will be successful.

And this brings us to the next risk: oppression. For a variety of reasons (including a desire to maintain artificial scarcity), established government and business interests

will want to retain control of molecular manufacturing and its effects. A technology that lets each individual design and build whatever they want does not appear compatible with centralised control. But if the technology is too restricted, a thriving black market will arise to supply citizens with restricted and valuable products. Unfortunately, this would also supply the criminals and perhaps even lead to loss of military control, leading to a spiral of further repression.

Nearly free manufacturing could lead to an orgy of conspicuous consumption that could damage the ecosystem. And advanced products, up to and including brain and body modification, could shake up society.